Universitext

Universitext is a series of textbooks that presents material from a wide variety of mathematical disciplines at master's level and beyond. The books, often well class-tested by their author, may have an informal, personal, or even experimental approach to their subject matter. Some of the most successful and established books in the series have evolved through several editions, always following the evolution of teaching curricula, into very polished texts.

Thus as research topics trickle down into graduate-level teaching, first textbooks written for new, cutting-edge courses may find their way into *Universitext*.

Alejandro Illanes

Continuum Theory

Springer

Alejandro Illanes
Instituto de Matematicas
National Autonomous University of Mexico
Mexico City, Mexico

ISSN 0172-5939 ISSN 2191-6675 (electronic)
Universitext
ISBN 978-3-031-91010-4 ISBN 978-3-031-91011-1 (eBook)
https://doi.org/10.1007/978-3-031-91011-1

Mathematics Subject Classification: 54F15, 54F16

This Springer imprint is published by the registered company Springer Nature Switzerland AG
The registered company address is: Gewerbestrasse 11, 6330 Cham, Switzerland

This book is dedicated to Ana and Diego

Una lágrima espontánea, en nuestro primer encuentro

Paciencia infinitamente elástica

Amor incondicional y permanente

Viaje a mi propio pasado, al ver y recordar la fragilidad, los miedos, las alegrías, las ilusiones, las esperanzas, la fuerza. . . , de los niños y jóvenes

Viaje al presente para echar a la basura algunos valores, prejuicios, conceptos, conocimientos. . . , que se han vuelto anacrónicos

Tremenda resistencia interior a no dormir

Mis logros personales no son lo mejor que voy a dejar en este mundo

Aprender más y enseñar menos

Hay temores mayores al miedo de lo que me puede pasar personalmente

Como aeropuertos, ayudamos a despegar, pero tenemos que estar listos para los aterrizajes necesarios

Tardes de ballet, piano, natación, dibujo, guitarra, acuarela, costura, olimpiadas, teatro...

Preface

In the middle of the 1980s, a group of topologists held a seminar at the National Autonomous University of México (UNAM) to read the classic book "Hyperspaces of Sets" by Sam Bernard Nadler, Jr. [112]. We had only basic experience in Continuum Theory; however, the book was so motivating that we immediately fell in love with hyperspaces. Some months later, we started working on many of the open problems that the author generously included in his book.

For some time we learned about Continuum Theory from several sources. In fact, we started thinking about writing our own book on continua. This idea was put aside in 1992 with the appearance of Sam Nadler's excellent book "Continuum Theory: An Introduction" [113].

Again, we organized a seminar to read and solve the exercises of this book, reinforcing our knowledge in this area.

Since then, I have taught introductory courses on Continuum Theory and on hyperspaces, for undergraduate and graduate students. I developed the present book as an extension of the notes taken in my courses. It reflects the exact way I like to introduce my students to my favorite topics in mathematics.

By the way, I want to express here my admiration for Sam. In fact, I consider Sam as a mentor, since I learned a large part of what I know about Continuum Theory and hyperspaces through his work.

The first half of this book is devoted to presenting the basic facts about Continuum Theory (Chap. 1 to 8). Beginning with significative examples, it describes the simplest but most powerful method for constructing continua, namely: intersect a decreasing sequence of continua. In these chapters we travel from the properties of the most elementary continua—finite graphs, dendrites, and dendroids, to the most intriguing ones—indecomposable and hereditarily indecomposable continua, passing through the structure of locally connected continua and the Hahn–Mazurkiewicz Theorem.

Each of the remaining chapters contains a brief introduction to a topic of interest to those working on Continuum Theory. I have tried to include some gems of the area, meaning, some special examples and results that deserve to be better known.

So, in this book you will find:

- The characterization of finite graphs as those Peano continua which do not contain a topological copy of the dendrite F_{ω} or an extended null comb.
- The elegant proof that each compact metric space is a continuous image of the Cantor set.
- An extensive account of what is known on geometric models of hyperspaces.
- The classical theorem stating that for each irreducible continuum such that all its indecomposable subcontinua have empty interior, there exists a monotone continuous mapping onto the interval $[0, 1]$ with fibers having empty interior.
- A proof of the Mountain Climbing Theorem.
- Borsuk's Theorem stating that dendroids have the fixed point property.
- The characterization of dendrites as those dendroids not containing semi-combs.
- The proof that the cone over a spiral surrounding a circle does not have the fixed point property.
- The proof that the Hilbert cube is homogeneous.
- The proof of the equivalence of being an absolute retract and having a convex structure.
- The proof of the most impressive properties of the pseudo-arc P: P is homogeneous, P is homeomorphic to each of its non-degenerate subcontinua, and P is characterized both as the unique chainable hereditarily indecomposable continuum and as the unique chainable homogeneous continuum.

The minimal knowledge required for reading this book is a first course on General Topology. In particular, it is necessary to be familiar with the notions of compactness, connectedness, separation axioms, and the topology of metric spaces.

Before finishing this introduction, I would like to recommend to the reader the excellent article on the history of Continuum Theory written by Janusz Jerzy Charatonik in 1998 [22]. Some more notes on the history of Continuum Theory can be found in [72].

As I mentioned before, the present book has its origins in my courses. Some students helped me by taking and typing notes. For this, I am especially grateful to María Elena Aguilera Miranda, Claudia Guadalupe Domínguez López, Luis Miguel García Velázquez, Margarita Hernández Urbina and Alberto Carlos Mercado Saucedo.

I am also very grateful with those students that helped me develop a specific subject: Jimmy Anel Naranjo Murillo, for the chapter on irreducible continua; Mauricio Esteban Chacón Tirado, for the proof that the Hilbert cube is homogeneous; and Erick Iván Rodríguez Castro, for the chapter on the stronger properties of the pseudo-arc.

Of course, I am grateful to all the students in my courses on Continuum Theory and my courses on hyperspaces, their presence was very motivating and from reading their faces I learned which subjects were more difficult and which needed to be explained in more detail. I would like to include all their names here, but I realize that they are so many and my memory is so poor that this would be impossible.

However, I can mention that a good number of them are known in the field because of their contributions to Continuum Theory.
Sadly, some of my colleagues who shared with me part of my journey on Continuum Theory are now gone. As a small tribute, I remember them here:

Janusz Jerzy Charatonik, 1934–2004
Víctor Neumann Lara, 1933–2004
Frank H. Sturm, 1983–2014
Sam Bernard Nadler, Jr., 1939–2016
Adalberto García Máynez y Cervantes, 1941–2016
Włodzimierz J. Charatonik, 1957–2021
Piotr Minc, 1949–2023
María Elena Aguilera Miranda, 1978–2024

I am very much indebted to Diego Illanes Martínez-de-la-Vega and Verónica Martínez-de-la-Vega for their help in checking some parts of this book.

I greatly appreciate the work and constructive comments of the anonymous referees, who helped me improve the book significantly. I express my gratitude to Dr. Remi Lodh, editor from Springer, who was always patient and helpful. I also thank Fairle T. Thattil and Deepika Suresh from Springer for all their help. I thank Leonardo Espinosa Pérez for his expert advice about the use of TeX.

Finally, I acknowledge UNAM and, particularly, the Institute of Mathematics for being an ideal place to develop research. Besides the excellent conditions provided to the researchers, it is important to mention that this center is an infinite source of enthusiastic and talented students.

This book was partially supported by the projects "Sistemas dinámicos discretos y teoría de continuos I" (IN105624) of PAPIIT, DGAPA, UNAM and "Teoría de Continuos e Hiperespacios, dos" (AI-S-15492) of CONAHCYT.

Mexico City, Mexico Alejandro Illanes
October, 2024

Declarations

Competing Interests The author has no competing interests to declare that are relevant to the content of this manuscript.

Contents

About the Author

Alejandro Illanes is a researcher at the National Autonomous University of Mexico, where he has taught for over 45 years. He has over 150 research papers in international journals and has supervised 16 doctoral theses. He is the author of a number of books.

Chapter 1
Introduction

1.1 Basic Properties and Examples

A *continuum* is a compact connected metric space with more than one point (sets with more than one point are called *non-degenerate*). A *subcontinuum* of a continuum X is a nonempty closed connected subset of X, so a subcontinuum can be a one-point set and non-degenerate subcontinua are continua.

A *curve* is a one-dimensional continuum (equivalently, a continuum having a basis $\mathcal{B}$ such that every element in $\mathcal{B}$ has a totally disconnected boundary). A *mapping* is a continuous function.

Given a continuum X with metric d, $p \in X$, $\emptyset \neq A \subset X$ and $\varepsilon > 0$, let $\text{diameter}(A) = \sup\{d(x, y) : x, y \in A\}$, let $B(p, \varepsilon)$ be the ε-ball around p with radius ε, and let

$$N(A, \varepsilon) = \bigcup\{B(p, \varepsilon) \subset X : p \in A\}.$$

The simplest continuum is the unit interval $[0, 1]$ in the real line. An *arc* is a continuum homeomorphic to $[0, 1]$. If A is an arc and $h : [0, 1] \to A$ is a homeomorphism, the *end-points* of A are $h(0)$ and $h(1)$, and we say that A *joins* $h(0)$ and $h(1)$. A continuum X is *arcwise connected* if every pair of distinct points in X are the end-points of an arc in X.

Another simple and important continuum is the unit circle in the plane, which is denoted by S^1. A *simple closed curve* is a continuum homeomorphic to S^1. A *finite graph* is a continuum which is the union of a finite number of arcs such that each pair of them have finite intersection.

The family of finite graphs is an important family of continua. A finite graph is constructed by taking a finite family of arcs $\alpha_1, \ldots, \alpha_n$ in the Euclidean space $\mathbb{R}^3$ in such a way that for each $i \in \{2, \ldots, n\}$, α_i intersects $\alpha_1 \cup \cdots \cup \alpha_{i-1}$ in a nonempty finite set. Thus, $X = \alpha_1 \cup \cdots \cup \alpha_n$ is a finite graph.

Two prominent finite graphs are K_5 and $K_{3,3}$.

A. Illanes, *Continuum Theory*, Universitext,
https://doi.org/10.1007/978-3-031-91011-1_1

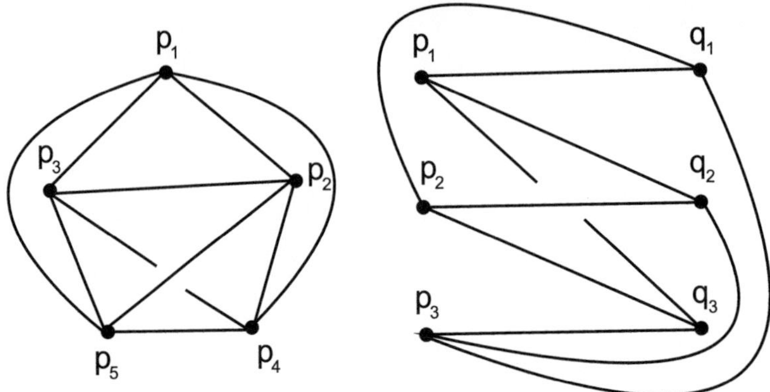

Fig. 1.1 K_5 and $K_{3,3}$

The finite graph K_5 is the *complete graph with* 5 *elements* and is constructed as follows: take 5 distinct points $p_1, \ldots, p_5$ in $\mathbb{R}^3$ and for each pair of indices $i, j \in \{1, \ldots, 5\}$, with $i \neq j$, take one arc $\alpha(\{i, j\})$ in $\mathbb{R}^3$ that joins p_i and p_j satisfying the following: if $\{i, j\} \neq \{k, l\}$, then $(\alpha(\{i, j\}) \setminus \{p_i, p_j\}) \cap (\alpha(\{k, l\}) \setminus \{p_k, p_l\}) = \emptyset$.

The finite graph $K_{3,3}$ is the *complete bipartite graph with* 3, 3 *elements* and it is constructed as follows: take two disjoint subsets $\{p_1, p_2, p_3\}$ and $\{q_1, q_2, q_3\}$ of $\mathbb{R}^3$, each with 3 distinct elements, and for each ordered pair of indices $i, j \in \{1, 2, 3\}^2$, take an arc $\alpha(i, j)$ in $\mathbb{R}^3$ that joins p_i and q_j in such a way that if $(i, j) \neq (k, l)$, then $(\alpha(i, j) \setminus \{p_i, q_j\}) \cap (\alpha(k, l) \setminus \{p_k, q_l\}) = \emptyset$.

In Fig. 1.1 we represent K_5 and $K_{3,3}$.

Finite graphs K_5 and $K_{3,3}$ cannot be embedded in the Euclidean plane $\mathbb{R}^2$ (this is a consequence of the Jordan Curve Theorem). The relevance of these graphs is the theorem of K. Kuratowski [89] which says that a finite graph X can be embedded in $\mathbb{R}^2$ if and only if X does not contain a homeomorphic copy of either K_5 or $K_{3,3}$.

Finite and countable products of continua are continua. Using this fact it is possible to produce a large family of continua. The products of intervals such as $[0, 1]^2, [0, 1]^3, \ldots$ have an important role in Continuum Theory. An *n-cell* is a continuum homeomorphic to $[0, 1]^n$. We can also consider the *Hilbert cube* $\mathbf{Q}$, that is defined as

$$\mathbf{Q} = [0, 1] \times [0, 1] \times \cdots .$$

Clearly, $\mathbf{Q}$ is compact and connected. This space can also be metrized with the metric (Exercise 1.17):

$$d_{\mathbf{Q}}((x_1, x_2, \ldots), (y_1, y_2, \ldots)) = \sum_{n=1}^{\infty} \frac{|x_n - y_n|}{2^n}. \tag{1.1}$$

1.2 Universality of the Hilbert Cube

The Hilbert cube is a universal continuum in the sense that every continuum can be embedded in $\mathbf{Q}$. This fact is proved in the following theorem.

Theorem 1.1 *Let X be a compact metric space. Then there is an embedding $\varphi : X \to \mathbf{Q}$.*

Proof Let d be a metric for X such that for all $p, q \in X$, $d(p,q) \leq 1$ (Exercise 1.18). By Exercise 1.19, X contains a countable dense subset $D = \{a_1, a_2, a_3, \ldots\}$ of X.

Let $\varphi : X \to [0,1] \times [0,1] \times \cdots$ be given by

$$\varphi(p) = (d(p, a_1), d(p, a_2), \ldots).$$

We check that φ is an embedding.

Claim 1. φ is continuous.

Since the range of φ is a product, to show the continuity of φ, it is enough to show that each of the coordinate functions of φ is continuous. So we need to check that for each $n \in \mathbb{N}$, the function that sends p to $d(p, a_n)$ is continuous. This continuity follows from the inequality $|d(p, a_n) - d(q, a_n)| \leq d(p, q)$, which is a direct consequence of the triangle inequality.

Claim 2. φ is one-to-one.

Let $p, q \in X$ be such that $p \neq q$. Let $\delta = \frac{d(p,q)}{2}$. Since $\delta > 0$ and $D = \{a_1, a_2, a_3, \ldots\}$ is dense in X, there is an $n \in \mathbb{N}$ such that $d(p, a_n) < \delta = \frac{d(p,q)}{2}$. If $d(q, a_n) < \delta$, then $d(p, q) \leq d(p, a_n) + d(a_n, q) < 2\delta = d(p, q)$, a contradiction. Hence $d(q, a_n) \geq \delta$. Thus $d(p, a_n) < d(q, a_n)$. We have shown that $\varphi(p) \neq \varphi(q)$. Therefore, φ is one-to-one.

Since φ is continuous, one-to-one, its domain is compact and its range is a Hausdorff space, we conclude that φ is an embedding. □

Manifolds, with and without boundary, comprise another important family of continua. So spheres, the torus, the Klein bottle and the Moebius strip are also examples of continua.

A *compactification of the ray* $[0, \infty)$ is a continuum X for which there is an embedding $h : [0, \infty) \to X$ such that $h([0, \infty))$ is dense in X. Since $h([0, \infty))$ is homeomorphic to $[0, \infty)$, we usually identify $[0, \infty)$ with $h([0, \infty))$. We often write $[0, \infty) \subset X$ and $[0, \infty)$ is dense in X. The set $R = X \setminus [0, \infty)$ is called *the remainder of* X. Since $[0, \infty)$ is homeomorphic to $[0, 1)$ and $(0, 1]$, these two sets are also called *rays*. The simplest compactification of $(0, 1]$ is the interval $[0, 1]$. The *topologist's curve* (or the $\sin(\frac{1}{x})$-*continuum*) is the continuum X defined as the closure, in the plane $\mathbb{R}^2$, of the graph of the mapping $\sin(\frac{1}{x})$, where $x \in (0, 1]$ (Fig. 1.2).

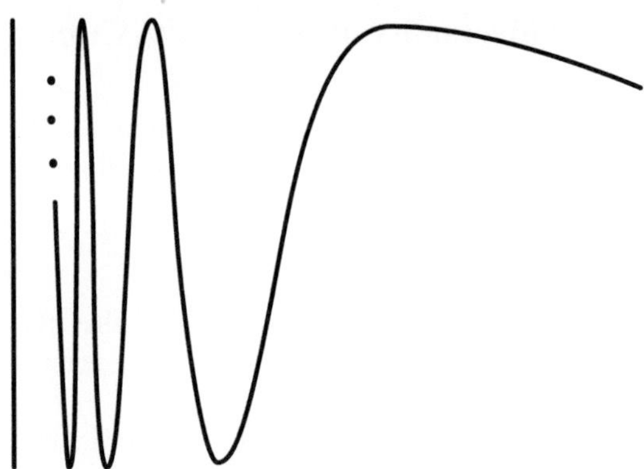

Fig. 1.2 The topologist's curve

1.3 Decreasing Sequences of Continua

The following theorem gives a very useful method for constructing continua.

Theorem 1.2 *Let X be a continuum and let $\{A_n\}_{n=1}^{\infty}$ be a sequence of subcontinua of X such that $A_1 \supset A_2 \supset \cdots$. Then $A = A_1 \cap A_2 \cap \cdots$ is a subcontinuum of X.*

Proof By Exercise 1.20, A is compact and nonempty.

In order to show that A is connected, suppose to the contrary that there exist disjoint nonempty closed subsets K and L such that $A = K \cup L$. Let U and V be disjoint open subsets of X such that $K \subset U$ and $L \subset V$.

Notice that

$$X \setminus (U \cup V) \subset X \setminus (K \cup L) = X \setminus A = X \setminus \bigcap \{A_n : n \in \mathbb{N}\} = \bigcup \{X \setminus A_n : n \in \mathbb{N}\}.$$

Thus, the family $\{X \setminus A_n : n \in \mathbb{N}\}$ is an open cover of the compact set $X \setminus (U \cup V)$. Hence there exist $k, n_1, \ldots, n_k \in \mathbb{N}$ such that

$$X \setminus (U \cup V) \subset (X \setminus A_{n_1}) \cup \cdots \cup (X \setminus A_{n_k}) = X \setminus (A_{n_1} \cap \cdots \cap A_{n_k}).$$

Let $m = \max\{n_1, \ldots, n_k\}$. Then

$$X \setminus (A_{n_1} \cap \cdots \cap A_{n_k}) = X \setminus A_m.$$

Thus

$$A_m \subset U \cup V.$$

Hence $A_m = (U \cap A_m) \cup (V \cap A_m)$, and moreover, $U \cap A_m$ and $V \cap A_m$ are disjoint open subsets of A_m. Since $K \subset U \cap A_m$ and $L \subset V \cap A_m$, $U \cap A_m$ and $V \cap A_m$ are nonempty. So we have a separation of A_m. This contradicts the connectedness of A_m and completes the proof that A is connected. □

1.4 Advanced Examples

Example 1.3 (The Sierpiński Triangle) The Sierpiński triangle is constructed in the following way. Consider a convex closed triangle T_1 in the plane $\mathbb{R}^2$. Consider the four triangles determined by joining the middle points of the edges of T_1, and let T_2 be the continuum obtained by removing the interior of the central triangle. In this way we have that T_2 is the union of three convex triangles. Repeat the procedure for each of the three triangles in T_2 to obtain a continuum T_3. Continue in this fashion to construct $T_3, T_4, \ldots$ The *Sierpiński triangle* is defined by

$$T = T_1 \cap T_2 \cap \cdots .$$

By Theorem 1.2, the Sierpiński triangle is a continuum.

In the second half of the nineteenth century, it seemed natural to define a plane curve as a continuous image of the interval [0, 1]. This idea came to an end in 1890, when G. Peano proved that the unit square is such an image [116]. After this, the definition of a plane curve as a continuum in the plane with empty interior was adopted, and it remains as such to the present day. However, in 1915, W. Sierpiński constructed Example 1.3 to show that we can obtain weird "curves" that satisfy this definition [36]. It is important to observe that the Hahn–Mazurkiewicz theorem (Theorem 2.9) implies that the Sierpiński triangle is also a continuous image of [0, 1] (Fig. 1.3).

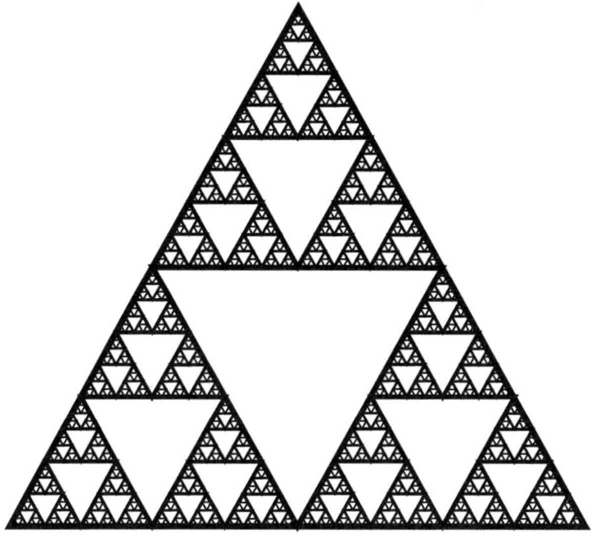

Fig. 1.3 The Sierpiński triangle

Example 1.4 (The Sierpiński Carpet) The Sierpiński carpet is constructed in a similar way to the Sierpiński triangle, but starting with the convex square $[0, 1]^2$ instead of a convex triangle.

Take $C_1 = [0, 1]^2$. Take the partition P on C_1 induced by the partition $0 < \frac{1}{3} < \frac{2}{3} < 1$ on $[0, 1]$. Then C_1 is decomposed into nine squares. We remove the interior of the middle square to obtain C_2. That is, $C_2 = C_1 \setminus ((\frac{1}{3}, \frac{2}{3}) \times (\frac{1}{3}, \frac{2}{3}))$.

So C_2 is the union of eight little squares, of area $(\frac{1}{3})^2$, induced by P.

Repeat this procedure for each of the eight squares to obtain C_3. Then C_3 is the union of 64 smaller squares of area $(\frac{1}{9})^2$.

Continue this procedure to obtain a sequence of subcontinua $\{C_n\}_{n=1}^{\infty}$ of the plane $\mathbb{R}^2$ such that $C_1 \supset C_2 \supset \cdots$. The *Sierpiński carpet* is defined as

$$C = C_1 \cap C_2 \cap \cdots .$$

Although the continuum C is called the Sierpiński carpet, the idea of the construction came from S. Mazurkiewicz (see page 709 of [22]). W. Sierpiński proved in [118] one of the most important properties of C: each one-dimensional (equivalently, each subcontinuum with empty interior) subcontinuum of the plane $\mathbb{R}^2$ can be embedded in C. So C is called a universal curve in the plane (Fig. 1.4).

Example 1.5 (The Menger Curve) The Menger curve has a similar construction as the Sierpiński carpet, but starting with $[0, 1]^3$ instead of $[0, 1]^2$. The construction is as follows. Let $E_1 = [0, 1]^3$. Divide E_1 by planes parallel to its faces into 27 equal cubes (in a similar manner to a Rubik's cube). Remove the central small cube D and all cubes of the subdivision sharing a two-dimensional face with D. In this way we obtain E_2, which consists of the remaining 20 small closed cubes. Repeating

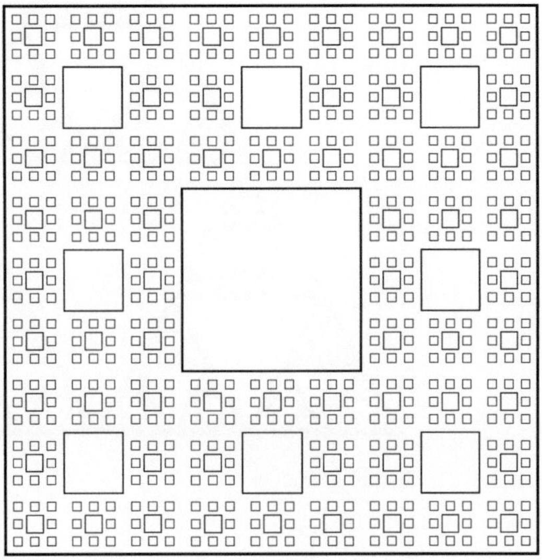

Fig. 1.4 The Sierpiński carpet

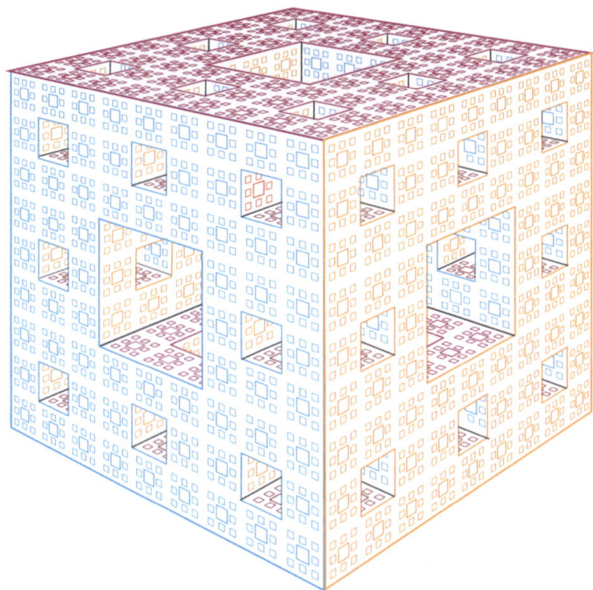

Fig. 1.5 The Menger curve

exactly the same procedure for each of the 20 cubes, we obtain a subcontinuum E_3 of E_2 which is the union of 400 smaller cubes. Continuing the process indefinitely, we obtain a sequence of continua $\{E_n\}_{n=1}^{\infty}$ such that $E_1 \supset E_2 \supset E_3 \supset \cdots$. Define

$$E = E_1 \cap E_2 \cap \cdots .$$

This continuum E is the *Menger curve*. It is known that the Menger curve is one-dimensional and every curve (one-dimensional continuum) can be embedded in E [107]. So E is called a universal curve. Surprisingly, R.D. Anderson proved that E is homogeneous (for every two points p and q in E there exists a homeomorphism $h : E \to E$ such that $h(p) = q$) [2] and [3] (Fig. 1.5).

Example 1.6 (Lakes of Wada) We construct this continuum in the square $X_0 = [0, 1]^2$. Imagine that X_1 is an island with two lakes, like in the first picture of Fig. 1.6. Suppose that the water of the sea around X_1 is blue, while the water of one island is red and the water of the other island is green.

Using the interval of time $[0, \frac{1}{2}]$, starting at 0, we dig a canal from the sea, which brings blue water to within a distance of 1 unit of every point of the land. Using the interval of time $[\frac{1}{2}, \frac{3}{4}]$, we dig a canal from the lake of red water, never intersecting the blue canal, which brings red water to within a distance of $\frac{1}{2}$ of every point of the land. Using the interval $[\frac{3}{4}, \frac{7}{8}]$, we dig a canal from the lake of green water, which brings green water to within a distance of $\frac{1}{4}$ of every point of the land. Using the

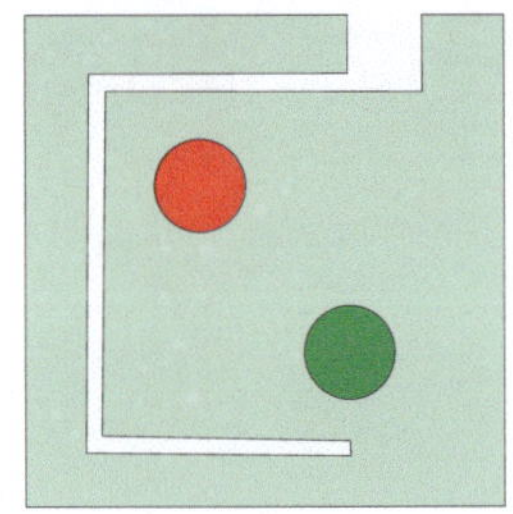
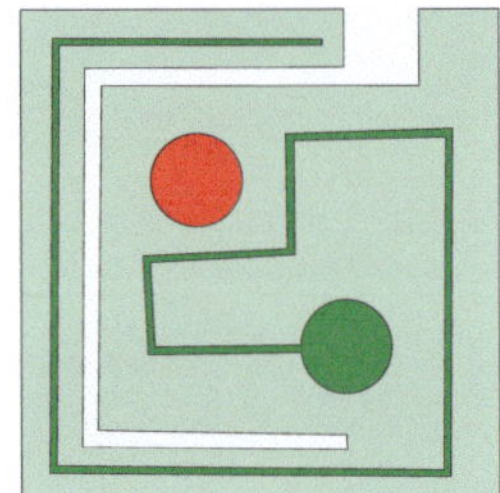

Fig. 1.6 The lakes of Wada

interval $[\frac{7}{8}, \frac{15}{16}]$, we dig a canal from the end of the first canal to bring blue water to within a distance $\frac{1}{8}$ of every point of the land, and we continue in this fashion. At the end of the process, the remaining land is a continuum $X \subset [0, 1]^2$ with the property that every point of X is an accumulation point of blue, red and green points.

The continuum X was described in 1917 by K. Yoneyama [127]. It is known by the name *the lakes of Wada*. Notice that X is the common boundary of three regions (open connected subsets of the plane), namely: the sets representing the blue water, red water and green water (see p. 716 and 717 of [22] for more information about continua with these properties).

Definition 1.7 A continuum X is *decomposable* if there exist proper subcontinua A and B of X such that $X = A \cup B$. If a continuum X is not decomposable, then X is *indecomposable*. If every subcontinuum of X is indecomposable, then X is *hereditarily indecomposable*.

Example 1.8 (The Buckethandle Continuum) The simplest indecomposable continuum was found by L.E.J. Brouwer in 1910. His construction was simplified by Z. Janiszewski, and B. Knaster also gave a nice description of it. For this reason, this continuum is known as the *Brouwer–Janiszewski–Knaster continuum*, it is also known by its less formal name: *the buckethandle continuum*.

Knaster's construction is as follows.

Consider the usual ternary Cantor set $C_0 \subset [0, 1]$.

The first step is to join any two points $(p, 0), (q, 0)$ in $C \times \{0\}$, which are symmetric with respect to $(\frac{1}{2}, 0)$, by a semi-circle in the upper half plane with its centre at $(\frac{1}{2}, 0)$.

The second step is to join any two points $(p, 0), (q, 0)$ of $C \times \{0\}$ in the interval $[\frac{2}{3}, 1] \times \{0\}$, which are symmetric with respect to $(\frac{5}{6}, 0)$, by a semi-circle in the lower half plane with its centre at $(\frac{5}{6}, 0)$.

In general, for each $n \in \mathbb{N}$ join any two points $(p, 0), (q, 0)$ of $C \times \{0\}$ in the interval $[\frac{2}{3^n}, \frac{3}{3^n}] \times \{0\}$, which are symmetric with respect to $(\frac{5}{2 \cdot 3^n}, 0)$, by a semi-circle in the lower half plane with its centre at $(\frac{5}{2 \cdot 3^n}, 0)$ (Fig. 1.7).

The buckethandle continuum is indecomposable and chainable (see Definition 1.9, and Exercises 1.34 and 1.39).

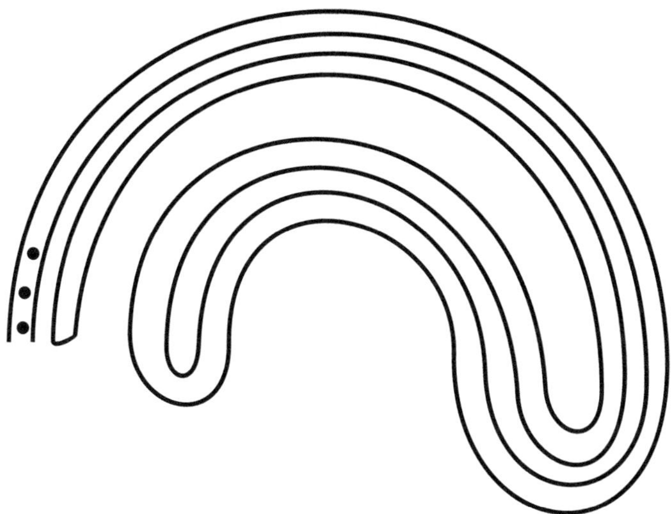

Fig. 1.7 The buckethandle continuum

Definition 1.9 Let X be a continuum. A *chain* in X is a finite sequence $\mathcal{U} = \{U_1, \ldots, U_n\}$ of subsets of X such that $U_i \cap U_j \neq \emptyset$ if and only if $|i - j| \leq 1$. The sets U_i are the *links* of the chain $\mathcal{U}$. The *first link* of $\mathcal{U}$ is U_1. The mesh of the chain $\mathcal{U}$ is denoted by mesh$(\mathcal{U})$ and defined by

$$\text{mesh}(\mathcal{U}) = \max\{\text{diameter}(U_i) : i \in \{1, \ldots, n\}\}.$$

If p and q are points of X such that $p \in U_1$ and $q \in U_n$, then we say that $\mathcal{U}$ *goes from* p *to* q. The union of the elements of $\mathcal{U}$ is denoted by $\bigcup \mathcal{U}$. Given $1 \leq i \leq j \leq n$, the union of the *subchain* of $\mathcal{U}$, from U_i to U_j, is denoted by $\mathcal{U}^*(i, j)$, that is, $\mathcal{U}^*(i, j) = U_i \cup \cdots \cup U_j$. The chain $\mathcal{U}$ is *open* if each U_i is open, and $\mathcal{U}$ *covers* X if $X = U_1 \cup \cdots \cup U_n$. The continuum X is *chainable* if for each $\varepsilon > 0$, there exists an open chain $\mathcal{U}$, covering X, such that mesh$(\mathcal{U}) < \varepsilon$.

A useful technique for constructing interesting continua is to take sequences of chains in $\mathbb{R}^2$, as in the following example (Fig. 1.8).

Example 1.10 (Another Indecomposable Continuum) Fix three distinct points a, b and c in the plane $\mathbb{R}^2$. Construct a chain $\mathcal{U}_1$ from a to c passing through b. This means that $\mathcal{U}_1$ goes from a to c and $b \in \bigcup \mathcal{U}_1$. We also ask that the links U_i are 2-cells and mesh$(\mathcal{U}_1) < \frac{1}{2^1}$. Inductively, construct a sequence of chains $\mathcal{U}_1, \mathcal{U}_2, \ldots$ such that the links of $\mathcal{U}_n$ are 2-cells, $\text{cl}_{\mathbb{R}^2}(\bigcup \mathcal{U}_{n+1}) \subset \bigcup \mathcal{U}_n$, mesh$(\mathcal{U}_n) < \frac{1}{2^n}$, and:

(1) if $n \equiv 1 \pmod 3$, then $\mathcal{U}_n$ goes from a to c passing through b,
(2) if $n \equiv 2 \pmod 3$, then $\mathcal{U}_n$ goes from a to b passing through c, and

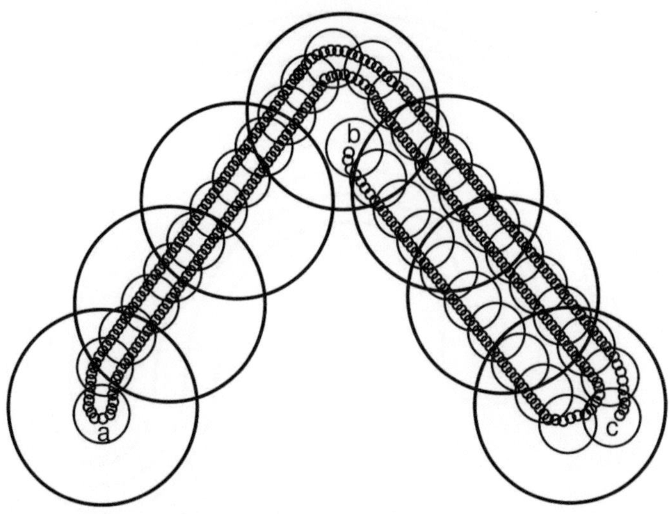

Fig. 1.8 Another indecomposable continuum

(3) if $n \equiv 3 \pmod 3$, then $\mathcal{U}_n$ goes from b to c passing through a.

By Theorem 1.2 the set

$$X = \mathrm{cl}_{\mathbb{R}^2}(\bigcup \mathcal{U}_1) \cap \mathrm{cl}_{\mathbb{R}^2}(\bigcup \mathcal{U}_2) \cap \cdots .$$

is a continuum in $\mathbb{R}^2$.

We check that X is indecomposable. Suppose to the contrary that there exist proper subcontinua A and B of X such that $X = A \cup B$. Then one of the sets A and B contains two of the points in the set $\{a, b, c\}$. Without loss of generality, we may assume that $a, b \in A$. Fix a point $p \in X \setminus A$ and $n \in \mathbb{N}$ such that $B(p, \frac{1}{n}) \cap A = \emptyset$ and $n \equiv 2 \pmod 3$. Let $\mathcal{U}_n = \{U_1, \ldots, U_m\}$. Then $a \in U_1$ and $b \in U_m$. Let $i \in \{1, \ldots, m\}$ be such that $p \in U_i$. Since $\mathrm{mesh}(\mathcal{U}_n) < \frac{1}{2^n}$, we have that $U_i \subset B(p, \frac{1}{n})$, so $U_i \cap A = \emptyset$. Then A intersects the sets $U_1 \cup \cdots \cup U_{i-1}$ and $U_{i+1} \cup \cdots \cup U_m$ (at least at the points a and b, respectively). Since these sets are open and disjoint, we obtain a contradiction with the connectedness of A. Therefore, X is indecomposable.

Example 1.11 (The Dyadic Solenoid) The *dyadic solenoid* is constructed as follows. Let T_1 be a solid torus (that means, a torus in the Euclidean space $\mathbb{R}^3$, including its interior). Inside T_1 take another solid torus T_2 wrapped around it longitudinally 2 times, in a smooth fashion without folding back, as shown in Fig. 1.9. Inside T_2, take another solid torus T_3 wrapped around it 2 times in the same way. Then T_3 is wrapped around the first torus T_1 4 times. Continue the procedure indefinitely. In this way we obtain a decreasing sequence of tori $T_1 \supset T_2 \supset T_3 \supset \cdots$. By Theorem 1.2, the set $S = \bigcap\{T_n : n \in \mathbb{N}\}$ is a continuum called the *dyadic solenoid*. Of course, a similar construction is possible by wrapping around 3 times, instead of 2 times, or any number of times, and then we obtain different solenoids.

Fig. 1.9 The dyadic solenoid

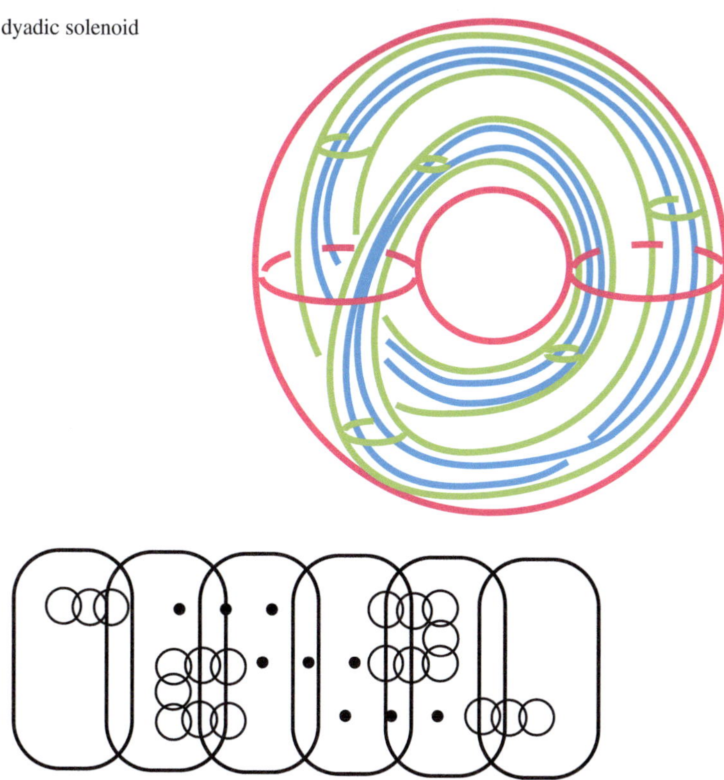

Fig. 1.10 Crooked chains

We finish this section by presenting the most interesting and important continuum: the *pseudo-arc*. The pseudo-arc is defined by taking a sequence of chains satisfying special properties. So, we start by establishing the appropriate definitions.

Example 1.12 (The Pseudo-Arc) Given chains $\mathcal{V}$ and $\mathcal{U}$ in a topological space, we say that $\mathcal{V}$ *refines* $\mathcal{U}$ if each link of $\mathcal{V}$ is contained in a link of $\mathcal{U}$. The chain $\mathcal{V} = \{V_1, \dots, V_m\}$ is *crooked* in $\mathcal{U} = \{U_1, \dots, U_n\}$ if:

- $\mathcal{V}$ refines $\mathcal{U}$, and
- if $k, l \in \{1, \dots, n\}$ with $k+3 \le l$ and $i, j \in \{1, \dots, m\}$ are such that $V_i \subset U_k$ and $V_j \subset U_l$, then there exist $r, s \in \{1, \dots, m\}$ such that $V_r \subset U_{l-1}$ and $V_s \subset U_{k+1}$ and either $i < r < s < j$ or $j < s < r < i$.

Figure 1.10 shows the behavior of crookedness. Chain $\mathcal{V}$ is represented by the small circles. The second condition of crookedness says that if two links V_i and V_j of $\mathcal{V}$ are contained in respective links U_l and U_k of $\mathcal{U}$ which are at least 3 links apart ($k + 3 \le l$), then in order that the links $\mathcal{V}$ advance from V_i to V_j, where it is possible that $j < i$, they **first** have to visit the adjacent link U_{k-1}, then they have to visit the link U_{l+1}, and after this it is possible to go to link U_k. Notice that if we

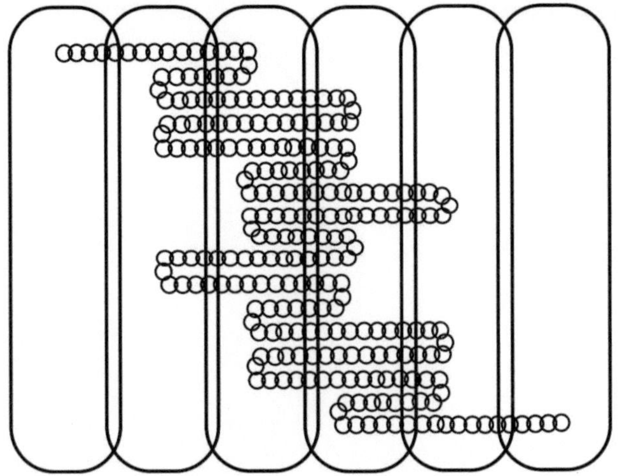

Fig. 1.11 Crooked in 6 links

want to go from link U_1 to link U_5, we first visit link U_4, and it is necessary to make a "zig-zag" from U_1 to U_4. Then we have to visit U_2, and we can go from U_2 to link U_5, making a "zig-zag", of course. In Fig. 1.11, we have pictured a crooked chain inside a chain of 6 links.

Now we know how to construct a crooked chain inside a chain with 6 links.

In order to construct a crooked chain inside a chain with 7 links, we can do the following: Inductively, construct a chain from link 1 to link 6 as in Fig. 1.11, then continue the chain with one subchain from link 6 to link 2 (we know how to do this), and finally continue the chain with one subchain from link 2 to link 7, with something similar to the chain in Fig. 1.11.

By a simple induction, it is now possible to see that if we take chains in the Euclidean plane in such a way that the links and the union of all its links are homeomorphic to an open disk, we can construct a chain crooked in a chain with any number of links. Thus, it is possible to construct a sequence of chains $\mathcal{U}_1, \mathcal{U}_2, \mathcal{U}_3, \ldots$ in the Euclidean plane such that for each $n \in \mathbb{N}$:

- $\mathcal{U}_{n+1}$ is crooked in $\mathcal{U}_n$,
- $\mathrm{mesh}(\mathcal{U}_n) < \frac{1}{2^n}$,
- the first element of $\mathcal{U}_{n+1}$ is contained in the first one of $\mathcal{U}_n$, and the last element of $\mathcal{U}_{n+1}$ is contained in the last one of $\mathcal{U}_n$.

We are ready to construct the pseudo-arc. Take the sequence $\mathcal{U}_1, \mathcal{U}_2, \mathcal{U}_3, \ldots$ as before.

For each $n \in \mathbb{N}$, let $A_n = \mathrm{cl}_{\mathbb{R}^2}(\bigcup \mathcal{U}_n)$. Then A_n is a subcontinuum of $\mathbb{R}^2$ and $A_{n+1} \subset A_n$. Then define the pseudo-arc P as

$$P = \bigcap_{n=1}^{\infty} A_n.$$

In 1921, Knaster and Kuratowski [85] asked if there exists a plane hereditarily indecomposable continuum. This question was answered in 1922 by Knaster in his doctoral dissertation [82]. He spent almost 40 pages describing the example and proving its properties. We have constructed the pseudo-arc in less than two pages and we will prove that it is hereditarily indecomposable (Theorem 1.13) in less than one page. This can be done thanks to the definition of a crooked chain, introduced by RH Bing, generalizing a concept due to E.E. Moise [108]. Chapter 17 is devoted to showing stronger properties of the pseudo-arc, namely, that the pseudo-arc is homogeneous, it is the only hereditarily indecomposable chainable continuum and it is homeomorphic to each of its non-degenerate subcontinua.

Theorem 1.13 *The pseudo-arc is hereditarily indecomposable.*

Proof Suppose to the contrary that P contains a decomposable subcontinuum X. Then there exist two proper subcontinua A and B of X such that $X = A \cup B$.

Fix points $a \in X \setminus B \subset A$ and $b \in X \setminus A \subset B$. Let $\varepsilon > 0$ be such that $B(a, \varepsilon) \cap B(b, \varepsilon) = \emptyset$, $B(a, \varepsilon) \cap B = \emptyset$ and $B(b, \varepsilon) \cap A = \emptyset$. Let $N \in \mathbb{N}$ be such that $N > 3$ and $\frac{1}{2^{N-1}} < \varepsilon$. We consider the chains

$$\mathcal{U}_N = \{U_1^{(N)}, U_2^{(N)}, \ldots, U_{m_N}^{(N)}\} \text{ and } \mathcal{U}_{N+1} = \{U_1^{(N+1)}, U_2^{(N+1)}, \ldots, U_{m_{N+1}}^{(N+1)}\}.$$

Since $X = A \cup B \subset U_1^{(N+1)} \cup \cdots \cup U_{m_{N+1}}^{(N+1)}$, there are $r, s \in \{1, \ldots, m_{N+1}\}$ such that

$$a \in U_r^{(N+1)} \text{ and } b \in U_s^{(N+1)}.$$

Let $i, j \in \{1, \ldots, m_N\}$ be such that $U_r^{(N+1)} \subset U_i^{(N)}$ and $U_s^{(N+1)} \subset U_j^{(N)}$. Then $a \in U_i^{(N)}$ and $b \in U_j^{(N)}$.

Suppose for example that $i \leq j$. Consider the subchain $\{U_i^{(N)}, \ldots, U_j^{(N)}\}$. We know that $U_i^{(N)} \cap U_{i+1}^{(N)} \neq \emptyset$ and the diameters of the sets $U_i^{(N)}$ and $U_{i+1}^{(N)}$ are less than $\frac{1}{2^N}$. Thus, diameter$(U_i^{(N)} \cup U_{i+1}^{(N)}) < \varepsilon$ and $U_i^{(N)} \cup U_{i+1}^{(N)} \subset B\,(a, \varepsilon)$. Similarly we have that $U_{j-1}^{(N)} \cup U_j^{(N)} \subset B\,(b, \varepsilon)$. Hence $U_i^{(N)}, U_{i+1}^{(N)} \notin \{U_{j-1}^{(N)}, U_j^{(N)}\}$. This implies that $i < i + 1 < j - 1 < j$. Therefore $3 \leq j - i$.

We may assume that $r < s$. The opposite case is similar.

Since $\mathcal{U}_{N+1}$ is crooked in $\mathcal{U}_N$, there exist $u, v \in \{1, \ldots, m_{N+1}\}$ such that

$$r < u < v < s,$$

$U_u^{(N+1)} \subset U_{j-1}^{(N)}$ and $U_v^{(N+1)} \subset U_{i+1}^{(N)}$. Then $U_v^{(N+1)} \subset B\,(a, \varepsilon)$ and $U_u^{(N+1)} \subset B\,(b, \varepsilon)$. Thus $U_v^{(N+1)} \cap B = \emptyset$ and $U_u^{(N+1)} \cap A = \emptyset$.

Since $X \cap U_r^{(N+1)} \neq \emptyset$, $X \cap U_s^{(N+1)} \neq \emptyset$, X is connected and $r < v < s$, we have that $X \cap U_v^{(N+1)} \neq \emptyset$. Fix a point $p \in X \cap U_v^{(N+1)}$. Then $p \notin B$, which implies that $p \in A$. Hence $A \cap U_v^{(N+1)} \neq \emptyset$. Thus, $A \cap U_v^{(N+1)} \neq \emptyset$, $r < u < v$

and A is connected. This implies that $A \cap U_u^{(N+1)} \neq \emptyset$. We observed before that this does not hold. We have obtained a contradiction. Therefore, P is hereditarily indecomposable. □

1.5 Decompositions

In this section we discuss another way to construct continua. Suppose that we have a continuum X, and a partition $\mathcal{D}$ of nonempty closed subsets of X. Such a partition is called a *decomposition* of X. In this situation, we can consider the quotient space $X/\mathcal{D}$ which is obtained by shrinking each element of $\mathcal{D}$ to a one-point set. Equivalently, the space $X/\mathcal{D}$ is the quotient space of the equivalence relation defined by: p is equivalent to q if and only if both points belong to the same element of $\mathcal{D}$.

We denote the quotient mapping by $\pi : X \to \mathcal{D}$. So, $X/\mathcal{D}$ has the identification topology given by: $\mathcal{U} \subset X/\mathcal{D}$ is open if and only if $\pi^{-1}(\mathcal{U})$ is open in X. As usual, a subset R of X is *$\mathcal{D}$-saturated* if it is the union of some elements of $\mathcal{D}$. Notice that the elements of $\mathcal{D}$ are $\mathcal{D}$-saturated.

Since π is onto, we have that $X/\mathcal{D}$ is compact. By Exercise 1.23, $X/\mathcal{D}$ is a continuum if and only if $X/\mathcal{D}$ is a Hausdorff space. The following theorem gives a necessary and sufficient condition in order that $X/\mathcal{D}$ be a Hausdorff space.

Theorem 1.14 *Let $\mathcal{D}$ be a decomposition of a continuum X. Then $X/\mathcal{D}$ is a continuum if and only if for each pair of distinct elements A, B of $\mathcal{D}$, there exist two disjoint $\mathcal{D}$-saturated open subsets U and V such that $A \subset U$ and $B \subset V$.*

Proof (Necessity) Let A and B be two distinct elements of $\mathcal{D}$. Then $\pi(A)$ and $\pi(B)$ are distinct one-point sets of $X/\mathcal{D}$. Since $X/\mathcal{D}$ is a Hausdorff space, there exist two disjoint open subsets W and Z of $X/\mathcal{D}$ such that $\pi(A) \in W$ and $\pi(B) \in Z$. Let $U = \pi^{-1}(W)$ and $V = \pi^{-1}(Z)$. Then U and V are disjoint open subsets of X, $A \subset U$ and $B \subset V$. By Exercise 1.40, U and V are $\mathcal{D}$-saturated.

(Sufficiency) By Exercise 1.23, we only have to show that $X/\mathcal{D}$ is a Hausdorff space. Let $\pi(A)$ and $\pi(B)$ be two distinct elements of $X/\mathcal{D}$, where $A, B \in \mathcal{D}$. By hypothesis, there exist two disjoint, $\mathcal{D}$-saturated open subsets U and V of X such that $A \subset U$ and $B \subset V$. By Exercise 1.40, $U = \pi^{-1}(\pi(U))$. Since $X/\mathcal{D}$ is endowed with the quotient topology, we have that $\pi(U)$ is open in $X/\mathcal{D}$. Similarly, $\pi(V)$ is open in $X/\mathcal{D}$. Notice that $\pi(A) \in \pi(U)$ and $\pi(B) \in \pi(V)$. Finally, if there exists an element $p \in \pi(U) \cap \pi(V)$, then there exist points $u \in U$ and $v \in V$ such that $\pi(u) = p = \pi(v)$. Then $v \in \pi^{-1}(\pi(u)) = U$. Thus $v \in U \cap V$, a contradiction. This proves that $\pi(U) \cap \pi(V) = \emptyset$. Therefore, $X/\mathcal{D}$ is a Hausdorff space. □

Given a continuum X, the decompositions $\mathcal{D}$ of X for which $X/\mathcal{D}$ is a Hausdorff space are called *upper semi-continuous decompositions of* X. In Theorem 9.7 we will reveal another way to see these decompositions.

1.6 The Brouwer Reduction Theorem

In many fields of mathematics, there are numerous existence theorems that can be proved using Zorn's Lemma. In Continuum Theory, the use of this lemma seems to be excessive because for continua it is possible to use the Brouwer Reduction Theorem; this result asserts the existence of minimal elements, working only with countable families. In Exercise 1.51 we ask the reader to prove the version of the Brouwer Reduction Theorem for the existence of maximal elements.

Theorem 1.15 (Brouwer Reduction Theorem) *Let Z be a second countable space. Let $\mathcal{A}$ be a nonempty family of closed subsets of X with the following property: for each sequence $\{A_n\}_{n=1}^{\infty}$ in $\mathcal{A}$ such that $A_1 \supset A_2 \supset \cdots$, there exists an $A \in \mathcal{A}$ such that $A \subset A_n$ for every $n \in \mathbb{N}$.*

Then there exists an $A \in \mathcal{A}$ which is minimal with respect to inclusion.

Proof We proceed by contradiction, so suppose that $\mathcal{A}$ contains no minimal elements.

Let $\{U_n : n \in \mathbb{N}\}$ be a countable basis of nonempty open subsets of Z.

Take an element $A_1 \in \mathcal{A}$. Since A_1 is not a minimal element of $\mathcal{A}$, there exists an $A \in \mathcal{A}$ such that $A \subsetneq A_1$. Take a point $p \in A_1 \setminus A$ and $r \in \mathbb{N}$ such that $p \in U_r$ and $U_r \cap A = \emptyset$. Then it is possible to define

$$m_1 = \min\{n \in \mathbb{N} : \text{there exists a } B \in \mathcal{A} \text{ such that } B \subset A_1,$$
$$U_n \cap A_1 \neq \emptyset \text{ and } U_n \cap B = \emptyset\}.$$

Let $A_2 \in \mathcal{A}$ be such that $A_2 \subset A_1$, $U_{m_1} \cap A_1 \neq \emptyset$ and $A_2 \cap U_{m_1} = \emptyset$. Since A_2 is not a minimal element of $\mathcal{A}$, there exists an $A \in \mathcal{A}$ such that $A \subsetneq A_2$. Proceeding as before, it is possible to define

$$m_2 = \min\{n \in \mathbb{N} : \text{there exists a } B \in \mathcal{A} \text{ such that } B \subset A_2,$$
$$U_n \cap A_2 \neq \emptyset \text{ and } U_n \cap B = \emptyset\}.$$

Let $A_3 \in \mathcal{A}$ be such that $A_3 \subset A_2$, $U_{m_2} \cap A_2 \neq \emptyset$ and $A_3 \cap U_{m_2} = \emptyset$.

Continuing in this way, it is possible to define sequences $\{m_k\}_{k=1}^{\infty}$, $\{A_k\}_{k=1}^{\infty}$ and $\{U_{m_k}\}_{k=1}^{\infty}$ of positive integers, elements of $\mathcal{A}$ and open subsets of X, respectively, such that for each $k \in \mathbb{N}$,

$$m_k = \min\{n \in \mathbb{N} : \text{there exists a } B \in \mathcal{A} \text{ such that } B \subset A_k,$$
$$U_n \cap A_k \neq \emptyset \text{ and } U_n \cap B = \emptyset\},$$

$A_{k+1} \subset A_k$, $U_{m_k} \cap A_k \neq \emptyset$ and $A_{k+1} \cap U_{m_k} = \emptyset$.

Given $k \in \mathbb{N}$, since $A_{k+2} \subset A_{k+1} \subset A_k$, $\emptyset \neq U_{m_{k+1}} \cap A_{k+1} \subset U_{m_{k+1}} \cap A_k$ and $U_{m_{k+1}} \cap A_{k+2} = \emptyset$, we have that m_{k+1} belongs to the set for which m_k is the minimum, so $m_k \leq m_{k+1}$. Since $A_{k+1} \cap U_{m_k} = \emptyset$ and $A_{k+1} \cap U_{m_{k+1}} \neq \emptyset$, we have that $m_k < m_{k+1}$. Therefore, the sequence $\{m_k\}_{k=1}^{\infty}$ is not bounded.

By hypothesis, there exists a $B_0 \in \mathcal{A}$ such that $B_0 \subset A_n$ for every $n \in \mathbb{N}$. Since B_0 is not a minimal element of $\mathcal{A}$, there exists an $A_0 \in \mathcal{A}$ such that $A_0 \subsetneq B_0$. Let $q \in B_0 \setminus A_0$ and $n \in \mathbb{N}$ be such that $q \in U_n$ and $A_0 \cap U_n = \emptyset$. Given $k \in \mathbb{N}$, by the definition of m_k, since $A_0 \subsetneq A_k$, $q \in U_n \cap A_k$ and $U_n \cap A_0 = \emptyset$, we have that $m_k \leq n$. This contradicts what we proved in the previous paragraph and finishes the proof that $\mathcal{A}$ has a minimal element. □

1.7 Exercises

Exercise 1.16 The finite graph in Fig. 1.12 does not contain any subgraph isomorphic (as a graph) to either of the graphs K_5 or $K_{3,3}$. However it contains a topological copy of one of them.

Exercise 1.17 Let $d_{\mathbf{Q}}$ be the metric defined on the Hilbert cube $\mathbf{Q}$ by the Eq. (1.1). Prove that the topology in $\mathbf{Q}$ induced by this metric is the same as the product topology.

Exercise 1.18 Every metric space admits a compatible metric bounded by 1.

Exercise 1.19 Every compact metric space contains a countable dense subset.

Exercise 1.20 Let X be a compact metric space, and let $\{A_n\}_{n=1}^{\infty}$ be a sequence of nonempty closed subsets of X such that $A_1 \supset A_2 \supset \cdots$. Prove that $A = A_1 \cap A_2 \cap \cdots$ is nonempty.

Exercise 1.21 Find a metric space which cannot be embedded in the Hilbert cube.

Exercise 1.22 Let X be a compact Hausdorff space with a countable basis. Prove that X can be embedded in the Hilbert cube. (Hint: use Urysohn's Lemma.)

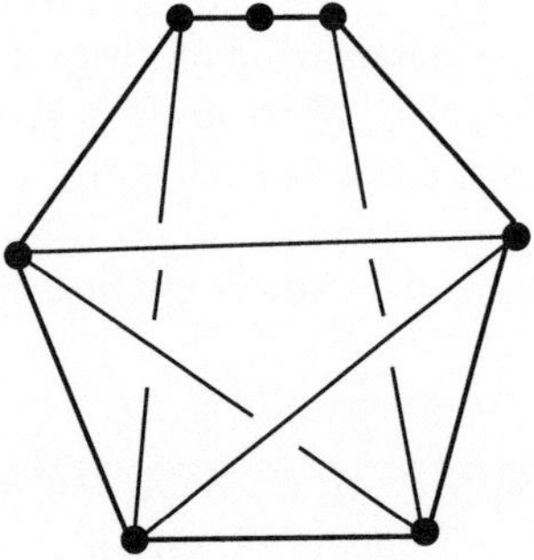

Fig. 1.12 A non-planar graph

Exercise 1.23 Let X be a continuum and let $f : X \to Y$ be an onto mapping, where Y is a Hausdorff space. Prove that Y is a continuum. (Hint: use a countable basis for X to construct a countable basis for Y.)

Exercise 1.24 Suppose that $X = \prod_{j \in J}[0, 1]_j$ and J is uncountable. Show that X is not metrizable. (Hint: let p be the point in X having all its coordinates equal to 0. Prove that there does not exist a sequence of open subsets $U_1, U_2, \ldots$ of X such that $\{p\} = U_1 \cap U_2 \cap \cdots$.)

Exercise 1.25 In a metric compact space, every sequence has a convergent subsequence.

Exercise 1.26 Let X be a metric space such that every sequence in X has a convergent subsequence in X. Prove that:

(a) for each $\varepsilon > 0$, the open cover $\{B(p, \varepsilon) : p \in X\}$ has a finite subcover.
(b) if $\mathcal{U}$ is an open cover of X, then there exists an $\varepsilon > 0$ (called a *Lebesgue number*) such that for each $p \in X$, there exists a $U \in \mathcal{U}$ such that $B(p, \varepsilon) \subset U$.
(c) X is compact.

Exercise 1.27 Let Z be a compact metric space, with metric d_Z. Then Z is connected if and only if for every $p, q \in Z$ and each $\varepsilon > 0$, there exists a finite sequence $z_0, z_1, \ldots, z_m$ in Z such that $z_0 = p$, $z_m = q$ and $d_Z(z_{i-1}, z_i) < \varepsilon$ for each $i \in \{1, \ldots, m\}$.

Exercise 1.28 Let X be a continuum. Suppose that $\mathcal{A}$ is a nonempty family of subcontinua of X with the property that for each finite number $A_1, \ldots, A_n$ of elements of $\mathcal{A}$, there exists an $A \in \mathcal{A}$ such that $A \subset A_1 \cap \cdots \cap A_n$. Prove that $\bigcap\{A : A \in \mathcal{A}\}$ is a subcontinuum of X.

Exercise 1.29 Let X be a compact metric space and let $f : X \to Y$ be a mapping, where Y is a metric space. Show that f is uniformly continuous. (That is, for each $\varepsilon > 0$, there exists a $\delta > 0$ such that if $p, x \in X$ and $d_X(p, x) < \delta$, then $d_Y(f(p), f(x)) < \varepsilon$, where d_X and d_Y are metrics for X and Y, respectively.)

Exercise 1.30 Let X be a compactification of $[0, \infty)$, with remainder R. Prove the following properties:

- if $0 \leq a < b$, then $[0, b)$ and (a, b) are open in X (Hint: let W be open in X such that $W \cap [0, \infty) = (a, b)$. Using that $[a, b]$ is compact, prove that $W = (a, b)$.);
- if $0 < a$, then $X \setminus \{a\}$ is disconnected and $[a, \infty) \cup R$ is homeomorphic to X;
- if A is a subcontinuum of X and $A \cap R \neq \emptyset$, then $A \subset R$ or $R \subset A$;
- if A is a subcontinuum of X, then A is of one of the following forms: (i) A is a subcontinuum of R, (ii) $A = [a, b]$, for some $0 \leq a \leq b \leq \infty$, or (iii) $A = [a, \infty) \cup R$ for some $a \in [0, \infty)$;
- there exists an onto mapping $f : X \to [0, 1]$ such that $f^{-1}(t)$ is a subcontinuum of X, for every $t \in [0, 1]$;
- if R is a one-point set, then X is an arc;

- if R is non-degenerate, then X is not pathwise connected (Hint: use Exercise 1.29.).

Exercise 1.31 Show that there exist two non-homeomorphic compactifications of $[0, \infty)$, each with an arc as remainder.

Exercise 1.32 Show that there exists a compactification of $[0, \infty)$ with remainder $[0, 1]^2$.

Exercise 1.33 Let R be a continuum. Prove that there exists a compactification of $[0, \infty)$ with remainder R. (Hint: by Theorem 1.1, we may assume that $R \subset \mathbf{Q}$. Show that for each $n \in \mathbb{N}$, $N(R, \frac{1}{n})$ is pathwise connected and construct X in $\mathbf{Q} \times [0, 1]$.)

Exercise 1.34 Show that the interval $[0, 1]$, the topologist's curve and the buckethandle continuum are chainable.

Exercise 1.35 Prove that the number of components of a topological space Z is greater than or equal to n if and only if there exist nonempty subsets $E_1, \ldots, E_n$ of Z such that $Z = E_1 \cup \cdots \cup E_n$ and if $i \neq j$, then $E_i \cap \mathrm{cl}_Z(E_j) = \emptyset$.

Exercise 1.36 Let X be a chainable continuum and let A, B and C be subcontinua of X such that $A \cap B \cap C \neq \emptyset$. Show that one of the sets A, B or C is contained in the union of the other two.

Exercise 1.37 Suppose that X is a chainable continuum. Let A,B be subcontinua of X such that $A \subset B$. Prove that $B \setminus A$ has at most two components. (Hint: use Exercises 1.35, 1.36 and 4.12.)

Exercise 1.38 Prove that the dyadic solenoid has a basis of neighborhoods $\mathcal{B}$ such that each element of $\mathcal{B}$ is homeomorphic to the product of the Cantor set and the interval $[0, 1]$.

Exercise 1.39 Give geometric arguments to show that the dyadic solenoid and the buckethandle continuum are indecomposable.

Exercise 1.40 Let $\mathcal{D}$ be a decomposition of a continuum X. Show that a subset R of X is $\mathcal{D}$-saturated if and only if $R = \pi^{-1}(\pi(R))$.

Exercise 1.41 Find a decomposition $\mathcal{D}$ of the interval $[0, 1]$ such that $[0, 1]/\mathcal{D}$ is not a Hausdorff space.

Exercise 1.42 Let $\mathcal{D}$ be a decomposition of a continuum X. Prove that if $\mathcal{D}$ has only a finite number of non-degenerate elements, then $X/\mathcal{D}$ is a continuum.

Exercise 1.43 Prove that the cone and the suspension of a continuum are continua.

Exercise 1.44 Show that the cone over the topological space $\{\frac{1}{n} : n \in \mathbb{N}\}$ is not metrizable and therefore is not homeomorphic to the set $\bigcup\{(\frac{1}{2}, 1)(\frac{1}{n}, 0) : n \in \mathbb{N}\}$, where for each $n \in \mathbb{N}$, $(\frac{1}{2}, 1)(\frac{1}{n}, 0)$ denotes the convex segment joining the point $(\frac{1}{2}, 1)$ and $(\frac{1}{n}, 0)$ in the plane. (Hint: show that there does not exist a local countable basis of neighborhoods at the vertex of the cone.)

Exercise 1.45 Let X be a compact metric space. Let X' be a subspace of the Hilbert cube $\mathbf{Q}$ with X' homeomorphic to X. Let $\theta = (0, 0, \ldots) \in \mathbf{Q}$. Define the *geometric cone* of X by $\bigcup\{(\theta, 1)(p, 0) \subset \mathbf{Q} \times [0, 1] : p \in X'\}$, where $(\theta, 1)(p, 0)$ is the convex segment in $\mathbf{Q} \times [0, 1]$ joining the points $(\theta, 1)$ and $(p, 0)$. Prove that the geometric cone of X is homeomorphic to the (topological) cone over the space X. (Hint: use the Transgression Theorem [33, Theorem 3.2, Chapter VI].)

Exercise 1.46 Let X be a continuum and let $f : X \to X$ be a mapping. Prove that there exists a minimal subcontinuum A of X such that $A = f(A)$. (Hint: for each $n \in \mathbb{N}$, let $B_n = f^n(X) = (f \circ \cdots \circ f)(X)$, n times, and let $B = B_1 \cap B_2 \cap \cdots$, and prove that $f(B) = B$.)

Exercise 1.47 Let X be a continuum and let B be a nonempty subset of X. Prove that there exists a minimal subcontinuum A of X such that $B \subset A$.

Exercise 1.48 Let X be a continuum and let B and C be nonempty closed subsets of X. Prove that there exists a minimal subcontinuum A of X such that $A \cap B \neq \emptyset$ and $A \cap C \neq \emptyset$.

Exercise 1.49 Let Z be a connected space such that $Z = A_1 \cup \cdots \cup A_n$, where the sets $A_1, \ldots, A_n$ are nonempty and closed in Z. Prove that there exists a permutation $\sigma : \{1, \ldots, n\} \to \{1, \ldots, n\}$ such that $A_{\sigma(1)} \cap A_{\sigma(2)} \neq \emptyset$; $(A_{\sigma(1)} \cup A_{\sigma(2)}) \cap A_{\sigma(3)} \neq \emptyset$; $(A_{\sigma(1)} \cup A_{\sigma(2)} \cup A_{\sigma(3)}) \cap A_{\sigma(4)} \neq \emptyset$; etc.

Exercise 1.50 Let $f : X \to Y$ be a function between continua. Prove that f is continuous if and only if given a sequence $\{x_n\}_{n=1}^{\infty}$ in X satisfying $\lim_{n\to\infty} x_n = x \in X$ and $\lim_{n\to\infty} f(x_n) = y \in Y$, we have that $f(x) = y$.

Exercise 1.51 Prove the Brouwer Reduction Theorem for the existence of maximal elements. That is: suppose that Z is a second countable space. Suppose that $\mathcal{A}$ is a nonempty family of closed subsets of Z with the following property:

For each sequence $\{A_n\}_{n=1}^{\infty}$ of elements of $\mathcal{A}$ such that $A_1 \subset A_2 \subset \cdots$, there exists an $A \in \mathcal{A}$ such that $A_n \subset A$ for every $n \in \mathbb{N}$. Then there exists an $A \in \mathcal{A}$ which is maximal with respect to inclusion.

Exercise 1.52 (Baire Category Theorem) Suppose that Z is a complete metric space. Prove that Z cannot be a countable union of closed subsets with empty interior. Conclude that the Cantor set has uncountably many points.

Exercise 1.53 Let X be a continuum, $\varepsilon > 0$ and $f : X \to [0, 1]$ an onto mapping such that for each $t \in [0, 1]$, diameter$(f^{-1}(t)) < \varepsilon$. Prove that there exists a $\delta > 0$ such that if $J \subset [0, 1]$ and diameter$(J) < \delta$, then diameter$(f^{-1}(J)) < \varepsilon$.

Exercise 1.54 Suppose that X is a compact metric space, U is an open subset of X, $0 \le a \le b \le 1$ and $f : X \to [0, 1]$ is a mapping. Suppose that $f^{-1}([a, b]) \subset U$. Then there exists a $\delta > 0$ such that $f^{-1}([a - \delta, b + \delta]) \subset U$.

Exercise 1.55 Prove Urysohn's Lemma for metric spaces, that is, suppose that X is a metric space and A, B are nonempty disjoint closed subsets of X, then prove that there exists a mapping $f : X \to [0, 1]$ such that $A = f^{-1}(0)$ and $B = f^{-1}(1)$. (Hint: consider the function $f(p) = \frac{\text{dist}(p, A)}{\text{dist}(p, A)+\text{dist}(p, B)}$).

Chapter 2
Locally Connected Continua

A topological space X is *locally connected at a point* $p \in X$ if for every open subset U of X with $p \in U$, there exists an open connected subset V of X such that $p \in V \subset U$. The space X is *locally connected* or *Peano* if it is locally connected at p for each $p \in X$.

A topological space X is *connected im kleinen* at a point p in X if for every open subset U of X with $p \in U$, there exists a connected subset M of X such that $p \in \text{int}_X(M) \subset M \subset U$. The space X is *connected im kleinen* if it is connected im kleinen at each $p \in X$.

Clearly, if a space X is locally connected at a point p, then X is connected im kleinen at p. The converse implication does not hold (see Exercise 2.12).

Recall that a space X is *arcwise connected* if for every pair of distinct points p and q of X, there exists an arc joining p and q in X, and recall that X is *pathwise connected* provided that for each pair of points p and q of X, there exists a mapping $g : [0, 1] \to X$ such that $g(0) = p$ and $g(1) = q$.

As we will see in this chapter, Peano continua have deep properties, for instance they are images of the interval $[0, 1]$, they are arcwise connected and their open connected subsets are arcwise connected. There are several ways to prove these properties. For example, we can use the results of this chapter to prove the characterization of the arcs in Chap. 5. On the other hand, it is possible to first prove the characterization of the arcs and use this to show that Peano continua are arcwise connected, etc.

The proofs of this chapter are based in the concept of property S (defined by W. Sierpiński [119]).

A. Illanes, *Continuum Theory*, Universitext,
https://doi.org/10.1007/978-3-031-91011-1_2

2.1 Property S

Definition 2.1 Let (X, d) be a metric space. A nonempty subspace Y of X has property S if for each $\varepsilon > 0$, there exist $n \in \mathbb{N}$ and connected subsets $A_1, \ldots, A_n$ of Y such that $Y = A_1 \cup \cdots \cup A_n$ and diameter$(A_i) < \varepsilon$ for every $i \in \{1, \ldots, n\}$.

Definition 2.2 Given a metric space X and $\varepsilon > 0$, an $S(\varepsilon)$-chain is a finite sequence $\mathcal{A} = \{A_1, \ldots, A_n\}$ of subsets of X satisfying the following.

- $A_i \cap A_{i+1} \neq \emptyset$ for each $i \in \{1, \ldots, n-1\}$,
- A_i is connected for each $i \in \{1, \ldots, n\}$,
- diameter$(A_i) < \frac{\varepsilon}{2^i}$ for each $i \in \{1, \ldots, n\}$.

When $\mathcal{A} = \{A_1, \ldots, A_n\}$ is an $S(\varepsilon)$-chain, the sets A_i are the *links* of $\mathcal{A}$. If $p \in A_1$ and $q \in A_n$, we say that $\mathcal{A}$ is a *$S(\varepsilon)$-chain from p to q*.

Given $A \subset X$, define $S(A, \varepsilon)$ as:

$$S(A, \varepsilon) = \{q \in X : \text{ there exists an } S(\varepsilon) - \text{chain from some point in } A \text{ to } q\}.$$

Theorem 2.3 *Let X be a metric space having property S. If A is a nonempty subset of X and $\varepsilon > 0$, then the set $S(A, \varepsilon)$ has property S.*

Proof Let $\delta > 0$. Let $m \in \mathbb{N}$ be such that $m \geq 4$ and $\frac{\varepsilon}{2^{m-3}} < \delta$.

Let

$$B = \{q \in S(A, \varepsilon) : \text{there exists an } S(\varepsilon) - \text{chain, with at most } m \text{ links from some point in } A \text{ to } q\}.$$

Given $x \in A$, $\{\{x\}\}$ is an $S(\varepsilon)$-chain from x to x. This shows that $A \subset B$.

Since X has property S, there exists a finite cover C of X such that each element of C is connected and has diameter less than $\frac{\varepsilon}{2^{m+1}}$. Let $C_1, \ldots, C_n$ be the elements of C intersecting B. Since C covers X, $B \subset C_1 \cup \cdots \cup C_n$. Let $i \in \{1, \ldots, n\}$. We claim that $C_i \subset S(A, \varepsilon)$. By the choice of C_i, we can choose a point $p \in C_i \cap B$. Thus there exists a point $a \in A$ and an $S(\varepsilon)$-chain $\{D_1, \ldots, D_k\}$ from a to p, where $k \leq m$. Given $x \in C_i$, since diameter$(C_i) < \frac{\varepsilon}{2^{m+1}} \leq \frac{\varepsilon}{2^{k+1}}$, we have that $\{D_1, \ldots, D_k, C_i\}$ is an $S(\varepsilon)$-chain from a to x. Hence $x \in S(A, \varepsilon)$. This proves that $C_i \subset S(A, \varepsilon)$.

Given $i \in \{1, \ldots, n\}$, let

$$\mathcal{F}_i = \{G \subset X : G \subset S(A, \varepsilon),\ G \cap C_i \neq \emptyset, G \text{ is connected and diameter}(G) < \frac{\delta}{4}\}.$$

Let $F_i = \bigcup\{G : G \in \mathcal{F}_i\}$.

We claim that $S(A, \varepsilon) = F_1 \cup \cdots \cup F_n$ and for each $i \in \{1, \ldots, n\}$, F_i is connected and diameter$(F_i) < \delta$. This will finish the proof that $S(A, \varepsilon)$ has property S.

Let $i \in \{1, \ldots, n\}$. We know that $C_i \subset S(A, \varepsilon)$. Given $p \in C_i$, taking $G = \{p\}$, we can conclude that $\{p\} \in \mathcal{F}_i$. We have shown that $C_i \subset F_i$.

Since C_i is connected and each element of $\mathcal{F}_i$ is connected and intersects C_i, we obtain that F_i is connected.

Since C_i has diameter less than $\frac{\varepsilon}{2^{m+1}} < \frac{\delta}{4}$ and the elements of $\mathcal{F}_i$ also have diameter less than $\frac{\delta}{4}$, we obtain that F_i has diameter less than $\frac{3}{4}\delta < \delta$.

The elements of $\mathcal{F}_i$ are contained in $S(A, \varepsilon)$, so $F_1 \cup \cdots \cup F_n \subset S(A, \varepsilon)$.

In order to show the other inclusion, take $p \in S(A, \varepsilon)$. Since $B \subset C_1 \cup \cdots \cup C_n \subset F_1 \cup \cdots \cup F_n$, we may assume that $p \notin B$. Since $p \in S(A, \varepsilon)$, there exists an $S(\varepsilon)$-chain $\mathcal{G} = \{G_1, \ldots, G_k\}$ from a point a in A to p. Since $p \notin B$, we have $m < k$. Set $H = G_m \cup \cdots \cup G_k$. By the definition of B, we have $G_m \subset B$, so $G_m \cap C_i \neq \emptyset$ for some $i \in \{1, \ldots, n\}$.

We check that $H \in \mathcal{F}_i$. By definition of $S(A, \varepsilon)$, we have $G_1 \cup \cdots \cup G_k \subset S(A, \varepsilon)$, so $H \subset S(A, \varepsilon)$. Notice that $H \cap C_i \neq \emptyset$ and H is connected. Finally, $\text{diameter}(H) \leq \text{diameter}(G_m) + \cdots + \text{diameter}(G_k) < \frac{\varepsilon}{2^m} + \cdots + \frac{\varepsilon}{2^k} < \frac{\varepsilon}{2^{m-1}} < \frac{\delta}{4}$. This completes the proof that $H \in \mathcal{F}_i$. Therefore, $p \in H \subset F_i$.

We have shown that $S(A, \varepsilon) = F_1 \cup \cdots \cup F_n$. Hence, $S(A, \varepsilon)$ has property S. □

Theorem 2.4 *Let X be a metric space with property S. Then for every $\varepsilon > 0$, X is a finite union of connected subsets with diameter less than ε each having property S. Moreover, these sets can be taken to be open or closed in X.*

Proof Let $\varepsilon > 0$. Since X has property S, there exist $n \in \mathbb{N}$ and connected subsets $A_1, \ldots, A_n$ of X such that $X = A_1 \cup \cdots \cup A_n$ and $\text{diameter}(A_i) < \frac{\varepsilon}{3}$ for each $i \in \{1, \ldots, n\}$. By Exercise 2.19, the sets $S(A_1, \frac{\varepsilon}{3}), \ldots, S(A_n, \frac{\varepsilon}{3})$ are open, connected, have diameter less than ε and cover X, and by Theorem 2.3, they also have property S.

By Exercise 2.17 the sets $\text{cl}_X(S(A_i, \frac{\varepsilon}{3}))$ also have property S. Thus, the sets $\text{cl}_X(S(A_i, \frac{\varepsilon}{3}))$ satisfy properties similar to those of the sets $S(A_i, \frac{\varepsilon}{3})$. □

Theorem 2.5 *If X is a Peano continuum, then for each $\varepsilon > 0$, X is the union of a finite family of Peano subcontinua with diameter less than ε.*

Proof Let $\varepsilon > 0$. By Theorem 2.4, there exist $n \in \mathbb{N}$ and closed connected subsets $A_1, \ldots, A_n$ of X such that $X = A_1 \cup \cdots \cup A_n$, $\text{diameter}(A_i) < \varepsilon$ and A_i has property S for every $i \in \{1, \ldots, n\}$. By Exercise 2.18 each A_i is a Peano continuum. □

2.2 An Auxiliary Mapping

Definition 2.6 Let X be a metric space. A finite family of subsets $\mathcal{A} = \{A_1, \ldots, A_n\}$ is a *weak chain* if for each $i \in \{1, \ldots, n-1\}$, $A_i \cap A_{i+1} \neq \emptyset$. The elements of $\mathcal{A}$ are called *links* and the number n is the *size of* $\mathcal{A}$. In the case when each of the links of $\mathcal{A}$ has diameter less than ε ($\varepsilon > 0$), $\mathcal{A}$ is a *weak ε-chain*. The *first link* (respectively, the *last link*) of $\mathcal{A}$ is A_1 (respectively, A_n) of $\mathcal{A}$. If the points p and q satisfy $p \in A_1$ and $q \in A_n$, then we say that $\mathcal{A}$ *goes from* p *to* q. If $\mathcal{A} = \{A_1, \dots, A_n\}$ and $\mathcal{B} = \{B_1, \dots, B_m\}$ are weak chains and $A_n \cap B_1 \neq \emptyset$, then the weak chain $\{A_1, \dots, A_n, B_1, \dots, B_m\}$ is denoted by $\mathcal{A} \vee \mathcal{B}$. This notion is extended in the natural way to define $\mathcal{B}_1 \vee \mathcal{B}_2 \cdots \vee \mathcal{B}_k$, where $\mathcal{B}_1, \dots, \mathcal{B}_1$ are weak chains.

Suppose that we have real numbers $a_0 < a_1 < \cdots < a_k$ and for each $i \in \{1, \dots, k\}$, we have a partition P_i of the interval $[a_{i-1}, a_i]$. Then $P_1 \sqcup \cdots \sqcup P_k$ denotes the partition of the interval $[a_0, a_k]$ formed by taking the elements of $P_1 \cup \cdots \cup P_k$ as an ordered sequence. A partition of a closed interval $J \subset \mathbb{R}$ is *uniform* if it decomposes J into intervals of the same length.

Theorem 2.7 *Let X be a connected space, $p, q \in X$ and $\mathcal{A}$ a finite cover of nonempty closed subsets of X. Then there exist $n \in \mathbb{N}$ and a weak chain $\mathcal{B} = \{B_1, \dots, B_n\}$ such that $\mathcal{B} \subset \mathcal{A}$, $\mathcal{B}$ covers X and $\mathcal{B}$ goes from p to q (and if we repeat B_n we get as many links as we wish).*

Proof Take an element $A_1 \in \mathcal{A}$ such that $p \in A_1$. Take the weak chain $\{A_1, \dots, A_m\}$ of elements of $\mathcal{A}$ such that the cardinality r of the set $\mathcal{A}_0 = \{A_i : i \in \{1, \dots, m\}\}$ reaches its maximum, in other words, r is the maximum number of pairwise distinct elements of $\mathcal{A}$ (observe that $\mathcal{A}$ is a set) that can be used to construct a weak chain. We denote by $\mathcal{A}_0$ the set of elements of $\mathcal{A}$ that we are using to construct the weak chain. Observe that the weak chain $\{A_1, \dots, A_m\}$ is not unique and $r \leq m$ since some elements can be repeated in the sequence $\{A_1, \dots, A_m\}$.

In the case when $\mathcal{A}_0 \neq \mathcal{A}$, let $D = \bigcup\{A : A \in \mathcal{A}_0\}$ and $E = \bigcup\{A : A \in \mathcal{A} \setminus \mathcal{A}_0\}$. Then D and E are nonempty closed subsets of X such that $X = D \cup E$. By the connectedness of X, $D \cap E \neq \emptyset$. Then there exist $A \in \mathcal{A} \setminus \mathcal{A}_0$ and $i \in \{1, \dots, m\}$ such that $A \cap A_i \neq \emptyset$. Hence the sequence $\{A_1, \dots, A_{i-1}, A_i, A, A_i, A_{i+1}, \dots, A_m\}$ is a weak chain and the set $\mathcal{A}_0 \cup \{A\}$ has $r + 1$ elements. This contradicts the choice of r and proves that $\mathcal{A}_0 = \mathcal{A}$. Therefore $\{A_1, \dots, A_m\}$ is a weak chain of elements of $\mathcal{A}$ whose elements cover X.

Choose $j \in \{1, \dots, m\}$ such that $q \in A_j$ and define

$$\mathcal{B} = \{A_1, \dots, A_m, A_{m-1}, \dots, A_j\}.$$

Clearly, $\mathcal{B}$ satisfies the required conditions. □

Theorem 2.8 *Let X be a continuum and $p, q \in X$. Let $\{\mathcal{B}_m\}_{m=1}^{\infty}$ be a sequence of weak chains of subcontinua of X such that for each $m \in \mathbb{N}$ the following properties hold.*

1. *The chain $\mathcal{B}_m = \{B_1^{(m)}, \dots, B_{k_m}^{(m)}\}$ goes from p to q,*
2. *$k_1 \geq 2$,*
3. *$\mathcal{B}_m$ is a weak $\frac{1}{2^m}$-chain,*
4. *$\mathcal{B}_{m+1} = \mathcal{C}_1^{(m+1)} \vee \cdots \vee \mathcal{C}_{k_m}^{(m+1)}$, where for each $i \in \{1, \dots, k_m\}$, $\mathcal{C}_i^{(m+1)}$ is a weak chain with size ≥ 2 and $\bigcup \mathcal{C}_i^{(m+1)} \subset B_i^{(m)}$.*

Then there exists an onto mapping $f : [0, 1] \to \bigcap\{\bigcup \mathcal{B}_m : m \in \mathbb{N}\}$ such that $f(0) = p$ and $f(1) = q$. Moreover, if each $\mathcal{B}_m$ is a chain, then f is one-to-one.

Proof Let d be a metric for X. For each $m \in \mathbb{N}$, we consider a partition

$$P_m : 0 = t_0^{(m)} < t_1^{(m)} < \cdots < t_{k_m}^{(m)} = 1$$

of the interval $[0, 1]$, satisfying the following properties.

(a) P_1 is a uniform partition of $[0, 1]$ with $k_1 + 1$ elements. That is, $t_0^{(1)} = \frac{0}{k_1}$, $t_1^{(1)} = \frac{1}{k_1}, t_2^{(1)} = \frac{2}{k_1}, \ldots, t_{k_1}^{(1)} = \frac{k_1}{k_1} = 1$;

(b) for each $m \in \mathbb{N}$, $P_{m+1} = Q_1 \sqcup \cdots \sqcup Q_{k_m}$, where for each $i \in \{1, \ldots, k_m\}$, Q_i has as many elements as the size of $\mathcal{C}_i^{(m+1)}$ plus one and it is a uniform partition of the interval $[t_{i-1}^{(m)}, t_i^{(m)}]$. In particular, P_{m+1} is a refinement of P_m.

It is easy to construct the partitions P_m inductively: (a) gives partition P_1. Once we have partition P_m, for each $i \in \{1, \ldots, k_m\}$ we have the interval $[t_{i-1}^{(m)}, t_i^{(m)}]$ and the size of $\mathcal{C}_i^{(m+1)}$ is given, so we can construct the uniform partition Q_i.

Since $k_1 \geq 2$, P_1 has at least three points, so the intervals determined by P_1 have length at most $\frac{1}{2}$. Inductively, suppose that the intervals determined by P_m have length at most $\frac{1}{2^m}$. Observe that each interval J determined by the partition P_{m+1} is an interval determined for some Q_i, so $J \subset [t_{i-1}^{(m)}, t_i^{(m)}]$. Since $|t_i^{(m)} - t_{i-1}^{(m)}| \leq \frac{1}{2^m}$ and the fact that $\mathcal{C}_i^{(m+1)}$ has size at least 2 implies that Q_i has at least 3 elements, the length of J is at most $\frac{1}{2^{m+1}}$. This completes the induction. Observe that we have shown that each interval determined by the partition P_{m+1} has length at most $\frac{1}{2^{m+1}}$ and it is contained in one interval determined by P_m.

Claim. Suppose that $i \in \{1, \ldots, k_m\}$ and $j \in \{1, \ldots, k_{m+1}\}$ are such that $[t_{j-1}^{(m+1)}, t_j^{(m+1)}] \subset [t_{i-1}^{(m)}, t_i^{(m)}]$. Then $B_j^{(m+1)} \subset B_i^{(m)}$.

We prove this claim. First, we consider the case $i = 1$. Suppose that $\mathcal{C}_1^{(m+1)} = \{C_1, \ldots, C_k\}$. By (b), Q_1 is a uniform partition of the interval $[t_0^{(m)}, t_1^{(m)}]$ with $k+1$ elements. Thus

$$[t_0^{(m)}, t_1^{(m)}] = [t_0^{(m+1)}, t_1^{(m+1)}] \cup \cdots \cup [t_{k-1}^{(m+1)}, t_k^{(m+1)}].$$

So $j \in \{1, \ldots, k\}$. By 4, the first k elements of $\mathcal{B}_{m+1}$ are $C_1, \ldots, C_k$. Therefore $B_1^{(m+1)} = C_1, \ldots, B_k^{(m+1)} = C_k$ and $B_j^{(m+1)} = C_j \in \mathcal{C}_1^{m+1}$. By 4, $B_j^{(m+1)} \subset B_1^{(m)} = B_i^{(m)}$. This finishes the proof of the case $i = 1$.

Now, we consider the case $i = 2$. Suppose $\mathcal{C}_2^{(m+1)} = \{D_1, \ldots, D_l\}$. By (b), Q_2 is a uniform partition of the interval $[t_1^{(m)}, t_2^{(m)}]$ with $l + 1$ elements. Thus

$$[t_1^{(m)}, t_2^{(m)}] = [t_k^{(m+1)}, t_{k+1}^{(m+1)}] \cup \cdots \cup [t_{k+l-1}^{(m+1)}, t_{k+l}^{(m+1)}].$$

So $j \in \{k+1, \ldots, k+l\}$. By 4, the elements of $\mathcal{B}_{m+1}$ between $B_{k+1}^{(m+1)}$ and $B_{k+l}^{(m+1)}$ are $D_1, \ldots, D_l$. Therefore $B_{k+1}^{(m+1)} = D_1, \ldots, B_{k+l}^{(m+1)} = D_l$ and $B_j^{(m+1)} = D_j \in C_2^{m+1}$. By 4, $B_j^{(m+1)} \subset B_2^{(m)} = B_i^{(m)}$. This finishes the proof of the case $i = 2$.

Proceeding in this way, we can prove the claim by a simple induction.

We are ready to define f.

Let $t \in [0, 1]$. For each $m \in \mathbb{N}$, define

$$J(m, t) = \{i \in \{1, \ldots, k_m\} : t \in [t_{i-1}^{(m)}, t_i^{(m)}]\}.$$

Observe that $J(m, t)$ has exactly one or two elements. Define

$$F(m, t) = \bigcup\{B_i^{(m)} : i \in J(m, t)\}.$$

In the case when $J(m, t)$ has two elements, we have $J(m, t) = \{i, i+1\}$ for some $i \in \{1, \ldots, k_m - 1\}$. Then $F(m, t) = B_i^{(m)} \cup B_{i+1}^{(m)}$. Thus $F(m, t)$ is a subcontinuum of X, and by 3, diameter$(F(m, t)) < \frac{2}{2^m}$. In the case when $J(m, t)$ has exactly one element, it is clearer that $F(m, t)$ is a subcontinuum of X with diameter less than $\frac{2}{2^m}$.

Given $m \in \mathbb{N}$, $i \in \{1, \ldots, k_m\}$ and $j \in \{1, \ldots, k_{m+1}\}$ such that $t \in [t_{j-1}^{(m+1)}, t_j^{(m+1)}] \cap [t_{i-1}^{(m)}, t_i^{(m)}]$, since P_{m+1} refines P_m, we conclude that either $[t_{j-1}^{(m+1)}, t_j^{(m+1)}] \subset [t_{i-1}^{(m)}, t_i^{(m)}]$ or $(t = t_{i-1}^{(m)} = t_j^{(m+1)}$ and $[t_{j-1}^{(m+1)}, t_j^{(m+1)}] \subset [t_{i-2}^{(m)}, t_{i-1}^{(m)}])$ or $(t = t_i^{(m)} = t_{j-1}^{(m+1)}$ and $[t_{j-1}^{(m+1)}, t_j^{(m+1)}] \subset [t_i^{(m)}, t_{i+1}^{(m)}])$. In the three cases there exists a $k \in J(m, t)$ such that $[t_{j-1}^{(m+1)}, t_j^{(m+1)}] \subset [t_{k-1}^{(m)}, t_k^{(m)}]$. By the claim, $B_j^{(m+1)} \subset B_k^{(m)} \subset F(m, t)$. With this, we have proved that $F(m+1, t) \subset F(m, t)$.

We have shown that for each $t \in [0, 1]$, the sequence $\{F(m, t)\}_{m=1}^{\infty}$ is a decreasing sequence of subcontinua of X such that for each $m \in \mathbb{N}$, diameter$(F(m, t)) < \frac{2}{2^m}$. By Theorem 1.2 we can define $f(t)$ by the equality:

$$\{f(t)\} = \bigcap\{F(m, t) \subset X : m \in \mathbb{N}\}.$$

Given $m \in \mathbb{N}$, since $0 \in [0, t_1^{(m)}]$, we have $f(0) \in B_1^{(m)}$. Since $p \in B_1^{(m)}$ and diameter$(B_1^{(m)}) < \frac{1}{2^m}$, we conclude that $f(0) = p$. Similarly, $f(1) = q$.

We check that f is continuous. Let $t \in [0, 1]$ and $\varepsilon > 0$. Let $m_0 \in \mathbb{N}$ be such that $\frac{4}{2^{m_0}} < \varepsilon$. Let $J = \bigcup\{[t_{i-1}^{(m_0)}, t_i^{(m_0)}] : i \in J(m_0, t)\}$. Then J is the union of the intervals of the form $[t_{i-1}^{(m_0)}, t_i^{(m_0)}]$ containing t. Then J is a neighborhood of t in $[0, 1]$. Given $u \in J$, there exists an $i \in J(m_0, t)$ such that $u \in [t_{i-1}^{(m_0)}, t_i^{(m_0)}]$. Then $i \in J(m_0, u) \cap J(m_0, t)$ and $B_i^{(m_0)} \subset F(m_0, u) \cap F(m_0, t)$.

Then diameter$(F(m_0, u) \cup F(m_0, t)) < \frac{4}{2^{m_0}}$. Since $f(u) \in F(m_0, u)$ and $f(t) \in F(m_0, t)$, we have that $d(f(u), f(t)) < \varepsilon$. Therefore f is continuous.

By definition, $f([0, 1]) \subset \bigcap\{\bigcup \mathcal{B}_m : m \in \mathbb{N}\}$. We check that the other inclusion also holds.

Let $p \in \bigcap\{\bigcup \mathcal{B}_m : m \in \mathbb{N}\}$. Let $\varepsilon > 0$. Since $f([0, 1])$ is closed in X, it is enough to show that $B(p, \varepsilon)$ intersects $f([0, 1])$. Let $m_0 \in \mathbb{N}$ be such that $\frac{2}{2^{m_0}} < \varepsilon$. Let $i \in \{1, \ldots, k_{m_0}\}$ be such that $p \in B_i^{(m_0)}$. Fix a point $t \in [t_{i-1}^{(m_0)}, t_i^{(m_0)}]$. Then $i \in J(m_0, t)$, so $B_i^{(m_0)} \subset F(m_0, t)$. Thus $f(t), p \in F(m_0, t)$, $d(f(t), p) < \frac{2}{2^{m_0}} < \varepsilon$ and $f(t) \in f([0, t]) \cap B(p, \varepsilon)$. This completes the proof that $f([0, 1]) = \bigcap\{\bigcup \mathcal{B}_m : m \in \mathbb{N}\}$.

Finally, we show that if each $\mathcal{B}_m$ is a chain, then f is one-to-one. Let $t, u \in [0, 1]$ be such that $t < u$. Let $m_0 \in \mathbb{N}$ be such that $\frac{1}{2^{m_0}} < \frac{u-t}{2}$.

Choose and fix $i, j \in \{1, \ldots, k_{m_0}\}$ such that $t \in [t_{i-1}^{(m_0)}, t_i^{(m_0)}]$ and $u \in [t_{j-1}^{(m_0)}, t_j^{(m_0)}]$. Since $[t_{i-1}^{(m_0)}, t_i^{(m_0)}]$ and $[t_{j-1}^{(m_0)}, t_j^{(m_0)}]$ have diameter less than $\frac{1}{2^{m_0}} < \frac{u-t}{2}$, we have that $[t_{i-1}^{(m_0)}, t_i^{(m_0)}] \cap [t_{j-1}^{(m_0)}, t_j^{(m_0)}] = \emptyset$. This implies that $i + 1 < j$. Since $\mathcal{B}_{m_0}$ is a chain, we have $B_i^{(m_0)} \cap B_j^{(m_0)} = \emptyset$. We have shown that $F(m_0, t) \cap F(m_0, u) = \emptyset$. Hence $f(t) \neq f(u)$. Therefore, f is one-to-one. □

2.3 The Hahn–Mazurkiewicz Theorem

Theorem 2.9 (Hahn–Mazurkiewicz) *Every Peano continuum is a continuous image of the interval* [0, 1].

Proof Let X be a Peano continuum. Let d be a metric for X. Choose two points $p, q \in X$. We will construct, inductively, a sequence $\{\mathcal{B}_m\}_{m=1}^{\infty}$ of weak chains such that for each $m \in \mathbb{N}$ the following hold:

1. $k_1 \geq 2$;
2. the weak chain $\mathcal{B}_m = \{B_1^{(m)}, \ldots, B_{k_m}^{(m)}\}$ goes from p to q;
3. $\mathcal{B}_m$ is a weak $\frac{1}{2^m}$-chain and for each $i \in \{1, \ldots, k_m\}$, $B_i^{(m)}$ is a Peano continuum;
4. $\mathcal{B}_{m+1} = \mathcal{C}_1^{(m+1)} \vee \cdots \vee \mathcal{C}_{k_m}^{(m+1)}$, where for each $i \in \{1, \ldots, k_m\}$, $\mathcal{C}_i^{(m+1)}$ is a weak chain with size ≥ 2 and $\bigcup \mathcal{C}_i^{(m+1)} = B_i^{(m)}$;
5. $\mathcal{B}_m$ covers X.

We construct $\mathcal{B}_1$. By Theorem 2.5, there exists a finite open cover $\mathcal{A}$ of X consisting of Peano continua of diameter less than $\frac{1}{2}$. By Theorem 2.7, there exists a weak chain $\mathcal{B}_1 = \{B_1^{(1)}, \ldots, B_{k_1}^{(1)}\}$, with $k_1 \geq 2$, such that $\mathcal{B}_1 \subset \mathcal{A}$, $\mathcal{B}_1$ covers X and $\mathcal{B}_1$ goes from p to q.

Suppose we have constructed $\mathcal{B}_1, \ldots, \mathcal{B}_m$ satisfying the required properties.

For each $i \in \{1, \ldots, k_m - 1\}$ choose a point $p_i \in B_i^{(m)} \cap B_{i+1}^{(m)}$. Set $p_0 = p$ and $p_{k_m} = q$.

Applying again Theorems 2.5 and 2.7, we obtain that for each $i \in \{1, \ldots, k_m\}$, there exists a weak $\frac{1}{2^{m+1}}$-chain

$$C_i^{(m+1)} = \{C_1^{(i)}, C_2^{(i)}, \ldots, C_{l_i}^{(i)}\},$$

where $l_i \geq 2$, such that $C_i^{(m+1)}$ covers $B_i^{(m)}$, goes from p_{i-1} to p_i and its elements are Peano subcontinua of $B_i^{(m)}$

Note that for each $i < k_m$, p_i is in the last link of $C_i^{(m+1)}$ and in the first one of $C_{i+1}^{(m+1)}$. Then we construct the weak $\frac{1}{2^{m+1}}$-chain $\mathcal{B}_{m+1}$ by setting:

$$\mathcal{B}_{m+1} = C_1^{(m+1)} \vee \cdots \vee C_{k_m}^{(m+1)}.$$

That is

$$\mathcal{B}_{m+1} = \{C_1^{(1)}, \ldots, C_{l_1}^{(1)}, C_1^{(2)}, \ldots, C_{l_2}^{(2)}; \ldots; C_1^{(k_m)}, \ldots, C_{l_{k_m}}^{(k_m)}\}.$$

Since $\mathcal{B}_m$ covers X and each $B_i^{(m)}$ is covered by the links of $C_i^{(m+1)}$, we conclude that $\mathcal{B}_{m+1}$ covers X. Since $p \in C_1^{(1)}$ and $q \in C_{l_{k_m}}^{(k_m)}$, $\mathcal{B}_{m+1}$ goes from p to q.

This finishes the inductive construction of the sequence $\{\mathcal{B}_m\}_{m=1}^{\infty}$.

Theorem 2.8 implies that there exists an onto mapping $f : [0, 1] \to X$ such that $f(0) = p$ and $f(1) = q$. □

Given a chain $C = \{C_1, \ldots, C_m\}$ of subsets of a set X, C is a *pearl necklace* if for each $i \in \{2, \ldots, m\}$, $C_{i-1} \cap C_i$ is a one-point set.

Lemma 2.10 *Let X be a Peano continuum, $\varepsilon > 0$ and $p, q \in X$, with $p \neq q$. Then there exists a pearl necklace $C = \{C_1, \ldots, C_m\}$ of Peano subcontinua of X such that C_1 is the only link containing p, C_m is the only link containing q, $m \geq 3$ and for each $i \in \{1, \ldots, m\}$,* diameter$(C_i) < \varepsilon$.

Proof Let d be a metric for X. By Theorem 2.5, there exists a finite covering $\mathcal{A}$ of X by Peano subcontinua with diameter less than $\min\{\varepsilon, \frac{d(p,q)}{3}\}$. By Theorem 2.7, there exists a weak chain $\mathcal{D} = \{D_1, \ldots, D_m\}$ of elements of $\mathcal{A}$ from p to q. We may assume that m is the minimal positive integer for which there exists such a weak chain. The minimality of m implies that $\mathcal{D}$ is a chain, the only link containing p is D_1 and the only link containing q is D_m. Since for each $i \in \{1, \ldots, m\}$, diameter$(D_i) < \frac{d(p,q)}{3}$, $m \geq 3$.

By Theorem 2.9, there exists an onto mapping $f : [0, 1] \to D_1$. Fix $s, t \in [0, 1]$ such that $f(s) = p$ and $f(t) \in D_2$. Since $s \neq t$, we may assume that $s < t$. Let $r = \min\{u \in [s, t] : f(u) \in D_2\}$. Set $C_1 = f([s, r])$. By Exercise 2.14, C_1 is a Peano continuum. Since $C_1 \subset D_1$, we have that diameter(C_1) is less than ε. Moreover $p \in C_1$ and the minimality of r implies that $C_1 \cap D_2 = \{f(r)\}$. We can repeat this argument, using the Peano continuum D_2, the point $f(r)$ and the compact set D_3, to obtain a Peano subcontinuum C_2 of D_2 such that $f(r) \in C_2$ and

$C_2 \cap D_3$ is a one-point set. Proceeding in this way the required pearl necklace C can be constructed. □

Theorem 2.11 *Every Peano continuum is arcwise connected.*

Proof Let X be a Peano continuum. Fix two different points $p, q \in X$. We will inductively construct a sequence of pearl necklaces $\{\mathcal{B}_m\}_{m=1}^{\infty}$ of subcontinua of X satisfying the hypothesis of Theorem 2.8. By Lemma 2.10, there exists a pearl necklace $\mathcal{B}_1 = \{B_1^{(1)}, \ldots, B_{k_1}^{(1)}\}$ such that $B_1^{(1)}$ is the only link containing p and $B_{k_1}^{(1)}$ is the only link containing q, $k_1 \geq 3$ and for each $i \in \{1, \ldots, k_1\}$, $\text{diameter}(B_i^{(1)}) < \frac{1}{2^1}$.

Inductively, suppose pearl necklaces $\mathcal{B}_1, \ldots, \mathcal{B}_m$ satisfying the conditions in the hypothesis of Theorem 2.8 have been constructed.

For each $i \in \{1, \ldots, k_m - 1\}$, let p_i be the only point in $B_i^{(m)} \cap B_{i+1}^{(m)}$. Let $p_0 = p$ and $p_{k_m} = q$. For each $i \in \{1, \ldots, k_m\}$, by Lemma 2.10, there exists a pearl necklace $C_i = \{C_1^{(i)}, \ldots, C_{r_i}^{(i)}\}$ in the Peano continuum $B_i^{(m)}$ such that $C_1^{(i)}$ is the only link containing p_{i-1}, $C_{r_i}^{(i)}$ is the only link containing p_i, $r_i \geq 3$ and for each $j \in \{1, \ldots, r_i\}$, $\text{diameter}(C_j^{(i)}) < \frac{1}{2^{m+1}}$. Since $\bigcup C_1 \subset B_1^{(m)}$, $\bigcup C_2 \subset B_2^{(m)}$ and $B_1^{(m)} \cap B_2^{(m)} = \{p_1\}$, we have that $(\bigcup C_1) \cap (\bigcup C_2) = \{p_1\}$. Since the only link of C_1 (respectively, C_2) containing p_1 is $C_{r_1}^{(1)}$ (respectively, $C_1^{(2)}$), we conclude that the sequence $\{C_1^{(1)}, \ldots, C_{r_1}^{(1)}, C_1^{(2)}, \ldots, C_{r_2}^{(2)}\}$ is a pearl necklace. Repeating this argument we conclude that the sequence

$$\mathcal{B}_{m+1} = \{C_1^{(1)}, \ldots, C_{r_1}^{(1)}; C_1^{(2)}, \ldots, C_{r_2}^{(2)}; \ldots; C_1^{(k_m)}, \ldots, C_{r_{k_m}}^{(k_m)}\} = C_1 \vee \cdots \vee C_{k_m}.$$

is a pearl necklace.

Notice that $\mathcal{B}_{m+1}$ satisfies the requirements in Theorem 2.8.

This completes the inductive construction. Now, we can apply Theorem 2.8 to conclude that there exists a one-to-one mapping $f : [0, 1] \to X$ such that $f(0) = p$ and $f(1) = q$. Therefore, X is arcwise connected. □

2.4 Exercises

Exercise 2.12 Show that the continuum in Fig. 2.1 is connected im kleinen at p, but it is not locally connected at p. The example is the union of a sequence of harmonic fans in the plane, approaching a point p.

Exercise 2.13 Show that in a topological space the following are equivalent:

- X is a Peano space;
- the components of the open subsets of X are open;
- X is connected im kleinen.

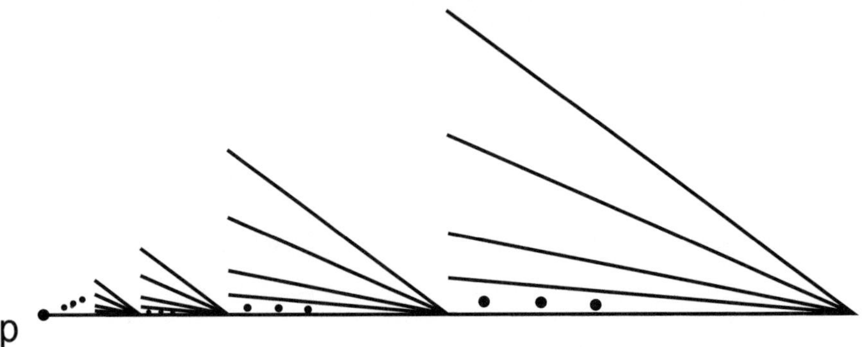

Fig. 2.1 Connectedness im kleinen does not imply local connectedness

Exercise 2.14 Let X be a Peano continuum and let $f : X \to Y$ be an onto mapping, where Y is a Hausdorff space. Prove that Y is a Peano continuum. (Hint: use Exercise 1.23.)

Exercise 2.15 Let X be a Peano continuum and $p \in X$ such that $X \setminus \{p\} = U \cup V$, where U and V are disjoint nonempty open subsets of X. Prove that each of the sets $X \setminus U$ and $X \setminus V$ is a Peano continuum. (Hint: use Exercise 4.2.)

Exercise 2.16 Let X be a continuum with metric d. Determine which of the following statements are equivalent to the fact that X is locally connected:

- X is a finite union of locally connected continua;
- each nonempty open subset of X contains a nonempty open connected subset of X;
- for each $\varepsilon > 0$ there exists a $\delta > 0$ such that if $x, y \in X$ and $d(x, y) < \delta$, then there exists a connected subset A of X such that $x, y \in A$ and diameter$(A) < \varepsilon$.

Exercise 2.17 Let Y be a nonempty subspace of a metric space X and let $Z \subset X$ be such that $Y \subset Z \subset \mathrm{cl}_X(Y)$. Prove that if Y has property S, then Z also has property S, but the converse implication does not hold.

Exercise 2.18 Prove that if a metric space X has property S, then X is a Peano space, but the converse implication does not hold. Show that if X is a compact metric space, then X is a Peano space if and only if X has property S.

Exercise 2.19 Let X be a metric space, let A be a nonempty subset of X and let $\varepsilon > 0$. Show that the following hold:

- diameter$(S(A, \varepsilon)) \leq$ diameter$(A) + 2\varepsilon$;
- if A is connected, then $S(A, \varepsilon)$ is connected;
- if X has property S, then $S(A, \varepsilon)$ is open in X.

Exercise 2.20 Let X be a nonempty topological space. Show that X is connected if and only if for every open cover $\mathcal{U}$ of X and every pair of points p and q of X, there exists a weak chain that goes from p to q, with elements of $\mathcal{U}$. Show that this equivalence does not hold if the word "open" is changed to "closed". Show that this equivalence also holds when "weak chain" is changed to "chain".

Exercise 2.21 Show that for each $\varepsilon > 0$, there exists a weak ε-chain covering the cube $[0, 1]^3$.

Exercise 2.22 Show that a continuum X is a Peano continuum if and only if it is a continuous image of $[0, 1]$.

Exercise 2.23 Prove that if X and Y are Peano continua, $x_1, \ldots, x_n$ are n distinct points of X and $y_1, \ldots, y_n$ are n distinct points of Y, then there exists an onto mapping $f : X \to Y$ such that $f(x_i) = y_i$ for every $i \in \{1, \ldots, n\}$. (Hint: use the Tietze Extension Theorem.)

Exercise 2.24 Prove that the set $\{0, 1\}$ with the Sierpiński topology is pathwise connected but it is not arcwise connected.

Exercise 2.25 Prove that if a Hausdorff topological space is pathwise connected, then it is arcwise connected.

Exercise 2.26 Let X be an arcwise connected continuum. Show that for every pair of distinct points p and q in X, there exists a unique arc joining them if and only if X does not contain simple closed curves.

Exercise 2.27 Let Y be a Peano continuum and let X be a continuum. Show that Y is a continuous image of X.

Exercise 2.28 Show that in Peano continua, ε-balls are not necessarily connected.

Exercise 2.29 A metric d for a continuum X is *convex* if for every two points p and q in X, there exists a point $x \in X$ such that $d(p, x) = d(q, x) = \frac{1}{2}d(p, q)$. Prove that if the metric for the continuum X is convex, then every ε-ball is connected, and then X is a Peano continuum.

Exercise 2.30 Show that in a Peano continuum, the closure of an open connected subset is not necessarily locally connected.

Exercise 2.31 Prove that if A and B are Peano subcontinua of a continuum X and $A \cap B \neq \emptyset$, then $A \cup B$ is a Peano continuum. Prove that if a continuum X is a finite union of Peano continua, then X is a Peano continuum.

Exercise 2.32 Let X be a Peano continuum. Prove that for each $p \in X$ and each open subset U of X with $p \in U$, there exists a Peano subcontinuum M of X such that $p \in \mathrm{int}_X(M) \subset M \subset U$.

Exercise 2.33 Prove that the open connected subsets of a Peano continuum are arcwise connected.

Exercise 2.34 Show that if U is an open connected subset of a Peano continuum and A is a nonempty compact subset of U, then there exists a Peano continuum Y such that $A \subset Y \subset U$.

Exercise 2.35 Let U be an open subset of a Peano continuum X. Let $p \in \mathrm{Fr}_X(U)$. Prove that for each open subset V of X, with $p \in V$, there exists an arc α, with end-points a and b such that $\alpha \setminus \{b\} \subset U$ and $b \in V \cap \mathrm{Fr}_X(U)$. Show that if $\mathrm{Fr}_X(U)$ is finite, then it is possible to ask that $p = b$, but in general this is not always possible.

Exercise 2.36 Let X be a Peano continuum and let A be a closed subset of $[0, 1]$. Show that every mapping from A to X has an extension to a mapping from $[0, 1]$ to X.

Exercise 2.37 Let X be a Peano space. Let C be a component of a subset D of X. Prove that $\mathrm{Fr}_X(C) \subset \mathrm{Fr}_X(D)$.

Exercise 2.38 Let X be a Peano continuum, $p \in X$ and let $\{a_n\}_{n=1}^{\infty}$ and $\{b_n\}_{n=1}^{\infty}$ be sequences in X such that $\lim_{n\to\infty} a_n = \lim_{n\to\infty} b_n = p$. Prove that there exists a sequence $\{J_n\}_{n=1}^{\infty}$ such that for each $n \in \mathbb{N}$, J_n is either an arc joining a_n and b_n or $J_n = \{a_n\} = \{b_n\}$ and $\lim_{n\to\infty} \mathrm{diameter}(J_n) = 0$. (Hint: if $a_n \neq b_n$, choose an arc J_n such that $\mathrm{diameter}(J_n) < \inf\{J : J \text{ is an arc in } X \text{ joining } a_n \text{ and } b_n\} + \frac{1}{n}$.)

Chapter 3
Cutting Wires and Bumping Boundaries

3.1 The Cut Wire Theorem

In this chapter we prove two important theorems: the Cut Wire Theorem and the Boundary Bumping Theorem. Both have many applications in Continuum Theory. Among other things, the first one allows us to prove that if two continua A and B satisfy $A \subsetneq B$, then there exists a subcontinuum C (in fact uncountably many continua, see Corollary 9.18) such that $A \subsetneq C \subsetneq B$, and the second one helps to find all the possible disconnections in compact metric spaces. Observe that the hypothesis for both theorems does not include metrizability, instead we only ask the spaces to be Hausdorff.

Definition 3.1 Let Z be a topological space. Given $p \in Z$ denote the component of p in Z by C_p. The *quasi-component* of p in Z is defined as

$$Q_p = \bigcap \{A \subset Z : A \text{ is open and closed in } Z \text{ and } p \in A\}.$$

Theorem 3.2 *Let Z be a compact Hausdorff space and $p \in Z$. Then $C_p = Q_p$.*

Proof The most important step in this proof is to show that Q_p is connected. Suppose to the contrary that Q_p is not connected. Since Q_p is an intersection of closed sets, Q_p is closed in Z. Thus, there exist disjoint nonempty closed subsets K and L of Z such that $Q_p = K \cup L$.

By Exercise 3.7, $C_p \subset Q_p \subset K \cup L$. By the connectedness of C_p, we may assume that $C_p \subset K$. Let U and V be disjoint open subsets of Z such that $K \subset U$ and $L \subset V$. Let $E = Z \setminus (U \cup V)$. Then E is compact and $p \notin E$.

Given $e \in E$, we have that $e \notin Q_p$, so there exists an open and closed subset A_e of Z such that $p \in A_e$ and $e \notin A_e$. The compactness of E implies that there exist $n \in \mathbb{N}$ and $e_1, \ldots, e_n$ such that

$$E \subset (X \setminus A_{e_1}) \cup \cdots \cup (X \setminus A_{e_n}) = X \setminus (A_{e_1} \cap \cdots \cap A_{e_n}).$$

A. Illanes, *Continuum Theory*, Universitext,
https://doi.org/10.1007/978-3-031-91011-1_3

Let $A = A_{e_1} \cap \cdots \cap A_{e_n}$. Notice that A is open and closed in Z, $p \in A$ and $E \subset X \setminus A$. Then $A \subset U \cup V$. By definition of Q_p, $Q_p \subset A$.

Notice that $A \cap U = A \setminus V$. This shows that $A \cap U$ is open and closed in Z, and since $p \in A \cap U$, we obtain that $Q_p \subset A \cap U$. This is a contradiction since $\emptyset \neq L \subset Q_p \cap V \subset A \cap U \cap V$.

With this contradiction we finish the proof that Q_p is connected. Then $Q_p \subset C_p$. By Exercise 3.7, we conclude that $Q_p = C_p$. □

Theorem 3.3 (Cut Wire Theorem) *Let Z be a compact Hausdorff space and A, B closed subsets of Z such that no connected subset of Z intersects both sets A and B. Then there exist disjoint closed subsets K and L of Z such that $A \subset K$, $B \subset L$ and $Z = K \cup L$.*

Proof Clearly, we may assume that A and B are nonempty. Take $a \in A$. By hypothesis, the component C_a of a in Z does not intersect B. Let Q_a be the quasi-component of a in Z. By Theorem 3.2, Q_a does not intersect B. For each $b \in B$, $b \notin Q_a$. Then there exists an open and closed subset E_b of Z such that $a \in E_b$ and $b \in Z \setminus E_b$. By the compactness of B, there exist $n \in \mathbb{N}$ and $b_1, \ldots, b_n \in B$ such that

$$B \subset (Z \setminus E_{b_1}) \cup \cdots \cup (Z \setminus E_{b_n}) = Z \setminus (E_{b_1} \cap \cdots \cap E_{b_n}).$$

Set $F_a = E_{b_1} \cap \cdots \cap E_{b_n}$.

We have shown that, for each $a \in A$, there exists an open and closed subset F_a of Z such that $a \in F_a$ and $B \cap F_a = \emptyset$.

The compactness of A implies that there exist $m \in \mathbb{N}$ and $a_1, \ldots, a_m \in A$ such that $A \subset F_{a_1} \cup \cdots \cup F_{a_m}$. Let $K = F_{a_1} \cup \cdots \cup F_{a_m}$ and $L = Z \setminus K$. Then K and L are disjoint closed subsets of Z such that $Z = K \cup L$, $A \subset K$ and $B \subset L$. □

3.2 The Boundary Bumping Theorem

Theorem 3.4 (Boundary Bumping Theorem) *Let X be a compact connected Hausdorff spaces, let A be a nonempty proper subset of X and let C be a component of A. Then* $\mathrm{cl}_X(C) \cap \mathrm{Fr}_X(A) \neq \emptyset$.

Proof Suppose to the contrary that $\mathrm{cl}_X(C) \cap \mathrm{Fr}_X(A) = \emptyset$. Let $Z = \mathrm{cl}_X(A)$. Then Z is a compact metric space. Given $p \in C$, let D be the component of the point p in Z. Since $C \subset A$, we have that $\mathrm{cl}_X(C) \subset \mathrm{cl}_X(A) = Z$, so $\mathrm{cl}_X(C)$ is a connected subset of Z containing p. Thus, $\mathrm{cl}_X(C) \subset D$. We analyze two cases.

First Case. A is closed.

In this case $Z = A$ and C is a component of Z. By Theorem 3.2, the component C_p coincides with the quasi-component Q_p of p in Z. Given $q \in \mathrm{Fr}_X(Z)$, we have that $q \notin C = Q_p$. This implies that there exists an open and closed subset A_q of Z such that $p \in A_q$ and $q \notin A_q$. Then $Z \setminus A_q$ is open in Z and $q \in Z \setminus A_q$. Since

$\text{Fr}_X(Z)$ is compact, there exist $n \in \mathbb{N}$ and $q_1, \ldots, q_n \in \text{Fr}_X(Z)$ such that

$$\text{Fr}_X(Z) \subset (Z \setminus A_{q_1}) \cup \cdots \cup (Z \setminus A_{q_n}) = Z \setminus (A_{q_1} \cap \cdots \cap A_{q_n}).$$

Let $E = A_{q_1} \cap \cdots \cap A_{q_n}$. Then E is open and closed in Z, $p \in E$ and $\text{Fr}_X(Z) \subset Z \setminus E$. Now, we see that E is open and closed in X. We know that E is closed in Z and Z is closed in X, so E is closed in X. Since E is open in Z, there exists an open subset U of X such that $E = U \cap Z$. Since $E \cap \text{Fr}_X(Z) = \emptyset$, we have that $E = U \cap Z = U \cap (\text{int}_X(Z) \cup \text{Fr}_X(Z)) = U \cap \text{int}_X(Z)$. Thus, E is open in X. Therefore, E is open and closed in X.

The connectedness of X implies that either $E = \emptyset$ or $E = X$. However, $p \in E$ and $E \subset Z = A \neq X$. This contradiction proves that this case is not possible.

Second Case. A is not necessarily closed.

We are supposing that $\text{cl}_X(C) \cap \text{Fr}_X(A) = \emptyset$. Since $\text{cl}_X(C) \subset \text{cl}_X(A)$, we have that $\text{cl}_X(C) \subset \text{int}_X(A)$. Then there exists an open subset U of X such that $\text{cl}_X(C) \subset U \subset \text{cl}_X(U) \subset \text{int}_X(A)$. We check that C is a component of $\text{cl}_X(U)$. Let F be the component of $\text{cl}_X(U)$ containing C. Then $C \subset F \subset \text{cl}_X(U) \subset \text{int}_X(A) \subset A$. Since C is a component of A, $C = F$. Hence C is a component of $\text{cl}_X(U)$. By the first case, $\text{cl}_X(C) \cap \text{Fr}_X(U) \neq \emptyset$. But $\text{cl}_X(C) \subset U$ and U is open, a contradiction.

Since in each case we obtain a contradiction, we conclude that $\text{cl}_X(C) \cap \text{Fr}_X(A) \neq \emptyset$. □

Theorem 3.5 *Let X be a non-locally connected continuum. Then there exist: a sequence of pairwise disjoint subcontinua $\{A_n\}_{n=1}^{\infty}$; two distinct points p and q of X; two sequences $\{p_n\}_{n=1}^{\infty}$ and $\{q_n\}_{n=1}^{\infty}$ of points of X; open subsets V, U of X; a sequence $\{C_n\}_{n=1}^{\infty}$ of pairwise distinct components of U and a closed subset D of X such that $\lim_{n\to\infty} q_n = q \in D = \text{cl}_X(V) \subset U$, $\lim_{n\to\infty} p_n = p \in V$, for every $n \in \mathbb{N}$, $\{p_n, q_n\} \subset A_n \subset C_n \cap D \setminus C_0$, where C_0 is the component of D containing p, and $q \in C_0$.*

Proof Since X is non-locally connected, by Exercise 2.13, there exists a point $p \in X$ such that X is not connected im kleinen at p. Then there exists an open subset U of X such that $p \in U$ and if C is the component of p in U, then $p \notin \text{int}_X(C)$. Note that $U \neq X$.

Fix an open set V of X such that $p \in V \subset \text{cl}_X(V) \subset U$. Let $\varepsilon > 0$ be such that $B(p, \varepsilon) \subset V$.

Choose a point $p_1 \in B(p, \varepsilon) \setminus C$ and let C_1 be the component of p_1 in U. Notice that $C_1 \neq C$.

Since C_1 is closed in U and $p \in U \setminus C_1$, $p \notin \text{cl}_X(C_1)$, we have that there exists $\varepsilon_1 > 0$ such that $\varepsilon_1 < \frac{\varepsilon}{2}$ and $B(p, \varepsilon_1) \cap C_1 = \emptyset$. Choose $p_2 \in B(p, \varepsilon_1) \setminus C$. Let C_2 be the component of p_2 in U. Note that $C_1 \neq C_2 \neq C$.

Proceeding in this way, it is possible to choose sequences: $\{p_n\}_{n=1}^{\infty}$ of points of V; and $\{C_n\}_{n=1}^{\infty}$ of components of U such that for each $m \in \mathbb{N}$, $p_m \in C_m$, $C, C_1, C_2, \ldots$ are pairwise distinct and $\lim_{n\to\infty} p_n = p$.

For each $n \in \mathbb{N}$, let A_n be the component of p_n in $\mathrm{cl}_X(V)$. Then for each $n \in \mathbb{N}$, A_n is a subcontinuum of X and $A_n \subset C_n$. By Theorem 3.4, $\emptyset \neq A_n \cap \mathrm{Fr}_X(\mathrm{cl}_X(V)) \subset A_n \cap (X \setminus V)$; and $C, A_1, A_2, A_3, \ldots$ are pairwise disjoint. Let $D = \mathrm{cl}_X(V)$ and C_0 be the component of p in D. Note that $C_0 \subset C$.

For each $n \in \mathbb{N}$, choose a point $q_n \in A_n \cap (X \setminus V)$. Taking a subsequence, if necessary, we may suppose that $\lim_{n\to\infty} q_n = q$ for some $q \in X$.

Since $p \in V$ and $q \notin V$, we have that $p \neq q$. Finally, we prove that $q \in C_0$. Suppose to the contrary that $q \notin C_0$. By Theorem 3.2, there exists an open and closed subset L of D such that $p \in L$ and $q \in D \setminus L$. Then there exists an $n \in \mathbb{N}$ such that $p_n \in L$ and $q_n \in D \setminus L$. Thus A_n intersects both sets L and $D \setminus L$ and $A_n \subset D$. This contradicts the connectedness of A_n and proves that $q \in C_0$. □

3.3 Exercises

Exercise 3.6 Prove that a compact metric space Z is totally disconnected (every component of Z is a one-point set) if and only if Z is 0-dimensional (Z has a basis of open and closed subsets).

Exercise 3.7 Prove that, in every topological space, the component of a point p is contained in the quasi-component of p.

Exercise 3.8 Show that the quasi-component of a point p is not always contained in the component of p.

Exercise 3.9 Prove that in a locally connected space Z, for each $p \in Z$ the component of p is equal to the quasi-component of p in Z.

Exercise 3.10 Show that a compact Hausdorff space is totally disconnected if and only if Z has a basis of open sets with empty boundary.

Exercise 3.11 Show that compactness is necessary in the Cut Wire Theorem.

Exercise 3.12 Show that compactness is necessary in the Boundary Bumping Theorem.

Exercise 3.13 Prove that if A and B are subcontinua of a continuum X and $A \subsetneq B$, then there exists a subcontinuum E of X such that $A \subsetneq E \subsetneq B$.

Exercise 3.14 Prove that if a continuum X has a basis of closed neighborhoods whose boundaries have a finite number of components, then X is a Peano continuum.

Exercise 3.15 A continuum X is *semi-locally connected* if it has a basis of neighborhoods whose complement has a finite number of components. Prove that Peano continua are semi-locally connected, but the converse implication does not hold.

Exercise 3.16 Show that a continuum X is semi-locally connected if and only if for each pair of distinct points p and q in X, there exists a subcontinuum M of X such that $p \in \text{int}_X(M)$ and $q \notin M$.

Exercise 3.17 A topological space Z is σ-connected if Z cannot be expressed as the union of countably many nonempty pairwise disjoint closed subsets. Prove that every continuum X is σ-connected. (Hint: suppose that $X = A_1 \cup A_2 \cup \cdots$, where the sets A_n are nonempty, closed and pairwise disjoint. Construct a continuum B_1, disjoint from A_1, intersecting A_2 but not contained in A_2. Let $m > 2$ be the minimum integer such that B_1 intersects A_m. Inside B_1 construct a subcontinuum B_2 intersecting A_m, not intersecting A_2 and not contained in A_m. Continue constructing the decreasing sequence $\cdots \subset B_3 \subset B_2 \subset B_1$. What happens with $\bigcap\{B_n : n \in \mathbb{N}\}$?)

Exercise 3.18 Prove that if a compact metric space Z can be expressed as a union of nonempty, countably many, pairwise disjoint continua, then each one of these sets is a component of Z.

Exercise 3.19 Prove that if X is a continuum, then there exists a sequence of pairwise disjoint non-degenerate subcontinua $\{A_n\}_{n=1}^{\infty}$ and there exists a point $p \in X$ such that $p \notin \bigcup\{A_n : n \in \mathbb{N}\}$, $\{p\} \cup (\bigcup\{A_n : n \in \mathbb{N}\})$ is compact and for each $\varepsilon > 0$, there exists an $N \in \mathbb{N}$ such that for each $n \geq N$, $A_n \subset B(p, \varepsilon)$.

Chapter 4
Indecomposable Continua

During the first developments of Continuum Theory, indecomposable continua were mostly considered as a curiosity until the outstanding discovery of the stronger properties of the pseudo-arc by E.E. Moise [108] and RH Bing [9]. These properties will be discussed in Chap. 17. Now, indecomposable continua, and in particular, the pseudo-arc play a very important role in this area. In this chapter, we develop the basic facts about indecomposability. To the interested reader, we recommend F.L. Jones' doctoral dissertation "A history and development of indecomposable continua theory" [75], which is a detailed and excellent survey of the history of indecomposable continua. In this direction, the paper "A brief history of indecomposable continua" by J.A. Kennedy [80] is also highly appealing. Both works have been translated into Spanish [76].

4.1 Composants

Theorem 4.1 *A continuum X is indecomposable if and only if every proper subcontinuum of X has empty interior.*

Proof (Sufficiency) Suppose that X is not indecomposable. Then there exist proper subcontinua A and B of X such that $X = A \cup B$. Notice that $X \setminus B$ is a nonempty open subset of X contained in A. Thus, $\text{int}_X(A) \neq \emptyset$. So the sufficiency is proved.

(Necessity) Suppose that there exists a proper subcontinuum D of X such that $\text{int}_X(D) \neq \emptyset$. Consider two cases.

Case 1. $X \setminus D$ is connected. Set $E = \text{cl}_X(X \setminus D)$. Then E and D are subcontinua of X such that $X = D \cup E$. Since $X \setminus D$ is contained in the closed set $X \setminus \text{int}_X(D)$, we have that $E \subset X \setminus \text{int}_X(D)$. Thus, $E \neq X$ and since $D \neq X$, we conclude that X is decomposable.

A. Illanes, *Continuum Theory*, Universitext,
https://doi.org/10.1007/978-3-031-91011-1_4

Case 2. $X \setminus D$ is not connected. In this case there exist nonempty subsets K and L of X such that $X \setminus D = K \cup L$, $\text{cl}_X(K) \cap L = \emptyset$ and $K \cap \text{cl}_X(L) = \emptyset$.

By Exercise 4.12, we have that $K \cup D$ and $L \cup D$ are connected. Notice that $X = D \cup L \cup K$. Since $K \cap \text{cl}_X(L) = \emptyset$, we have that $\text{cl}_X(L) \subset L \cup D$. This implies that $\text{cl}_X(L \cup D) = L \cup D$. Thus, $L \cup D$ is a subcontinuum of X. Since $(L \cup D) \cap H = \emptyset$, we conclude that $L \cup D$ is a proper subcontinuum of X. Similarly, $K \cup D$ is a proper subcontinuum of X. Finally, since $X = (K \cup D) \cup (L \cup D)$, we conclude that X is decomposable. □

Definition 4.2 Let X be a continuum and $p \in X$. Define the *composant* κ_p of p in X as

$$\kappa_p = \{q \in X : \text{ there is a proper subcontinuum } A \text{ of } X \text{ such that } p, q \in A\}.$$

Definition 4.3 Let X be a continuum. Define, for $p, q \in X$ the relation $p \sim q$ if and only if there exists a proper subcontinuum A of X such that $p, q \in A$.

Theorem 4.4 *If X is an indecomposable continuum, then $\sim$ is an equivalence relation.*

Proof Clearly, $\sim$ is a symmetric relation.

Given $p \in X$, the set $\{p\}$ is a proper subcontinuum of X containing p. Hence, $p \sim p$ and $\sim$ is reflexive.

Let $p, q, r \in X$ be such that $p \sim q$ and $q \sim r$. Then there exist proper subcontinua A and B of X such that $p, q \in A$ and $q, r \in B$. The union $A \cup B$ is a subcontinuum of X. Since A and B are proper subcontinua of X and X is indecomposable, we have that $X \neq A \cup B$. Thus $A \cup B$ is a proper subcontinuum of X containing the points p and r. Then $p \sim r$. Hence $\sim$ is transitive. We have shown that $\sim$ is an equivalence relation. □

Lemma 4.5 *Let X be a continuum and $p \in X$. Then the composant κ_p of p in X is a union of countably many proper subcontinua of X.*

Proof Let $\mathcal{U} = \{U_1, U_2, \ldots\}$ be a countable basis of nonempty open subsets of X. Let $J = \{m \in \mathbb{N} : p \notin U_m\}$. Given $m \in J$, let C_m be the component of p in $X \setminus U_m$.

We check that $\bigcup\{C_m : m \in J\} = \kappa_p$.

Let $q \in \bigcup\{C_m : m \in J\}$. Then $q \in C_m$ for some $m \in J$. Since $p \in C_m$ and $C_m \neq X$, we have that $q \in \kappa_p$. Thus $\bigcup\{C_m : m \in J\} \subset \kappa_p$.

To show the other inclusion, take $q \in \kappa_p$. Then there exists a proper subcontinuum A of X such that $p, q \in A$. Since $A \neq X$ and A is closed, there exists an $m \in \mathbb{N}$ such that $U_m \cap A = \emptyset$. Hence, $A \subset C_m$. Therefore, $q \in C_m$. □

Theorem 4.6 *Every indecomposable continuum X has uncountably many composants.*

Proof Suppose that X contains at most countably many composants. By Theorem 4.4, X is the union of its composants. By Lemma 4.5, each composant is the union of countably many proper subcontinua of X. Then X is a countable union of proper subcontinua. By Theorem 4.1 each proper subcontinuum of X has empty interior in X. Therefore, X is a countable union of closed subsets with empty interior. This contradicts the Baire Category Theorem (Exercise 1.52). Thus, X has uncountably many composants. □

Theorem 4.7 *If X is a continuum and $p \in X$, then the composant κ_p is dense.*

Proof Let U be a nonempty open subset of X. Fix a nonempty open subset V of X such that $\mathrm{cl}_X(V) \subset U$. If $p \in V$, we obtain that κ_p intersects U. Now, suppose that $p \notin V$, let C be the component of p in $X \setminus V$. By the Boundary Bumping Theorem (Theorem 3.4), we have that $\emptyset \neq \mathrm{cl}_X(C) \cap \mathrm{Fr}_X(X \setminus V) = C \cap \mathrm{Fr}_X(V) \subset U$. Then C is a proper subcontinuum of X that contains p and intersects U. Hence, $C \subset \kappa_p$. Therefore κ_p is dense in X. □

Theorem 4.8 *If X is a decomposable continuum, then X has one or three composants.*

Proof By hypothesis, there exist proper subcontinua A and B of X such that $X = A \cup B$.

By the connectedness of X, there exists a point $x \in A \cap B$. Then $A \subset \kappa_x$ and $B \subset \kappa_x$. Thus $X = A \cup B = \kappa_x$. Therefore, X is one of the composants of X.

If X does not have more composants, we are done. Suppose then that there exists a $p \in X$ such that $\kappa_p \neq X$. This implies that there exists a $q \in X$ such that no proper subcontinuum of X contains the set $\{p, q\}$. Notice that $p \notin \kappa_q$ and $q \notin \kappa_p$. Then $\kappa_q \neq \kappa_p$. Since $\kappa_q \neq X$, at the moment we have three distinct composants κ_p, κ_q and X.

We may assume that $p \in A$. Then $p \notin B$ (if $p \in B$, as we saw with the point x, we have that $\kappa_p = X$, a contradiction). Also notice that $q \notin A$, so $q \in B$.

Take $u \in X$. We will see that $\kappa_u \in \{X, \kappa_p, \kappa_q\}$. We consider three cases.

Case 1. $u \in \kappa_p \cap \kappa_q$.

In this case there exist proper subcontinua C and D of X such that $u, p \in C$ and $u, q \in D$. Thus, $C \cup D$ is a subcontinuum of X containing p and q. By the choice of q, this implies that $C \cup D = X$. Since $C \subset \kappa_u$ and $D \subset \kappa_u$, we conclude that $X = \kappa_u$.

Case 2. $u \notin \kappa_p$.

In this case, we will see that $\kappa_u = \kappa_q$. Since $u \notin \kappa_p$, $u \in B \setminus A$. Given $w \in \kappa_u$, there exists a proper subcontinuum E of X such that $u, w \in E$. Then $p \notin E$, so $E \cup B$ is a proper subcontinuum of X containing q and w. Hence $w \in \kappa_q$. We have shown that $\kappa_u \subset \kappa_q$.

Now, take $z \in \kappa_q$. Then there exists a proper subcontinuum F of X such that $z, q \in F$. Notice that $p \notin F$. Then $F \cup B$ is a proper subcontinuum of X and $z, u \in F \cup B$. Therefore, $z \in \kappa_u$. We have shown that $\kappa_q \subset \kappa_u$. This completes the proof that $\kappa_q = \kappa_u$.

Case 3. $u \notin \kappa_q$.

In a similar way as in Case 2, it can be proved that $\kappa_u = \kappa_p$. □

In our definition of continuum we included the property of being a metric space. A topological space is a *Hausdorff continuum* provided that X is a nondegenerate compact connected Hausdorff space. As we mentioned in Chap. 3, many of the results of this book are valid for Hausdorff continua. This is not the case for the theorems in this section. In [8], D.P. Bellamy constructed indecomposable Hausdorff continua with one and two composants.

4.2 Another Characterization

Theorem 4.9 *Let U and V be nonempty open disjoint subsets of the indecomposable continuum X. Then there exist closed subsets A and B of X such that $X = A \cup B$, $A \cap B \subset U$, $A \cap V \neq \emptyset$ and $B \cap V \neq \emptyset$.*

Proof Let W be a nonempty open subset of X such that $\mathrm{cl}_X(W) \subset U$. By Theorem 4.6, we can choose two distinct composants κ_1 and κ_2 of X. By Theorem 4.4, $\kappa_1 \cap \kappa_2 = \emptyset$. By Theorem 4.7, we can choose points $a \in \kappa_1 \cap V$ and $b \in \kappa_2 \cap V$. Let D be the component of a in $X \setminus W$. Since D is a proper subcontinuum of X, $b \notin D$. By the Cut Wire Theorem (Theorem 3.3), applied to the space $X \setminus W$, there exist closed disjoint subsets K and L of X such that $X \setminus W = K \cup L$, $a \in K$ and $b \in L$. Let $A = K \cup \mathrm{cl}_X(W)$ and $B = L \cup \mathrm{cl}_X(W)$. Then $A \cap B = \mathrm{cl}_X(W) \subset U$, $A \cup B = X$, $a \in A \cap V$ and $b \in B \cap V$. □

Theorem 4.10 *Let X be a continuum. Then X is hereditarily indecomposable if and only if for every pair of disjoint closed subsets A and B of X and each open subset U of X that intersects every component of A, there exist closed subsets K and L of X such that $X = K \cup L$, $A \subset K$, $B \subset L$ and $K \cap L \subset U \setminus (A \cup B)$.*

Proof (Necessity) Let $V = U \setminus B$. Notice that V intersects every component of A and $V \cap B = \emptyset$. We check that no component of $X \setminus V$ intersects both sets $A \setminus V$ and B. Suppose to the contrary that there exists a component D of $X \setminus V$ intersecting $A \setminus V$ and B. Let A_1 be a component of A such that $D \cap A_1 \neq \emptyset$. Since X is hereditarily indecomposable, by Exercise 4.19 we have that either $A_1 \subset D$ or $D \subset A_1$. Since $A_1 \cap V \neq \emptyset$ and $D \cap V = \emptyset$, we have that $D \subset A_1$. This is impossible since $D \cap B \neq \emptyset$ and $A \cap B = \emptyset$. This contradiction proves the impossibility of the existence of D. Thus, we can apply the Cut Wire Theorem (Theorem 3.3) to the space $X \setminus V$ and obtain that there exist closed disjoint subsets M and N of X such that $X \setminus V = M \cup N$, $A \setminus V \subset M$ and $B \subset N$. Notice that the sets $A \cup M$ and N are closed and disjoint. Let P and Q be open disjoint subsets of X such that $A \cup M \subset P$ and $N \subset Q$. Let $K = X \setminus Q$ and $L = X \setminus P$. Then K and L are closed in X. Since P and Q are disjoint, $X = K \cup L$. Notice that $A \subset P \subset X \setminus Q = K$, $B \subset Q \subset L$ and $K \cap L = X \setminus (P \cup Q) \subset X \setminus (A \cup M \cup N) = (X \setminus A) \cap V = U \setminus (A \cup B)$.

(Sufficiency) Suppose that X contains a decomposable subcontinuum D. Then D can be written as $D = A \cup C$, where A and C are proper subcontinua of D. Fix a point $b \in C \setminus A$ and let $B = \{b\}$. Let U be an open subset of X such that

$U \cap A \neq \emptyset$ and $U \cap C = \emptyset$. By hypothesis there exist closed subsets K and L of X such that $X = K \cup L$, $A \subset K$, $b \in L$ and $K \cap L \subset U \setminus (A \cup \{b\})$. Then $K \setminus U$ and $L \setminus U$ are closed and disjoint. Since $C \cap U = \emptyset$, C is connected and is contained in $(K \setminus U) \cup (L \setminus U)$, and $b \in L \cap C$, we obtain that $C \subset L \setminus U$ and $C \cap (K \setminus U) = \emptyset$. This implies that $C \cap K = \emptyset$, and then $C \cap A = \emptyset$. This contradicts the connectedness of D and finishes the proof that X is hereditarily indecomposable. □

Definition 4.11 A connected space X is *irreducible between* two points p and q of X if no proper closed connected subset of X contains the set $\{p, q\}$. A point $p \in X$ is a *point of irreducibility* in X if there exists a point $q \in X$ such that X is irreducible between p and q. The space X is *irreducible* if there exist two points p and q in X such that X is irreducible between p and q.

The notion of irreducibility plays a significant role in Continuum Theory. Related to this, we mention that indecomposable continua are irreducible (see Exercise 4.21), chainable continua are also irreducible [113, Theorem 12.5]. In Chap. 11 (Theorem 11.13) we include the important theorem that describes the structure of those continua with the property that each indecomposable subcontinuum has empty interior. Exercise 4.27 shows that each continuum contains many irreducible subcontinua. In Exercises 4.17, 4.20 and 4.21, we ask the reader to prove some relations between irreducibility and indecomposability.

4.3 Exercises

Exercise 4.12 (Ears Lemma) If Z is a connected space, A is a connected subset of Z and $Z \setminus A = K \cup L$, where $\mathrm{cl}_Z(K) \cap L = \emptyset = K \cap \mathrm{cl}_Z(L)$, then $A \cup K$ and $A \cup L$ are connected.

Exercise 4.13 Prove that an indecomposable continuum is not locally connected at any of its points.

Exercise 4.14 Find a decomposable continuum without points of local connectedness.

Exercise 4.15 Show that there exist a continuum X and a point $p \in X$ such that $X \setminus \kappa_p$ is the Hilbert cube.

Exercise 4.16 Show that there exist a decomposable continuum X and a point $p \in X$ such that κ_p is not open in X.

Exercise 4.17 Show that X is not irreducible if and only if for each $p \in X, \kappa_p = X$.

Exercise 4.18 A continuum X is *connected by continua* if for every pair of points $p, q \in X$ and every $\varepsilon > 0$, there exists $n \in \mathbb{N}$ and a chain of subcontinua $\{A_1, \ldots, A_n\}$ of X such that for each $i \in \{1, \ldots, n\}$, diameter$(A_i) < \varepsilon$, $p \in A_1$ and $q \in A_n$. Show that arcwise connected continua are connected by continua, but the converse implication does not hold. Also show that if a continuum X is connected by continua, then X is decomposable.

Exercise 4.19 Prove that a continuum X is hereditarily indecomposable if and only if for each pair of subcontinua A and B of X, we have that either $A \cap B = \emptyset$ or $A \subset B$ or $B \subset A$.

Exercise 4.20 Prove that a continuum X is indecomposable if and only if there exist three distinct points a, b and c in X such that X is irreducible between any pair of points of the set $\{a, b, c\}$.

Exercise 4.21 Show that a continuum X is indecomposable if and only if every point $p \in X$ is a point of irreducibility of X.

Exercise 4.22 Show that a continuum X is indecomposable if and only if there exists a point $p \in X$ such that the composant κ_p of p in X has empty interior.

Exercise 4.23 Prove that a continuum X is indecomposable if and only if X has two disjoint composants.

Exercise 4.24 Show that a continuum X is indecomposable if and only if each connected subset of X is either dense or has empty interior.

Exercise 4.25 Prove that a continuum X is hereditarily indecomposable if and only if no subcontinuum A of X contains a subcontinuum C satisfying $A \setminus C$ is disconnected.

Exercise 4.26 Suppose that Z is a continuum, $Z = X \cup Y$ and $X \cap Y = \{p\}$, where X and Y are indecomposable continua. Find the composants of Z.

Exercise 4.27 Prove that for every pair of points p and q in a continuum X, there exists a subcontinuum A of X such that A is irreducible between p and q.

Chapter 5
Characterizing Arcs and Circles

There are many characterizations of the continuum [0, 1]. Among the continuum specialists there is the idea (or joke) that given any topological property defined on continua, there is a way to use it to characterize [0, 1]. In this chapter, we present the simplest topological characterization of [0, 1], which makes use of cut points. According to J.J. Charatonik [22, p. 710]: "Topological characterizations of an arc as a continuum containing exactly two non-cut points (or expressed in similar terms) were obtained in 1916–1920 by W. Sierpiński, S. Straszewicz and R.L. Moore, who also characterized a simple closed curve as a continuum that is separated by any pair of its points."

5.1 Existence of Non-cut Points

Definition 5.1 A point p of a continuum X *cuts* X if $X \setminus \{p\}$ is not connected. The notation $X \setminus \{p\} = U|V$ means that $X \setminus \{p\} = U \cup V$, and U and V are disjoint nonempty open subsets of X.

Theorem 5.2 *Let X be a continuum and $p \in X$. Suppose that $X \setminus \{p\} = U|V$. Then U contains a non-cut point of X.*

Proof Suppose to the contrary that every point of U cuts X. For each $u \in U$, fix open sets U_u and V_u such that $X \setminus \{u\} = U_u|V_u$ and $p \in V_u$. By Exercise 4.12, $U_u \cup \{u\}$ is a connected subset of $X \setminus \{p\} = U \cup V$ and since $u \in (U_u \cup \{u\}) \cap U$, we obtain that

$$U_u \cup \{u\} \subset U.$$

Given $x, u \in U$, define $x \leq u$ if and only if $x \in U_u$ or $x = u$.

A. Illanes, *Continuum Theory*, Universitext,
https://doi.org/10.1007/978-3-031-91011-1_5

First we check that

$$x \leq u \text{ if and only if } U_x \subset U_u.$$

Take $x, u \in U$ such that $x \neq u$.

First suppose that $x \leq u$. Then $x \in U_u$. By Exercise 4.12, $V_u \cup \{u\}$ is connected. Since $V_u \cup \{u\} \subset X \setminus \{x\} = U_x \cup V_x$ and $p \in (V_u \cup \{u\}) \cap V_x$, we have that $V_u \cup \{u\} \subset V_x$. Taking complements we obtain that $U_x \cup \{x\} \subset U_u$, so $U_x \subset U_u$.

Now suppose that $U_x \subset U_u$. We need to show that $x \in U_u$. If $x \notin U_u$, since $x \neq u$, we have that $x \in V_u$. Thus, $U_x \cup \{x\} \subset U_u \cup V_u$, which contradicts the fact that $U_x \cup \{x\}$ is connected and intersects both sets U_u and V_u. This proves that $x \in U_u$. Hence, $x \leq u$.

Next, we check that for each $x \in U$,

$$\mathrm{cl}_X(U_x) = U_x \cup \{x\}.$$

Since $X \setminus V_x = U_x \cup \{x\}$, we have that $U_x \cup \{x\}$ is closed in X. Then $\mathrm{cl}_X(U_x) \subset U_x \cup \{x\}$. If $x \notin \mathrm{cl}_X(U_x)$, then $\mathrm{cl}_X(U_x) = U_x$. Thus, U_x is a nonempty proper open and closed subset of X, a contradiction. Hence, $x \in \mathrm{cl}_X(U_x)$ and $\mathrm{cl}_X(U_x) = U_x \cup \{x\}$.

Let $\mathcal{A} = \{U_u \cup \{u\} : u \in U\}$. We check that $\mathcal{A}$ satisfies the hypothesis of the Brouwer Reduction Theorem (Theorem 1.15). Take a sequence $\{x_n\}_{n=1}^{\infty}$ such that for each $n \in \mathbb{N}$, $U_{x_{n+1}} \cup \{x_{n+1}\} \subset U_{x_n} \cup \{x_n\}$, so $\mathrm{cl}_X(U_{x_{n+1}}) \subset \mathrm{cl}_X(U_{x_n})$. By Exercise 1.20, there exists a point $p \in \mathrm{cl}_X(U_{x_n)} = U_{x_n} \cup \{x_n\}$ for all $n \in \mathbb{N}$. By the first paragraph of the proof of this theorem, $p \in U$. Then for each $n \in \mathbb{N}$, $p \leq x_n$ and $\mathrm{cl}_X(U_p) = U_p \cup p \subset \mathrm{cl}_X(U_{x_n})$.

We have shown that $\mathcal{A}$ satisfies the hypothesis of Theorem 1.15. Then there exists a $q \in U$ such that $U_q \cup \{q\}$ is a minimal element of $\mathcal{A}$. Take an element $z \in U_q$. Then $z < q$, so $U_z \cup \{z\} \subset U_q$. Since $q \notin U_q$, we obtain that $U_z \cup \{z\}$ is properly contained in $U_q \cup \{q\}$. This contradicts the minimality of $U_q \cup \{q\}$ and ends the proof that U contains an element that is not a cut point of X □

Corollary 5.3 *Every continuum contains at least two non-cut points.*

Proof Let $x_1 \neq x_2$ be points in X. If they are non-cut points, we are done. If not, let's say x_1 is a cut point of X, then $X \setminus \{x_1\} = U|V$. By Theorem 5.2, each of the sets U and V has a non-cut point of X. □

5.2 Arcs

Theorem 5.4 *The unit interval* $[0, 1]$ *is (topologically) the only continuum having exactly two non-cut points.*

Proof Clearly, $[0, 1]$ has exactly two non-cut points. To prove the other implication, take a continuum X having exactly two non-cut points u and v.

Given $p \in X \setminus \{u, v\}$, p is a cut point of X, so we can fix two open subsets U_p and V_p such that $X \setminus \{p\} = U_p|V_p$. By Theorem 5.2, each set U_p and V_p has a non-cut point, so we may suppose that $u \in U_p$ and $v \in V_p$.

Given $p, q \in X \setminus \{u, v\}$, define $p \leq q$ if $U_p \subset U_q$. We show that $\leq$ is a linear order. Clearly, $\leq$ is transitive and reflexive.

We prove that $\leq$ is antisymmetric.

Suppose that $p \leq q$ and $q \leq p$. Then $U_p = U_q$. Suppose that $p \neq q$. Then $U_p \cup \{p\} \subset X \setminus \{q\} = U_q \cup V_q$. By Exercise 4.12, we have that $U_p \cup \{p\}$ is connected and since it contains the point u and $u \in U_q$, we obtain that $U_p \cup \{p\} \subset U_q$. Thus $p \in U_q = U_p$, a contradiction. We have shown that $p = q$. Therefore, $\leq$ is antisymmetric.

We have proved that $\leq$ is a partial order. In order to finish the proof that $\leq$ is a linear order, we check that if $p \neq q$, then $p < q$ or $q < p$. We analyze two cases.

Case 1. $p \in U_q$.

In this case, $p \notin V_q$. Then $V_q \cup \{q\} \subset X \setminus \{p\} = U_p \cup V_p$. Since $V_q \cup \{q\}$ is connected and $v \in (V_q \cup \{q\}) \cap V_p$, we conclude that $V_q \cup \{q\} \subset V_p$. Then $q \notin U_p$, so $U_p \cup \{p\} \subset X \setminus \{q\} = U_q \cup V_q$. Since $U_p \cup \{p\}$ is connected and $u \in (U_p \cup \{p\}) \cap U_q$, we have that $U_p \cup \{p\} \subset U_q$. Hence, $p \leq q$. Therefore, $p < q$.

Case 2. $p \in V_q$.

In this case $p \notin U_q$. Then $U_q \cup \{q\} \subset X \setminus \{p\} = U_p \cup V_p$. Since $U_q \cup \{q\}$ is connected and $u \in (U_q \cup \{q\}) \cap U_p$, we conclude that $U_q \cup \{q\} \subset U_p$. Thus $q \leq p$. Therefore, $q < p$.

We have shown that $\leq$ is a linear order.

We complete the definition of $\leq$ by making $u \leq p \leq v$ for all $p \in X$. Clearly, $\leq$ is a linear order defined on X.

We show that for every $q \in X \setminus \{u, v\}$, $\{p \in X : p < q\} = U_q$.

Take $p \in U_q$. Since $v \notin U_q$, we have that $p \neq v$. If $p = u$, by definition $p < q$. Finally, suppose that $p \in X \setminus \{u, v\}$. In Case 1 above, we proved that the condition $p \in U_q$ implies that $p < q$. We have shown that for each $p \in U_q$, $p < q$.

Now suppose that $p < q$. Then $p = u$ or $U_p \subset U_q$. In the case when $p = u$, by the choice of U_q, $p \in U_q$, and in the case when $p \neq u$ and $U_p \subset U_q$, we have that $U_p \cup \{p\}$ is a connected subset of $X \setminus \{q\} = U_q \cup V_q$ that intersects U_q, so $U_p \cup \{p\} \subset U_q$. Therefore, $p \in U_q$.

This completes the proof that $\{p \in X : p < q\} = U_q$.

Similarly, it is possible to show that $\{p \in X : q < p\} = V_q$ for each $q \in X \setminus \{u, v\}$.

In particular, we obtain that for every $q \in X \setminus \{u, v\}$, $\{p \in X : p < q\}$ and $\{p \in X : q < p\}$ are open in X. Moreover, since $\{p \in X : p < u\} = \emptyset$, $\{p \in X : u < p\} = X \setminus \{u\}$, $\{p \in X : p < v\} = X \setminus \{v\}$ and $\{p \in X : v < p\} = \emptyset$, we conclude that for each $q \in X$, $\{p \in X : p < q\}$ and $\{p \in X : q < p\}$ are open.

Consider the topology τ induced on X by taking as subbasis the following family.

$$\mathfrak{F} = \{\{p \in X : p < q\} : q \in X\} \cup \{\{p \in X : q < p\} : q \in X\}.$$

Clearly, (X, τ) is a Hausdorff space. Let τ_0 be the original topology for X. We have seen that the elements of $\mathfrak{F}$ belong to τ_0. Then $\tau \subset \tau_0$. Thus, the identity function $id_X : (X, \tau_0) \to (X, \tau)$ is continuous. Since (X, τ) is a Hausdorff space and (X, τ_0) is compact, we conclude that id_X is a homeomorphism. Therefore, $\tau_0 = \tau$. That is, the original topology of X is given by the order $\leq$.

Notice that a basis for the topology of X is the family

$$\mathfrak{B} = \{[u, q) : q \in X\} \cup \{(p, q) : p, q \in X\} \cup \{(q, v] : q \in X\},$$

where the intervals are defined in the natural way.

Now, we check that X is a Peano continuum. In order to show this, it is enough to see that each element of $\mathfrak{B}$ is connected. According to Exercise 5.6, it is enough to prove that each interval of the form $[p, q]$, where $p < q$, is connected.

Suppose to the contrary that $p < q$ and $[p, q] = K \cup L$, where K and L are nonempty closed disjoint subsets of X. Since K and L play symmetric roles, we only need to consider two cases.

Case 1. $p, q \in K$.

In this case, the sets $[u, p] \cup K \cup [q, v]$ and L are nonempty, closed and disjoint, and $X = ([u, p] \cup K \cup [q, v]) \cup L$. This contradicts the connectedness of X and proves that this case is impossible.

Case 2. $p \in K$ and $q \in L$.

In this case, the sets $[u, p] \cup K$ and $L \cup [q, v]$ are nonempty, closed and disjoint, and $X = ([u, p] \cup K) \cup (L \cup [q, v])$. This contradicts the connectedness of X and proves that this case is also impossible.

This finishes the proof that X is a Peano continuum.

We are ready to finish the proof of the theorem. By Theorem 2.11, there exists an arc α in X that joins u and v.

We check that $X = \alpha$. Suppose to the contrary that there exists a point $p \in X \setminus \alpha$. Then $p \notin \{u, v\}$. Since $X = U_p \cup \{p\} \cup V_p$, we have that $\alpha \subset U_p \cup V_p$, where U_p and V_p are open and disjoint. Since α intersects both sets U_p and V_p, we contradict the connectedness of α. This completes the proof that $X = \alpha$. Therefore, X is an arc. □

5.3 Simple Closed Curves

Theorem 5.5 *A continuum X is a simple closed curve if and only if for every pair of distinct points $x, y \in X$, $X \setminus \{x, y\}$ is disconnected.*

Proof Clearly S^1 has the mentioned property. Suppose then that X is a continuum such that for every pair of distinct points $x, y \in X$, $X \setminus \{x, y\}$ is disconnected.

First, we check that X does not have cut points. Suppose to the contrary that there exist $p \in X$ and open subsets U and V of X such that $X \setminus \{p\} = U|V$.

By Theorem 5.2, there exist $a \in U$ and $b \in V$ such that $X \setminus \{a\}$ and $X \setminus \{b\}$ are connected. Set $Z = X \setminus \{a\} = (U \setminus \{a\}) \cup \{p\} \cup V$. Then Z is connected and $Z \setminus \{p\} = (U \setminus \{a\}) \cup V$. Note that $U \setminus \{a\}$ and V are disjoint and open in Z. By Exercise 4.12, $(U \setminus \{a\}) \cup \{p\}$ is connected. Similarly, $(V \setminus \{b\}) \cup \{p\}$ is connected. Hence $X \setminus \{a, b\} = (U \setminus \{a\}) \cup \{p\} \cup (V \setminus \{b\})$ is connected, contradicting the assumption on X. This ends the proof that X does not have cut points.

Fix two points $x, y \in X$, with $x \neq y$. By hypothesis, $X \setminus \{x, y\} = W \cup Z$, where W and Z are nonempty disjoint open subsets of X. Set $K = W \cup \{x, y\}$ and $L = Z \cup \{x, y\}$. Note that $K \cup L = X$ and $K \cap L = \{x, y\}$.

We show that K and L are continua. Since $K = X \setminus Z$ and $L = X \setminus W$, we obtain that K and L are closed. Now we prove that K and L are connected. Since x is not a cut point of X, $X \setminus \{x\} = W \cup \{y\} \cup Z$ is connected. Then we can apply Exercise 4.12 to the connected space $W \cup \{y\} \cup Z$ to obtain that $W \cup \{y\}$ and $Z \cup \{y\}$ are connected. Similarly it is possible to prove that $W \cup \{x\}$ and $Z \cup \{x\}$ are connected. This implies that $K = (W \cup \{x\}) \cup (W \cup \{y\})$ is connected. Similarly, L is connected. Therefore K and L are continua.

Now, we check that K and L are arcs. By Corollary 5.3 and Theorem 5.4, it is enough to show that each element $w \in W$ is a cut point of K and each element $z \in Z$ is a cut point of L. If we suppose that this property does not hold, then we have three cases.

Case 1. There exists a non-cut point w_0 of K such that $w_0 \in W$ and there exists a non-cut point z_0 of L such that $z_0 \in Z$.

In this case $K \setminus \{w_0\}$ and $L \setminus \{z_0\}$ are connected and, since $x, y \in (K \setminus \{w_0\}) \cap (L \setminus \{z_0\})$, we have that $(K \setminus \{w_0\}) \cup (L \setminus \{z_0\})$ is connected. Since $(K \setminus \{w_0\}) \cup (L \setminus \{z_0\}) = X \setminus \{w_0, z_0\}$, we obtain that $X \setminus \{w_0, z_0\}$ is connected. This contradicts the hypothesis and shows that this case is not possible.

Case 2. There exists a non-cut point w_0 of K such that $w_0 \in W$ and every point $z \in Z$ is a cut-point of L.

In this case, by Theorem 5.4, L is an arc with end-points x and y. Fix a point $z_0 \in Z$. Then $L \setminus \{z_0\}$ is the union of two connected subsets A and B of X (in fact, A and B are subintervals of L) such that $L \setminus \{z_0\} = A \cup B$, where $x \in A$ and $y \in B$. Then $X \setminus \{w_0, z_0\} = (K \setminus \{w_0\}) \cup (L \setminus \{z_0\}) = (K \setminus \{w_0\}) \cup A \cup B$ is connected since $x \in (K \setminus \{w_0\}) \cap A$ and $y \in (K \setminus \{w_0\}) \cap B$. Hence $X \setminus \{w_0, z_0\}$ is connected. This contradicts the hypothesis and proves that this case is also impossible.

Case 3. There exists a non-cut-point z_0 of L such that $z_0 \in Z$ and every point $w \in W$ is a cut-point of K.

In this case, with a similar argument as in Case 2, we obtain a contradiction.

Cases 1, 2 and 3, show that each point $w \in W$ cuts K and each element $z \in Z$ cuts L. Therefore, K and L are arcs such that $K \cap L = \{x, y\}$. Since the end-points of K are x and y and the same happens with L, we conclude that X is a simple closed curve. □

5.4 Exercises

Exercise 5.6 Suppose that X is a set with a linear order $<$. Given $p \in X$, let $(\leftarrow, p) = \{x \in X : x < p\}$ and $(p, \rightarrow) = \{x \in X : p < x\}$. Consider X with the topology τ having as subbase the elements in the following family:

$$\{(\leftarrow, p) : p \in X\} \cup \{(p, \rightarrow) : p \in X\}.$$

Prove that if the closed intervals in X are connected, then all intervals in X are connected. Also show that connectedness of open intervals does not imply connectedness of all intervals.

Exercise 5.7 Let X be a set with a linear order $<$. Define on X a topology τ as in Exercise 5.6. Prove the following.

(a) (X, τ) is connected if and only if $(X, <)$, as an ordered set, has the supremum property and for every pair of points $p, q \in X$, with $p < q$, we have that $(p, q) \neq \emptyset$,
(b) (X, τ) is compact if and only if $(X, <)$, as an ordered set, has the supremum property and has a minimum and a maximum (that is, there exist $x_0, x_1 \in X$ such that $x_0 \leq p \leq x_1$ for all $p \in X$).

Exercise 5.8 A simple triod is a space homeomorphic to the cone over the discrete space with three elements. Let X be a Peano continuum. Suppose that X does not contain a simple triod. Prove that X is an arc or a simple closed curve.

Exercise 5.9 At the end of the proof of Theorem 5.4 we proved that if the topology of a continuum is given by the topology induced by a linear order, then X is an arc. This was made by proving that X is a Peano continuum, taking an arc α and showing that $X = \alpha$. Another way of proving this result, without using that Peano continua are arcwise connected, is as follows. Suppose that X is a continuum such that its topology is given by a linear order. By Exercise 5.7, X has a minimum x_0 and a maximum x_1. Fix a countable dense subset D of $X \setminus \{x_0, x_1\}$. Find a way to define a one-to-one onto order-preserving function f between D and the set of rational numbers in $(0, 1)$ and extend f to a one-to-one onto order-preserving function between X and $[0, 1]$.

Exercise 5.10 Let X be a continuum. Suppose that X is the union of two subcontinua A and B such that $A \cap B = \{p\}$ for some $p \in X$. Prove that either p is a cut point of X or $A = \{p\}$ or $B = \{p\}$.

Exercise 5.11 An onto mapping between continua $f : X \to Y$ is monotone if $f^{-1}(y)$ is connected for all $y \in Y$. Prove that f is monotone if and only if $f^{-1}(B)$ is connected for each connected subset of Y.

Exercise 5.12 Show that if a continuum X is a monotone image of $[0, 1]$, then X is an arc.

Exercise 5.13 Show that if a continuum X is a monotone image of a simple closed curve, then X is a simple closed curve.

Exercise 5.14 Prove that if X is a continuum with the property that the complement of any connected set is connected, then X is a simple closed curve. Also show that this statement is false if we change the complement of any connected set to the complement of any subcontinuum.

Exercise 5.15 Prove that if X is a continuum, then the set of non-cut points of X is not contained in a proper connected subset of X

Exercise 5.16 Prove that if X is a chainable Peano continuum, then X is an arc.

Exercise 5.17 Prove that every continuum can be expressed as the union of two connected proper and non-degenerate dense subspaces.

Chapter 6
Finite Graphs

6.1 Definition and Characterizations

Finite graphs were defined in Chap. 1 as those continua which are a finite union of arcs such that the intersection of every pair of them is finite. In order to have a deeper understanding of the nature of these continua, in this chapter we give a more detailed, but equivalent definition (see Exercise 6.15).

Definition 6.1 A *finite graph* is a continuum X for which there exist:

(a) a finite set V of points in X, called *vertices*, and
(b) a finite set $\mathcal{A}$ of arcs in X, called *edges*

such that:

(1) $X = \bigcup\{A : A \in \mathcal{A}\}$ and V is the set of end-points of the elements of $\mathcal{A}$,
(2) the end-points of each element $A \in \mathcal{A}$ are two distinct points $p, q \in V$, and $(A \setminus \{p, q\}) \cap (V \cup (\bigcup\{B : B \in \mathcal{A} \setminus \{A\}\})) = \emptyset$.

Definition 6.2 A *tree* is a finite graph without simple closed curves. A point x in a continuum X *separates* the points p and q if there exist disjoint open subsets U and V of X such that $X \setminus \{x\} = U \cup V$, $p \in U$ and $q \in V$.

Definition 6.3 Given $n \in \mathbb{N}$, a *simple n-od* is a continuum X homeomorphic to the cone over a discrete set with exactly n points. The point that corresponds to the vertex of the cone is called the *top* of the n-od. The arcs from the base of the cone to the top are called the *legs* of the n-od. A simple 3-od is called a *simple triod*. Note that a simple 1-od is an arc and the same holds for a simple 2-od. Given a finite graph X and a point $p \in X$, the *order* of p in X is 2 if $p \notin V$ (that is, if p is not a vertex). If $p \in V$, the order of p in X is the number of edges A for which p is an end-point of A. Note that it is possible to have vertices of order 2. The points of order ≥ 3 are called *ramification points*. The set of ramification points of X is denoted by $R(X)$.

A. Illanes, *Continuum Theory*, Universitext,
https://doi.org/10.1007/978-3-031-91011-1_6

Theorem 6.4 *Let X be a continuum. Then X is a finite graph if and only if every point p in X has a neighborhood homeomorphic to a simple n-od (for some $n \in \mathbb{N}$) for which p is its top.*

Proof (Necessity) Suppose that X is a finite graph, let $p \in X$. We consider two cases.

Case 1. p is not a vertex of X.

In this case, by (1) and (2) in Definition 6.1, there is only one edge A of X such that $p \in A$. Suppose that the end-points of A are a and b. Then $p \in A \setminus \{a, b\}$. By Exercise 6.11, A has p in its interior. Thus, A is a simple 2-od, A is a neighborhood of p and p is the top of A.

Case 2. p is a vertex of X.

Let $A_1, \ldots, A_n$ be the edges of X containing p. Given $i \in \{1, \ldots, n\}$, since the only vertices of X in A_i are the end-points of A_i, we have that p is an end-point of A_i. Let q_i be the other end-point of A_i. Fix a point $p_i \in A_i \setminus \{p, q_i\}$ and denote by pp_i and p_iq_i the subarcs of A_i joining the respective pairs of points p, p_i and p_i, q_i. Let

$$B = (\bigcup\{J : J \text{ is an edge of } X \text{ and } p \notin J\}) \cup (\bigcup\{p_iq_i : i \in \{1, \ldots, n\}\}).$$

Since B is a finite union of arcs, we have that B is closed in X.

Let $A = pp_1 \cup \cdots \cup pp_n$. Given $i \neq j$, by (2), $pp_i \cap pp_j = \{p\}$. This implies that A is a simple n-od and p is its top.

Since $p \in X \setminus B \subset A$, we have that p is an interior point of A.

(Sufficiency) Let d be a metric for X.

Claim 1. If G is a finite graph in X and L is an arc in X, then $L \setminus G$ has a finite number of components.

We prove Claim 1. Suppose to the contrary that $L \setminus G$ contains infinitely many components. Choose a sequence $\{C_n\}_{n=1}^{\infty}$ of pairwise distinct components of $L \setminus G = L \setminus (G \cap L)$. We may assume that $\bigcup\{C_n : n \in \mathbb{N}\}$ does not contain end-points of L. For each $n \in \mathbb{N}$, $\mathrm{cl}_X(C_n) = a_nb_n$, for some $a_n, b_n \in L \cap G$, where a_nb_n is the subarc of L joining a_n and b_n. We may assume that $\lim_{n\to\infty} a_n = p$ for some $p \in L \cap G$. Since $\{C_n\}_{n=1}^{\infty}$ is a sequence of open intervals in the arc L, we have that $\lim_{n\to\infty} \text{diameter}(C_n) = 0$. In particular, $\lim_{n\to\infty} d(a_n, b_n) = 0$. By Exercise 2.31, G is a Peano continuum. By Exercise 2.38 there exists a sequence $\{J_n\}_{n=1}^{\infty}$ of subarcs of G such that for each $n \in \mathbb{N}$, J_n joins a_n and b_n, and $\lim_{n\to\infty} \text{diameter}(J_n) = 0$. Then the set $S_n = a_nb_n \cup J_n$ is a simple closed curve and $\lim_{n\to\infty} \text{diameter}(S_n) = 0$.

By hypothesis, p has a neighborhood M in X which is a simple r-od for some $r \in \mathbb{N}$. Then there exists an $N \in \mathbb{N}$ such that for each $n \geq N$, we have that $S_n \subset M$. This is a contradiction since M does not contain simple closed curves. We have proved Claim 1.

Claim 2. If G is a finite graph in X and L is an arc in X such that $G \cap L \neq \emptyset$, then $L \cup G$ is a finite graph.

We prove Claim 2. If $L \subset G$, we are done. Suppose then that $L \not\subset G$. Let $C_1, \ldots, C_m$ be the components of $L \setminus G$ (there are only finitely many by Claim 1). For each $i \in \{1, \ldots, m\}$, $\mathrm{cl}_X(C_i)$ is a subarc of L, let p_i and q_i be the end-points of $\mathrm{cl}_X(C_i)$, we denote $\mathrm{cl}_X(C_i)$ by $p_i q_i$. Suppose that $V(G)$ is the set of vertices of G and $E(G)$ is the set of edges of G. Let $Y = G \cup L$. Define $V(Y) = V(G) \cup \{p_1, \ldots, p_m, q_1, \ldots, q_m\}$. For each edge A of G, Consider $A \cap V(Y)$ and let $\mathcal{E}_A$ be the set of subarcs into which A is divided by the points of $V(Y)$. Note that if no point of $\{p_1, \ldots, p_m, q_1, \ldots, q_m\}$ is in A, then $\mathcal{E}_A = \{A\}$. Note also that A is a finite graph with vertices $A \cap V(Y)$ and edges $\mathcal{E}_A$.

Define $E(Y) = (\bigcup\{\mathcal{E}_A : A \in E(G)\}) \cup \{p_1 q_1, \ldots, p_m q_m\}$. Note that $V(Y)$ and $E(Y)$ are finite, the elements of $E(Y)$ are subarcs of Y, Y is the union of the elements of $E(Y)$ and Y is a finite graph with set of vertices $V(Y)$ and set of edges $E(Y)$.

We can now finish the proof that X is a finite graph. By hypothesis every point p in X has a neighborhood which is a simple n-od (for some $n \in \mathbb{N}$). By compactness, it is possible to cover X with a finite number of these neighborhoods. Thus, X can be covered by a finite number of arcs $B_1, \ldots, B_s$. By Exercise 1.49, we may assume that $B_1 \cap B_2 \neq \emptyset$, $(B_1 \cup B_2) \cap B_3 \neq \emptyset$, $(B_1 \cup B_2 \cup B_3) \cap B_4 \neq \emptyset$, etc. By Claim 3, $B_1 \cup B_2$ is a finite graph. Applying again Claim 3, we obtain that $B_1 \cup B_2 \cup B_3$ is also a finite graph. Proceeding in this way we conclude that X is a finite graph. □

6.2 Order of a Subset

Definition 6.5 Let X be a continuum and $A \subset X$. A *base of neighborhoods of* A *in* X is a family $\mathcal{B}$ of open subsets of X such that for each open subset U of X that contains A, there exists an element $V \in \mathcal{B}$ such that $A \subset V \subset U$. Given a cardinal number α, we say that the *order of* A *in* X *is less than or equal to* α ($o(A, X) \leq \alpha$) if A has a base of neighborhoods $\mathcal{B}$ in X such that for each element of $\mathcal{B}$ its boundary in X has at most α points. The *order of* A *in* X *is equal to* β, written $\beta = o(A, X)$, if $o(A, X) \leq \beta$ and $o(A, X) \not\leq \alpha$ for any cardinal number $\alpha < \beta$. If $A = \{p\}$ for some $p \in X$, we write $o(p, X)$ instead of $o(\{p\}, X)$. In Exercise 6.20 we ask to show that, for a finite graph X and a vertex p of X, $o(p, X)$ coincides with the order defined in Definition 6.3.

Theorem 6.6 *A continuum* X *is a finite graph if and only if* $R(X)$ *is finite and for each* $p \in X$, $o(p, X)$ *is finite.*

Proof (Necessity) Let $p \in X$. If p is not a vertex of X, it is easy to show that $o(p, X) = 2$. If p is a vertex, by Exercise 6.20, $o(p, X)$ is the number of edges A for which p is an end-point of A. Thus, $o(p, X)$ is also finite. Moreover, since there is a finite number of vertices, there is a finite number of points p in X for which $o(p, X)$ is greater than 2, so $R(X)$ is finite.

(Sufficiency) We are going to use Theorem 6.4. So we need to check that each point $p \in X$ has a neighborhood M in X such that M is an m-od for some $m \in \mathbb{N}$ and p is the top of M.

Since we are assuming that for every $p \in X$, $o(p, X)$ is finite, by Exercise 6.23, X is a Peano continuum.

Let $p \in X$. Since X is a Peano continuum, we have that X is arcwise connected, so there are arcs in X having p as an end-point. Then p is the top of a simple k-od for some k (k could be equal to 1).

If $n \in \mathbb{N}$ is such that p is the top of a simple n-od, then we can choose n arcs $J_1, \ldots, J_n$ in X such that p is an end-point of each J_i and $J_i \cap J_j = \{p\}$ if $i \neq j$. Given a neighborhood U of p such that U does not contain any J_i, by the connectedness of J_i, we have that $J_i \cap \mathrm{Fr}_X(U) \neq \emptyset$. This shows that each small neighborhood of p in X contains at least n points in its boundary. Thus, $o(p, X) \geq n$.

Since we are assuming that $o(p, X)$ is finite, the number

$$m = \max\{n \in \mathbb{N} : p \text{ is the top of a simple } n - \text{od in } X\}$$

is well defined.

Let T be a simple m-od in X such that p is the top of T. We are going to show that T is a neighborhood of p.

Set $T = L_1 \cup \cdots \cup L_m$, where $L_1, \ldots, L_m$ are arcs in X such that p is an end-point of each L_i and $L_i \cap L_j = \{p\}$, if $i \neq j$. For each $i \in \{1, \ldots, m\}$, let x_i be the end-point of L_i such that $x_i \neq p_i$. Since $R(X)$ is finite we can take an open connected subset U of X such that $U \cap R(X) \subset \{p\}$. We can also ask that U does not contain any point x_i.

We claim that $U \subset T$. Suppose to the contrary that there exists a point $q \in U \setminus T$. By Exercise 2.33, there exists an arc $L \subset U$ joining p and q. Walking in L, from q to p, there exists a first point q_1 in $L \cap T$. Thus, there exists a subarc L_0 of L joining q and q_1, such that $L_0 \cap T = \{q_1\}$. We may assume that $q_1 \in L_1$.

If $q_1 \neq p$, by the choice of U, $q_1 \neq x_1$. This implies that $L_0 \cup L_1$ is a simple triod with q as its top, so $o(q_1, X) \geq 3$. This contradicts the fact that $U \cap R(X) \subset \{p\}$. Therefore $q_1 = p$. Note that for each $i \in \{1, \ldots, m\}$, $L_0 \cap L_i = \{p\}$, this implies that $L_0 \cup L_1 \cup \cdots \cup L_m$ is a simple $(m+1)$-od having p as its top. This contradicts the choice of m and proves that $U \subset T$.

We have shown that each point p of X has a neighborhood in X which is a simple m-od for some $m \in \mathbb{N}$. By Theorem 6.4, we conclude that X is a finite graph. □

Definition 6.7 A continuum X is a *null comb* if it is homeomorphic to the subcontinuum Y of the Euclidean plane defined by

$$Y = ([0, 1] \times \{0\}) \cup (\bigcup\{\{\frac{1}{n}\} \times [0, \frac{1}{n}] : n \in \mathbb{N}\}).$$

Fig. 6.1 Null Comb and F_ω

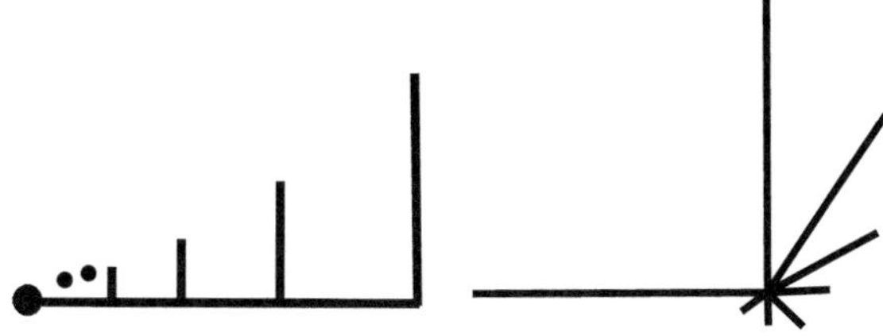

A continuum X is an F_ω if it is homeomorphic to the subcontinuum Z of the Euclidean plane defined by

$$Z = ([-1, 0] \times \{0\}) \cup (\bigcup \{\theta v_n : n \in \mathbb{N}\}),$$

where $\theta = (0, 0)$, $v_n = (\frac{1}{n}, \frac{1}{n^2})$ and θv_n denotes the convex segment joining θ and v_n (Fig. 6.1).

V. Martínez-de-la-Vega and N. Ordoñez [105, Theorem 5] found that finite graphs can be characterized as the Peano continua not containing either of two forbidden subcontinua: the null comb and F_ω.

Theorem 6.8 ([105, Theorem 5]) *A Peano continuum X is a finite graph if and only if X does not contain a null comb or an F_ω.*

Proof Let X be a Peano continuum with metric d. By Exercise 6.25, we only need to show that if X is not a finite graph, then X either contains a null comb or an F_ω. We consider two cases.

Case 1. There exists an arc α in X such that $\alpha \cap R(X)$ is infinite.

Let p and q be the end-points of α. We consider the arc α with the natural order in which $p < q$. By Exercise 6.26, we may assume that there exists a sequence $\{x_n\}_{n=1}^{\infty}$ in α such that $x_1 > x_2 > \cdots$ and for each $n \in \mathbb{N}$, $o(x_n, X) \geq 3$. Let $x = \lim_{n\to\infty} x_n$. Given points $u, v \in \alpha$ such that $u < v$, let $[u, v]$ be the subarc of α joining u and v. We also define $[u, u] = \{u\}$.

Let U_1 be a connected open subset of X such that diameter$(U_1) < 1$, $x_1 \in U_1$ and $\mathrm{cl}_X(U_1) \cap [p, x_2] = \emptyset$. Since $o(x_1, X) \geq 3$, by Exercise 6.27, $U_1 \not\subset \alpha$. Then, we can choose a point $w_1 \in U_1 \setminus \alpha$. By Exercise 2.33, we can choose an arc w_1z_1 with end-points w_1 and z_1 such that $w_1z_1 \subset U_1$ and $w_1z_1 \cap \alpha = \{z_1\}$. Note that $w_1z_1 \cap [p, x_2] = \emptyset$ and $x_2 < z_1$.

Let U_2 be a connected open subset of X such that diameter$(U_2) < \frac{1}{2}$, $x_2 \in U_2$ and $\mathrm{cl}_X(U_2) \cap ([p, x_3] \cup [z_1, q] \cup \mathrm{cl}_X(U_1)) = \emptyset$. Proceeding as we did with U_1, it is possible to choose a point $w_2 \in U_2 \setminus \alpha$ and an arc w_2z_2 with end-points w_2 and z_2 such that $w_2z_2 \subset U_2$ and $w_2z_2 \cap \alpha = \{z_2\}$. Note that $w_2z_2 \cap ([p, x_3] \cup [z_1, q] \cup w_1z_1) = \emptyset$ and $x_3 < z_2 < z_1$.

Proceeding in this way, it is possible to construct two sequences of points $\{w_n\}_{n=1}^{\infty}$ and $\{z_n\}_{n=1}^{\infty}$, and a sequence of pairwise disjoint arcs $\{w_nz_n\}_{n=1}^{\infty}$, such that for each $n \in \mathbb{N}$, $w_n z_n$ joins the points w_n and z_n, diameter$(w_n z_n) < \frac{1}{n}$, $d(x_n, z_n) < \frac{1}{n}$, $w_n z_n \cap \alpha = \{z_n\}$ and $z_1 > z_2 > \cdots$. Note that $\lim_{n\to\infty} z_n = x$.

By Exercise 6.28, the continuum $Y = [x, z_1] \cup (\bigcup\{w_n z_n : n \in \mathbb{N}\})$ is a null comb.

Case 2. Every arc in X contains only finitely many points of $R(X)$.

By Theorem 6.4, there exists a point $p \in X$ such that no neighborhood of p in X is a simple m-od ($m \in \mathbb{N}$).

Choose a connected open subset U_1 in X such that diameter$(U_1) < 1$ and $p \in U_1$. Choose a point $p_1 \in U_1 \setminus \{p\}$. By Exercise 2.33, there exists an arc $p_1 p$ contained in U and with end-points p_1 and p.

Since $p_1 p \cap R(X)$ is finite, we can choose a connected open subset U_2 of X such that $p \in U_2$, $p_1 \notin U_2$, diameter$(U_2) < \frac{1}{2}$ and $U_2 \cap R(X) \cap p_1 p \subset \{p\}$. By the choice of p, $p_1 p$ is not a neighborhood of p in X. Then we can choose a point $p_2 \in U_2 \setminus p_1 p$. By Exercise 2.33, we can choose an arc $p_2 q$ contained in U_2, with end-points p_2 and q, and satisfying $p_2 q \cap p_1 p = \{q\}$. If $q \in p_1 p \setminus \{p_1, p\}$, then $q \in R(X)$, this is impossible since $U_2 \cap R(X) \cap p_1 p \subset \{p\}$. Thus, $q = p_1$ or $q = p$. By the choice of U_2, $q \neq p_1$. Hence, $q = p$. This proves that $p_2 p \cap p_1 p = \{p\}$. In particular, $p_2 p \cup p_1 p$ is an arc.

Proceeding in this way it is possible to find a sequence of points $\{p_n\}_{n=1}^{\infty}$ and a sequence $\{p_n p\}_{n=1}^{\infty}$ of arcs such that for each $n \in \mathbb{N}$, the end-points of $p_n p$ are the points p_n and p, diameter$(p_n p) < \frac{1}{n}$ and if $n \neq m$, then $p_n p \cap p_m p = \{p\}$.

By Exercise 6.28, $Y = \bigcup\{p_n p : n \in \mathbb{N}\}$ is an F_ω. □

Theorem 6.9 *Let X be a continuum containing a non-locally connected subcontinuum Y. Then there exists an uncountable subset G of Y such that for each $x \in G$, $o(x, X)$ is infinite and $X \setminus \{x\}$ is connected.*

Proof By Theorem 3.5, there exist: a sequence of pairwise disjoint subcontinua $\{A_n\}_{n=1}^{\infty}$ of Y, two distinct points p and q in Y, and sequences $\{p_n\}_{n=1}^{\infty}$ and $\{q_n\}_{n=1}^{\infty}$ in Y such that $\lim_{n\to\infty} p_n = p$, $\lim_{n\to\infty} q_n = q$ and $\{p_n, q_n\} \subset A_n$ for all $n \in \mathbb{N}$. We may assume that the sets $A = \{p\} \cup \{p_n : n \in \mathbb{N}\}$ and $B = \{q\} \cup \{q_n : n \in \mathbb{N}\}$ are disjoint. By Urysohn's Lemma, there exists a mapping $f : X \to [0, 1]$ such that $f(A) = \{0\}$ and $f(B) = \{1\}$. Given $n \in \mathbb{N}$, since $\{p_n, q_n\} \subset A_n$ and A_n is connected, we have that $f(A_n) = [0, 1]$.

Given $t \in (0, 1)$, for each $n \in \mathbb{N}$, we choose a point $x_n^{(t)} \in A_n \cap f^{-1}(t)$. By the compactness of Y, we can choose an element $x^{(t)} \in Y$ such that for each neighborhood W of $x^{(t)}$ in Y, the set $\{n \in \mathbb{N} : x_n^{(t)} \in W\}$ is infinite. Note that $f(x^{(t)}) = t$. So, $x^{(t)} \in f^{-1}((0, 1))$. Moreover, $x^{(t)} \neq x^{(s)}$ if $t \neq s$.

Now, we show that for each $t \in (0, 1)$, $o(x^{(t)}, X)$ is infinite.

Let T be an arbitrary open subset of X such that $x^{(t)} \in T \subset f^{-1}((0, 1))$. We check that $\mathrm{Fr}_X(T)$ is infinite. By the choice of $x^{(t)}$, the set $J = \{n \in \mathbb{N} : x_n^{(t)} \in T \cap Y\}$ is infinite. Given $n \in J$, $x_n^{(t)} \in A_n \cap T \cap Y$. Since $p_n \in A_n \cap (Y \setminus f^{-1}((0, 1)))$ and $T \cap Y \subset f^{-1}((0, 1))$, we have that $p_n \notin T$. By the connectedness of A_n, we have that $\mathrm{Fr}_X(T) \cap A_n \neq \emptyset$. Since J is infinite and the sets $A_1, A_2, A_3, \ldots$ are pairwise disjoint, we conclude that $\mathrm{Fr}_X(T)$ is infinite. This finishes the proof that $o(x^{(t)}, X)$ is infinite.

Let $F = \{x^{(t)} : t \in (0, 1)$ and $x^{(t)}$ cut $X\}$ and $G = \{x^{(t)} : t \in (0, 1)$ and $x^{(t)}$ does not cut $X\}$. We are going to prove that F is at most countable, and hence that G is uncountable. Suppose to the contrary that F is uncountable.

Given $t \in F$, let U_t and V_t be nonempty disjoint open subsets of X such that $X \setminus \{x^{(t)}\} = U_t \cup V_t$. Since $f(p) = 0 < t = f(x^{(t)})$, $p \neq x^{(t)}$ and we may suppose that $p \in U_t$. Then there exists an $n \in \mathbb{N}$ such that $p_n \in U_t$ for all $n \geq N$. Given $n \geq N$, since the sets $A_1, A_2, A_3, \ldots$ are pairwise disjoint, we may assume that $x^{(t)} \notin A_n$. Then $A_n \cap U_t \neq \emptyset$, $A_n \subset U_t \cup V_t$ and A_n is connected. Thus $A_n \subset U_t \subset X \setminus V_t$.

Given $s \in F \setminus \{t\}$, by the choice of $x^{(s)}$, we have that

$$x^{(s)} \in \mathrm{cl}_X(\bigcup\{A_n : n \geq N\}) \subset X \setminus V_t,$$

and since $x^{(s)} \neq x^{(t)}$, we conclude that $x^{(s)} \in U_t$. By symmetry, $x^{(t)} \in U_s$. By Exercise 4.12, $\{x^{(t)}\} \cup V_t$ is a connected subset of X and it does not contain the point $x^{(s)}$. Thus, $\{x^{(t)}\} \cup V_t \subset U_s \cup V_s$. Since $x^{(t)} \in U_s$, we conclude that $\{x^{(t)}\} \cup V_t \subset U_s$. Therefore, $V_t \cap V_s = \emptyset$.

We have shown that $\{V_t : t \in F\}$ is an uncountable family of nonempty pairwise disjoint open subsets of X. Since this contradicts the separability of X, we conclude that F is at most countable. Therefore, G is uncountable. □

Theorem 6.10 *A continuum X is a finite graph if and only if $o(A, X) < \aleph_0$ for every subcontinuum A of X.*

Proof (Sufficiency) By hypothesis every point in X has finite order in X. By Theorem 6.9, X is a Peano continuum. If X is not a finite graph, by Theorem 6.8, X contains either a null comb or an F_ω. By Exercise 6.29, X contains a subcontinuum A such that $o(A, X) \geq \aleph_0$, a contradiction. Therefore, X is a finite graph.

The necessity follows from Exercise 6.13. □

6.3 Exercises

Exercise 6.11 Let X be a finite graph. Prove that if A is an edge with end-points p and q, then $A \setminus \{p, q\}$ is open and connected.

Exercise 6.12 Let X be a finite graph with set of vertices V. Prove that the components of $X \setminus V$ are the sets of the form $J \setminus \{p, q\}$, where J is an edge with end-points p and q.

Exercise 6.13 Let X be a finite graph and let A be a subcontinuum of X. Prove that:

(a) for each edge L of X, $A \cap L$ has at most two components,

(b) $\mathrm{Fr}_X(A)$ is finite,
(c) $o(A, X)$ is finite.

Exercise 6.14 Let X be a finite graph. Then X can be expressed as a union of edges in many ways. Suppose that we have described X in two ways: one with n vertices and c edges and another with m vertices and e edges. Prove that $n - c = m - e$.

Exercise 6.15 Prove that a continuum X is a finite graph if and only if there exist arcs $A_1, \dots, A_n$ in X such that $X = A_1 \cup \cdots \cup A_n$ and $A_i \cap A_j$ is finite if $i \neq j$.

Exercise 6.16 Let G and K be finite graphs in a continuum X such that $G \cap K$ is finite. Show that $G \cup K$ is a finite graph.

Exercise 6.17 Find two arcs in the Euclidean plane $\mathbb{R}^2$ whose intersection has uncountably many components.

Exercise 6.18 Let X be a finite graph. Prove that X is a tree if and only if the number of edges is equal to the number of vertices minus one.

Exercise 6.19 Prove that the non-degenerate subcontinua of finite graphs are finite graphs.

Exercise 6.20 Prove that for a finite graph, the order of the points given in Definition 6.3 coincides with the one given in Definition 6.5.

Exercise 6.21 Prove that a continuum X is an arc or a simple closed curve if and only if $o(p, X) \leq 2$ for all $p \in X$.

Exercise 6.22 Prove that a continuum X is a simple closed curve if and only if $o(p, X) = 2$ for all $p \in X$.

Exercise 6.23 Prove that if X is a finite graph, then there exists an $n \in \mathbb{N}$ such that $o(p, X) \leq n$ for all $p \in X$.

Exercise 6.24 Prove that if X is a continuum such that $o(p, X)$ is finite for every $p \in X$, then X is a Peano continuum.

Exercise 6.25 Prove that if X either contains a null comb or an F_w, then X is not a finite graph.

Exercise 6.26 Prove that every infinite subset of $[0, 1]$ contains a sequence that is either strictly increasing or strictly decreasing.

Exercise 6.27 Prove that if T is a simple n-od in a continuum X, $n \geq 2$ and $p \in \mathrm{int}_X(T)$, then $o(p, X) \leq n$.

Exercise 6.28 Let Y be a subcontinuum of a continuum X. Suppose that there exist sequences of points $\{p_n\}_{n=1}^{\infty}$ and $\{z_n\}_{n=1}^{\infty}$, and a sequence of arcs $\{p_n z_n\}_{n=1}^{\infty}$ in X such that for each $n \in \mathbb{N}$, the end-points of $p_n z_n$ are p_n and z_n, and $\mathrm{diameter}(p_n z_n) < \frac{1}{n}$. Then

(a) if there exists an arc β with end-points z_1 and x, where $x = \lim_{n\to\infty} z_n$, $\{z_n\}_{n=1}^{\infty}$ is a sequence either strictly increasing or strictly decreasing in β, the arcs $p_1 z_1$,

$p_2z_2, \ldots$ are pairwise disjoint and for every $n \in \mathbb{N}$, $p_nz_n \cap \beta = \{z_n\}$, then $Y = \beta \cup (\bigcup\{p_nz_n : n \in \mathbb{N}\})$ is a null comb.

(b) suppose that for every $n \neq m$, we have that $z_n = z_m$ and $p_nz_n \cap p_mz_m = \{z_m\}$. Then $Y = \bigcup\{p_nz_n : n \in \mathbb{N}\}$ is an F_ω.

Exercise 6.29 Let X be a continuum containing either a null comb or an F_ω. Prove that there exists a subcontinuum A of X such that $o(A, X) \geq \aleph_0$.

Exercise 6.30 Prove that a finite graph is a tree if and only if for every pair of distinct points p and q, there exists a point $x \in X$ such that p and q are in distinct components of $X \setminus \{x\}$

Exercise 6.31 Prove that if the set of non-cut points of a continuum X is at most countable, then X does not contain a simple closed curve.

Exercise 6.32 Prove that a continuum X is a tree if and only if the set of non-cut points of X is finite.

Exercise 6.33 Prove that a continuum is a simple triod if and only if X exactly contains three non-cut points.

Exercise 6.34 Find all the continua that contain exactly four non-cut points. Do the same for five points.

Exercise 6.35 Prove that every finite graph contains a finite number of simple closed curves.

Exercise 6.36 Let X be a Peano continuum. Prove that X is a simple closed curve if and only if the complement of every subcontinuum is connected. (Hint: prove that if $p, q \in X$ and $p \neq q$, then $X \setminus \{p, q\}$ is disconnected. In order to do this, take disjoint open subsets U and V of X such that $p \in U$ and $q \in V$. Cover $X \setminus (U \cup V)$ by a finite number of subcontinua non intersecting $\{p, q\}$ and connect these subcontinua.)

Exercise 6.37 Let X and Y be continua and let $f : X \to Y$ be an open mapping. Prove that:

(a) if U is open, then $\mathrm{Fr}_Y(f(U)) \subset f(\mathrm{Fr}_X(U))$,
(b) for each $p \in X$, $o(f(p), Y) \leq o(p, X)$,
(c) if X is a finite graph, then Y is a finite graph,
(d) if X is an arc, then Y is an arc, and
(e) if X is a simple closed curve, then Y is either an arc or a simple closed curve and both cases are possible.

Exercise 6.38 Find all the open continuous images of a simple n-od and the theta curve (the suspension of a discrete space with exactly three points).

Chapter 7
Dendroids

7.1 Definition and Problem

Definition 7.1 A continuum X is *unicoherent* if $A \cap B$ is connected for every pair of subcontinua A and B of X such that $X = A \cup B$, and X is *hereditarily unicoherent* if each subcontinuum of X is unicoherent. A *dendroid* is an arcwise connected hereditarily unicoherent continuum. A *dendrite* is a Peano dendroid. It is easy to show (Exercise 7.21) that dendroids are uniquely arcwise connected. Then, for each pair of points $p, q \in X$ we may define the unique arc pq in X connecting p and q, if $p \neq q$, and $pq = \{p\}$, if $p = q$. A point p in a dendroid X is an *end-point* if it is an end-point of any arc in X that contains p. The point p is a *ramification point* if there exists a simple triod T in X such that p is the top of T. Finally, the point p is *ordinary* if p is neither a ramification point nor an end-point. We denote by $E(X)$, $R(X)$ and $O(X)$, respectively, the set of end-points, ramification points and ordinary points of the dendroid X.

A *fan* is a dendroid with exactly one ramification point.

Two important examples of fans are:

The *harmonic fan*, defined as the cone over the *harmonic sequence* given by

$$\{0\} \cup \{\frac{1}{n} : n \in \mathbb{N}\},$$

and the *Cantor fan*, defined as the cone over the Cantor set (Fig. 7.1).

The definition of dendroid appeared for the first time in [83]. The students of B. Knaster used to say that he thought that dendroids have the following property: if X is a dendroid then for each $\varepsilon > 0$, there exist a tree $T \subset X$ and a retraction $r : X \to T$ such that for each $t \in T$, $\text{diameter}(r^{-1}(t)) < \varepsilon$.

Sixty years later the problem of determining if dendroids have this property remains open. J.B. Fugate [40] and [41] has given some positive partial answers.

A. Illanes, *Continuum Theory*, Universitext,
https://doi.org/10.1007/978-3-031-91011-1_7

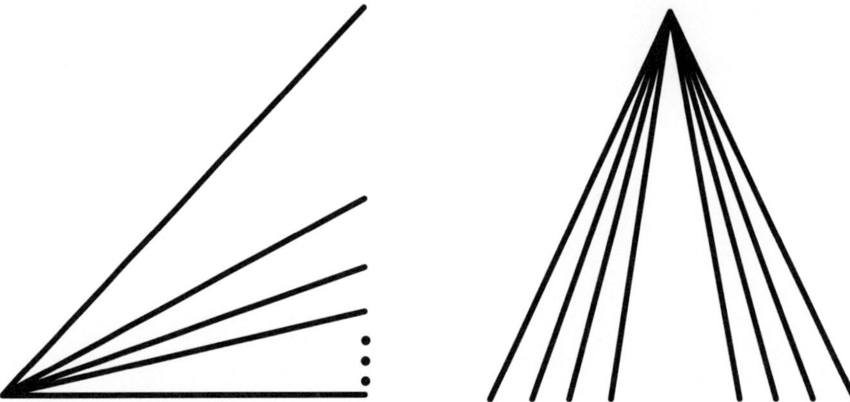

Fig. 7.1 Harmonic and Cantor fans

For locally connected dendroids, the solution of the problem is easy, and we include the proof in Theorem 7.16. On the other hand, R. Cauty claimed to have solved this problem in the positive and he wrote the paper: "Sur l'approximation interne des dendroïdes par des arbres" in 2007. This paper is available on the internet. It has not been published and the proof still contains many gaps, so nobody has been able to say if the presented proof is at least the correct way to solve the problem. Sadly, Cauty died in 2013.

7.2 Maximal Arcs

In [18], K. Borsuk proved that dendroids have the fixed point property (see Theorem 13.6). In his proof, he used the following two theorems which establish the existence of maximal arcs in dendroids.

Theorem 7.2 *Let X be a dendroid. Let $a, b_1, b_2, \ldots$ be elements of X such that $a \neq b_1$, $ab_1 \subset ab_2 \subset \cdots$. Then there exists a $b \in X$ such that $ab_n \subset ab$ for all $n \in \mathbb{N}$.*

Proof Given $m \in \mathbb{N}$, let

$$Y_m = \mathrm{cl}_X(\bigcup\{b_m b_n : m \leq n\}).$$

Given $r > n > m$, we have that ab_m is a subarc of ab_n and ab_n is a subarc of ab_r. Then, if we consider the natural order $<$ in the arc ab_r for which $a < b_r$, we have that $a \leq b_m \leq b_n \leq b_r$. Thus $b_m b_n \subset b_m b_r$.

By Exercise 7.20, Y_m is arcwise connected. By Exercise 4.18, Y_m is decomposable. Then we can fix two proper subcontinua A_m and B_m of Y_m such that

$Y_m = A_m \cup B_m$. Since $\{b_n : n > m\} \subset A_m \cup B_m$, we may suppose that $\{n > m : b_n \in B_m\}$ is infinite.

We claim that $b_m \in A_m \setminus B_m$. Suppose to the contrary that $b_m \in B_m$. Given $n \geq m$, we take $r > n$ such that $b_r \in B_m$. Then $b_m b_r \subset B_m$. We know that $b_m b_n \subset b_m b_r$. Thus, $b_m b_n \subset B_m$ for all $n > m$. Then $Y_m = \mathrm{cl}_X(\bigcup\{b_m b_n : m \leq n\}) \subset B_m$, and so $Y_m = B_m$. This contradicts the choice of B_m. We have shown that $b_m \notin B_m$. Hence $b_m \in A_m \setminus B_m$. By Exercise 7.29, there exists a unique point $u_m \in B_m$ such that $b_m u_m \cap B_m = \{u_m\}$.

Now we show that $Y_m = b_m u_m \cup B_m$. Let $Z_m = b_m u_m \cup B_m$. Since $b_m \in Y_m$ and $B_m \subset Y_m$, we have that $u_m \in Y_m$. Then $b_m u_m \subset Y_m$. This proves that $b_m u_m \cup B_m \subset Y_m$. Given $n > m$, we take $r > n$ such that $b_r \in B_m$. As we saw before, $b_m b_n \subset b_m b_r$. Since b_m, b_r belong to Z_m and Z_m is a continuum, we have that $b_m b_r \subset Z_m$. Thus $b_m b_n \subset Z_m$. With this, we conclude that $Y_m = \mathrm{cl}_X(\bigcup\{b_m b_n : m \leq n\}) \subset Z_m$. Therefore, $Y_m = Z_m$.

Now we check that $\{B_m : m \in \mathbb{N}\}$ has the finite intersection property. It is enough to show that for every $m \in \mathbb{N}$, there exists an $N_m \in \mathbb{N}$ such that $b_n \in B_m$ for all $n \geq N_m$. Let $m \in \mathbb{N}$. We choose $N_m \in \mathbb{N}$ such that $b_{N_m} \in B_m$. If $n > N_m$, then there exists an $r \in \mathbb{N}$ such that $r > n$ and $b_r \in B_m$. Then $b_{N_m} b_n \subset b_{N_m} b_r \subset B_m$ and then $b_n \in B_m$.

We are ready to finish the proof of the theorem. Fix $b \in \bigcap\{B_m : m \in \mathbb{N}\}$. Given $m \in \mathbb{N}$, we are going to see that $b_m \in ab$. Take $n > m$ such that $b_n \in B_m$. Then $ab_m \subset ab_n$. Since $ab \cup B_m$ is a continuum containing the points a and b_n, we have that $ab_m \subset ab_n \subset ab \cup B_m$. Thus $b_m \in ab \cup B_m$. Since we have checked before that $b_m \notin B_m$, we conclude that $b_m \in ab$. Therefore, $ab_m \subset ab$. □

Theorem 7.3 *Let X be a dendroid and $a, b \in X$. Then there exists a maximal arc in X containing the points a and b.*

Proof Clearly, it is enough to prove this theorem in the case that $a \neq b$.

Let $\mathcal{B} = \{ce \subset X : c, e \in X \text{ and } ab \subset ce\}$.

We will apply the Brouwer Reduction Theorem for the existence of maximal sets (Exercise 1.51) to obtain a maximal element of $\mathcal{B}$. In order to do this, take an increasing sequence $\{c_n e_n : n \in \mathbb{N}\}$ in $\mathcal{B}$, where $c_1 e_1 \subset c_2 e_2 \subset \cdots$. Choose $a_0 \in ab \setminus \{a, b\}$. Fix the natural order in the arc ab in such a way that $a < b$. Let $n \in \mathbb{N}$. Since the $c_n e_n$ is an arc containing ab, we can extend this order to a natural order $<_n$ for $c_n e_n$ and changing the names of the points c_n and e_n, if necessary, we may assume that $c_n <_n e_n$. Given $1 \leq m < n$, the order $<_n$ restricted to $c_m e_m$ is the only one in which $a <_n b$, so $<_n$ restricted to $c_m e_m$ coincides with $<_m$. Thus $c_m a_0 \subset c_n a_0$. Similarly, $a_0 e_m \subset a_0 e_n$. By Theorem 7.2, there exist $x, y \in X$ such that for every $n \in \mathbb{N}$, $c_n a_0 \subset x a_0$ and $a_0 e_n \subset a_0 y$.

Since the arc $x a_0$ (respectively, $a_0 y$) contains the point a (respectively, b) and $a_0 \in ab \setminus \{a, b\}$, it follows that $x a_0 \cap a_0 y = \{a_0\}$. Thus xy is an arc in X containing all the arcs $c_n e_n$.

We have shown that $\mathcal{B}$ satisfies the hypothesis of the Brouwer Reduction Theorem. Therefore, $\mathcal{B}$ contains a maximal element. □

7.3 Dendrites

In this section we prove a series of important characterizations and results related to dendrites. For a very complete survey on dendrites, we refer the reader to the paper by J.J. Charatonik and W.J. Charatonik [23].

Theorem 7.4 *Let X be a continuum. Then X is a dendrite if and only if X is a Peano continuum without simple closed curves.*

Proof If X is a dendrite, by definition, X is a Peano continuum and since simple closed curves are not unicoherent, we conclude that X does not contain one.

In order to prove the sufficiency, suppose that X is a Peano continuum containing no simple closed curves. Theorem 2.11 implies then that X is uniquely arcwise connected. So, we only need to prove that X is hereditarily unicoherent.

Suppose to the contrary that there exist two subcontinua A and B of X such that $A \cap B$ is not connected. Then there exist nonempty disjoint closed subsets K and L of X such that $A \cap B = K \cup L$. Let U and V be open disjoint subsets of X such that $K \subset U$ and $L \subset V$. Note that $A \setminus (U \cup V)$ and $B \setminus (U \cup V)$ are disjoint and closed in X, so there exist disjoint open subsets R and S of X such that $A \setminus (U \cup V) \subset R$ and $B \setminus (U \cup V) \subset S$.

Note that $A \subset R \cup U \cup V$ and $B \subset S \cup U \cup V$. Let P and Q be the components of $R \cup U \cup V$ and $S \cup U \cup V$ that contain A and B, respectively. By Exercise 2.13, P and Q are connected open subsets of X. By Exercise 2.33, P and Q are arcwise connected.

Choose points $p \in K$ and $q \in L$. Since $p, q \in A \subset P$, we have that the unique arc pq joining the points p and q in X is contained in P. Similarly, $pq \subset Q$.

Thus, $pq \subset P \cap Q \subset (R \cup U \cup V) \cap (S \cup U \cup V) = (R \cap S) \cup (U \cup V) = U \cup V$. Hence $pq \subset U \cup V$. This contradicts the connectedness of pq and proves that X is hereditarily unicoherent. □

Theorem 7.5 *Let X be a dendrite $p, q \in X$ and $x \in pq \setminus \{p, q\}$. Then x separates p and q in X.*

Proof First we check that p and q are in distinct components of $X \setminus \{x\}$. Let C be the component of $X \setminus \{x\}$ containing p. Suppose that $q \in C$. Since X is locally connected, by Exercise 2.13, C is open and connected. By Exercise 2.33, $pq \subset C$. This implies that $x \in C$, a contradiction. Hence, $q \notin C$.

Let $D = \bigcup\{E : E$ is a component of $X \setminus \{x\}$ and $E \neq C\}$. Then $q \in D$ and by Exercise 2.13, C and D disjoint open subsets of X and $X \setminus \{x\} = C \cup D$. □

Corollary 7.6 *If X is a dendrite, then every connected subset of X is arcwise connected.*

Proof Let A be a connected subset of X and let $p, q \in A$ be such that $p \neq q$. Suppose that $pq \not\subset A$. Let $x \in pq \setminus A$. Then $x \in pq \setminus \{p, q\}$. By Theorem 7.5, there exist disjoint open subsets U and V of X such that $p \in U$, $q \in V$ and

$X \setminus \{x\} = U \cup V$. Then A is a connected subset of $U \cup V$, $p \in A \cap U$ and $q \in A \cap V$, a contradiction. This shows that $pq \subset A$. Therefore, A is arcwise connected. □

Theorem 7.7 *A continuum X is a dendrite if and only if any pair of distinct points can be separated by a third point.*

Proof The necessity is immediate from Theorem 7.5.

In order to prove the sufficiency, we first show that X is locally connected. Suppose the contrary. By Theorem 3.5, there exists a sequence of pairwise disjoint subcontinua $\{A_n\}_{n=1}^{\infty}$ of X, sequences of points $\{p_n\}_{n=1}^{\infty}$, $\{q_n\}_{n=1}^{\infty}$ and two distinct points p and q in X such that $\lim_{n\to\infty} p_n = p$, $\lim_{n\to\infty} q_n = q$ and $\{p_n, q_n\} \subset A_n$ for every $n \in \mathbb{N}$.

By hypothesis there exists a point $x \in X$ separating p and q. Then there exist disjoint open subsets U and V of X such that $X \setminus \{x\} = U \cup V$, with $p \in U$ and $q \in V$. Take $N \in \mathbb{N}$ such that $p_n \in U$ and $q_n \in V$ for all $n \geq N$. Given $n \geq N$, $A_n \cap U \neq \emptyset$ and $A_n \cap V \neq \emptyset$. By connectedness of A_n, A_n is not contained in $U \cup V = X \setminus \{x\}$. Thus $x \in A_n$ for all $n \geq N$, this contradicts the fact that the sets A_n are pairwise disjoint. Therefore, X is locally connected.

Now, we show that X does not contain simple closed curves. Suppose to the contrary that S is a simple closed curve contained in X. Let $p, q \in S$ be such that $p \neq q$. By hypothesis there exists a point $x \in X$ and there exist two disjoint open subsets U and V of X such that $X \setminus \{x\} = U \cup V$, with $p \in U$ and $q \in V$.

Let α and β be the two distinct arcs in S that join p and q. Then $S = \alpha \cup \beta$ and $\alpha \cap \beta = \{p, q\}$. Since $\alpha \cap U \neq \emptyset$ and $\alpha \cap V \neq \emptyset$, the connectedness of α implies that α is not contained in $U \cup V$. Thus $x \in \alpha$. Similarly $x \in \beta$. This implies that $x \in \alpha \cap \beta = \{p, q\}$, a contradiction. Thus X does not contain simple closed curves. Therefore X is a dendrite. □

Theorem 7.8 *Let X be a dendrite and $p \in X$. Then $o(p, X) = 1$ if and only if p is an end-point of X.*

Proof The necessity is left as an exercise (Exercise 7.35). In order to prove the sufficiency, we suppose that p is an end-point and we need to show that $o(p, X) = 1$.

Fix a point $q \in X \setminus \{p\}$. Given a point $x \in pq \setminus \{p, q\}$, by Theorem 7.5, there exist open subsets U_x and V_x of X such that $X \setminus \{x\} = U_x \cup V_x$, $p \in U_x$ and $q \in V_x$. Since V_x is open, we have that $X \setminus V_x = U_x \cup \{x\}$ is closed. Thus, $\mathrm{Fr}_X(U_x) \subset \{x\}$. Since X is connected, the boundary of U_x in X is nonempty. Thus $\mathrm{Fr}_X(U_x) = \{x\}$.

We show that $\{U_x : x \in pq \setminus \{p, q\}\}$ is a base of neighborhoods of p in X. Let W be an open subset of X such that $p \in W$ and $q \notin W$. Given a point $z \in X \setminus W$, since X is locally connected, we can take an open connected subset T_z of X such that $z \in T_z$ and $p \notin \mathrm{cl}_X(T_z)$. By the compactness of $X \setminus W$, there exist $n \in \mathbb{N}$ and $z_1, \ldots, z_n \in X \setminus W$ such that $X \setminus W \subset T_{z_1} \cup \ldots \cup T_{z_n}$.

For each $i \in \{1, \ldots, n\}$, let q_i be the first point in the arc $z_i p$, going from z_i to p, in the arc qp. Then $z_i q_i \cap qp = \{q_i\}$. If $q_i = p$, then $z_i q = z_i p \cup pq$ is an arc containing p and p is not an end-point of $z_i q$. This contradicts that p is an end-point of X. Hence $q_i \neq p$.

Then $\{q_1, \ldots, q_n\}$ is a finite subset of $qp \setminus \{p\}$. Thus we can take a point $x \in qp \setminus \{p, q\}$ such that $(\mathrm{cl}_X(T_{z_1}) \cup \cdots \cup \mathrm{cl}_X(T_{z_n}) \cup \{q_1, \ldots, q_n\}) \cap xp = \emptyset$.

We claim that $U_x \subset W$. Let

$$B = (qx \setminus \{x\}) \cup (\mathrm{cl}_X(T_{z_1}) \cup z_1q_1) \cup \cdots \cup (\mathrm{cl}_X(T_{z_n}) \cup z_nq_n).$$

Since $\{q_1, \ldots, q_n\} \subset qx \setminus \{x\}$, we have that B is a connected subset of X, $q \in B \cap V_x$ and $x \notin B$. Then $B \subset U_x \cup V_x$ and the connectedness of B implies that $B \subset V_x$. Since $X \setminus W \subset B \subset V_x \subset X \setminus U_x$, we conclude that $U_x \subset W$.

We have shown that $\{U_x : x \in pq \setminus \{p, q\}\}$ is a base of neighborhoods of p in X such that the boundary of each element of the base is a one-point set. Therefore $o(p, X) = 1$. □

Theorem 7.9 *A continuum X is a dendrite if and only if each point of X is either a cut point of X or it has order* 1 *in X.*

Proof (Necessity) Suppose that X is a dendrite. Let $p \in X$. If p does not have order 1 in X, by Theorem 7.8, p is not an end-point of X, so there exist points a and b in X such that $p \in ab \setminus \{a, b\}$. By Theorem 7.5, p is a cut point of X. This ends the necessity.

(Sufficiency) If X is not locally connected, by Theorem 6.9, X contains a non-cut point x with infinite order, contrary to the hypothesis. Therefore, X is locally connected.

By Theorem 7.4, we only need to show that X does not contain simple closed curves. Suppose to the contrary that there exists a simple closed curve S in X.

Let $p \in S$. Clearly, p is not an end-point of X and its order in X is at least 2. By hypothesis, p cuts X. Then $X \setminus \{p\} = U_p \cup V_p$, where U_p and V_p are nonempty disjoint open subsets of X. Since $S \setminus \{p\}$ is connected, we may assume that $S \setminus \{p\} \subset U_p$.

We see that if $p, q \in S$ and $p \neq q$, then $V_p \cap V_q = \emptyset$. By Exercise 4.12, $V_q \cup \{q\}$ is connected. Since $q \in U_p$, we have that $(V_q \cup \{q\}) \cap U_p \neq \emptyset$. Since $p \in S \setminus \{q\} \subset U_q$, we have that $p \notin V_q \cup \{q\}$. Then $V_q \cup \{q\}$ is a connected subset of $U_p \cup V_p$ and intersects U_p. This implies that $V_q \cup \{q\} \subset U_p$. Thus $V_q \cap V_p = \emptyset$. Therefore $\{V_p : p \in S\}$ is an uncountable family of nonempty pairwise disjoint open subsets of X, contradicting that X is separable.

We have shown that X does not contain simple closed curves and so X is a dendrite. □

Definition 7.10 Given a continuum X and a point $p \in X$, the *complement index*, $c(p, X)$, is the cardinality of the set $\{C \subset X : C$ is component of $X \setminus \{p\}\}$. We also define $a(p, X)$ as the supremum of the cardinalities of the families of the form $\{\alpha_j : j \in J\}$, where each α_j is an arc, p is an end-point of α_j and $\alpha_i \cap \alpha_j = \{p\}$, if $i \neq j$.

Theorem 7.11 *Let X be a continuum such that for each $p \in X$, $c(p, X) = o(p, X)$. Then X is a dendrite.*

Proof Let $p \in X$. By Theorem 7.9, we only have to prove that either $o(p, X) = 1$ or p is a cut point of X. If $X \setminus \{p\}$ is connected, we have that $c(p, X) = 1$, so $o(p, X) = 1$; and if $X \setminus \{p\}$ is not connected, then p is a cut point of X. □

Theorem 7.12 *Let X be a dendrite. Then for every $p \in X$, $c(p, X) = o(p, X)$.*

Proof Let $p \in X$. First, we show that $o(p, X) \leq \aleph_0$. In order to do this, it is enough to show that p has a base of neighborhoods in X with finite boundary.

Let U be an open subset of X such that $p \in U$. By the local connectedness of X, it is possible to cover $X \setminus U$ with a finite number of subcontinua $A_1, \ldots, A_n$ of X such that $p \notin A_1 \cup \cdots \cup A_n$. Let $A = A_1 \cup \cdots \cup A_n$, let $W = X \setminus A$ and let V be the component of W containing p. By the local connectedness of X, the components of W are open in X (Exercise 2.13), so $W \setminus V$ is open in X and $\text{cl}_X(V) \cap (W \setminus V) = \emptyset$. Thus $\text{Fr}_X(V) = \text{cl}_X(V) \cap (X \setminus V) = \text{cl}_X(V) \cap ((X \setminus W) \cup (W \setminus V)) = \text{cl}_X(V) \cap A = (\text{cl}_X(V) \cap A_1) \cup \cdots \cup (\text{cl}_X(V) \cap A_n)$. By the hereditary unicoherence of X, each set of the form $\text{cl}_X(V) \cap A_i$ is a subcontinuum of X.

We check that in fact for each $i \in \{1, \ldots, n\}$, $\text{cl}_X(V) \cap A_i$ is a one-point set. Suppose to the contrary that $\text{cl}_X(V) \cap A_i$ is a non-degenerate continuum. Choose two distinct points $x, y \in \text{cl}_X(V) \cap A_i$. Let S and T be disjoint subcontinua of X such that $x \in \text{int}_X(S)$ and $y \in \text{int}_X(T)$. Then we can choose points $a \in S \cap V$ and $b \in T \cap V$. By Exercise 2.33, there exists an arc $\alpha \subset V$ joining a and b. Then $\alpha \cup S \cup T$ and A_i are subcontinua of X whose intersection is $S \cap A_i$ and $T \cap A_i$. Since these sets are closed and disjoint, and respectively contain x and y, this contradicts the hereditary unicoherence of X and proves that $\text{cl}_X(V) \cap A_i$ is a one-point set. Thus $\text{Fr}_X(V)$ is finite. Therefore $o(p, X) \leq \aleph_0$.

By Exercise 7.40, $c(p, X) \leq o(p, X)$ and we only need to consider two cases.

Case 1. $c(p, X) = \aleph_0$.

In this case $c(p, X) = \aleph_0 = o(p, X)$.

Case 2. $c(p, X) = n$ for some $n \in \mathbb{N}$.

By Exercise 7.40, $n \leq o(p, X)$. Then we only need to show that for every open subset U of X with $p \in U$, there exists an open subset V of X with exactly n points in the boundary.

Let $W_1, \ldots, W_n$ be the components of $X \setminus \{p\}$. Then for each $i \in \{1, \ldots, n\}$, the set $B_i = W_i \cup \{p\}$ is a subcontinuum of X and so it is a dendrite. Since $B_i \setminus \{p\} = W_i$ is connected, by Theorem 7.9, $o(p, B_i) = 1$. Hence there exists an open subset V_i of B_i such that $p \in V_i \subset B_i \cap U$ and $\text{Fr}_{B_i}(V_i) = 1$. Set $V = V_1 \cup \cdots \cup V_n$. Clearly, V is open in X, $p \in V \subset U$ and $\text{Fr}_X(V) = \text{Fr}_{B_1}(V_1) \cup \cdots \cup \text{Fr}_{B_n}(V_n)$. Therefore $\text{Fr}_X(V)$ contains exactly n elements. □

Corollary 7.13 *A continuum X is a dendrite if and only if for each $p \in X$, $c(p, X) = o(p, X)$.*

Theorem 7.14 *If X is a dendrite, then $R(X)$ is at most countable.*

Proof Suppose to the contrary that $R(X)$ is uncountable. Since each point of $R(X)$ has order at least 3 in X, by Theorem 7.13, each point of $R(X)$ cuts X.

Fix a dense countable dense set $D = \{p_n : n \in \mathbb{N}\}$ of X. Given $n, m \in \mathbb{N}$, let

$$C(n, m) = \{p \in R(X) : p \text{ separates } p_n \text{ and } p_m \text{ in } X\}.$$

Since each point of $R(X)$ cuts X, we have $R(X) = \bigcup\{C(n, m) : n, m \in \mathbb{N}\}$. Since we are assuming that $R(X)$ is uncountable, there exist $n, m \in \mathbb{N}$ such that $C(n, m)$ is uncountable.

Let $B = C(n, m)$. Given $p \in B$, since p separates p_n and p_m in X, we have that $p \in p_n p_m \setminus \{p_n, p_m\}$. Hence $B \subset p_n p_m \setminus \{p_n, p_m\}$.

Since for each $p \in B$, by Theorem 7.12, $X \setminus \{p\}$ has at least 3 components, there exist three nonempty pairwise disjoint open subsets U_p, V_p and W_p of X such that $X \setminus \{p\} = U_p \cup V_p \cup W_p$. We may assume that $p_n \in U_p$, $p_m \in V_p$. Since $p_n p \setminus \{p\}$ and $p_m p \setminus \{p\}$ are connected subsets of $X \setminus \{p\} = U_p \cup V_p \cup W_p$ and $U_p \cup V_p$ contains p_n and p_m, we have that $p_n p_m \setminus \{p\} \subset U_p \cup V_p$.

We claim that the sets W_p are pairwise disjoint. Take $p, q \in B$ such that $p \neq q$. By the previous paragraph, $q \in U_p \cup V_p$ and $p \in U_q \cup V_q$. By Exercise 4.12, $W_p \cup \{p\}$ is a connected subset of $X \setminus \{q\} = (U_q \cup V_q) \cup W_q$ and it intersects $U_q \cup V_q$. Hence $W_p \cup \{p\} \subset U_q \cup V_q$ and $W_p \cap W_q = \emptyset$.

We have shown that $\{W_p : p \in B\}$ is a family of nonempty pairwise disjoint open subsets of X. This contradicts the fact that X is separable and proves that $R(X)$ is at most countable. □

Example 7.15 There exists a dendroid X such that $R(X)$ is uncountable.

A dendroid X is shown in Fig. 7.2. To construct X, take the usual Cantor set C contained in the interval $[0, 1]$. Let $Z = C \times [0, 1]$. In the Cantor set $C \times \{0\}$ identify each pair of points which are the end-points of a component of $([0, 1] \setminus C) \times \{0\}$ to a point.

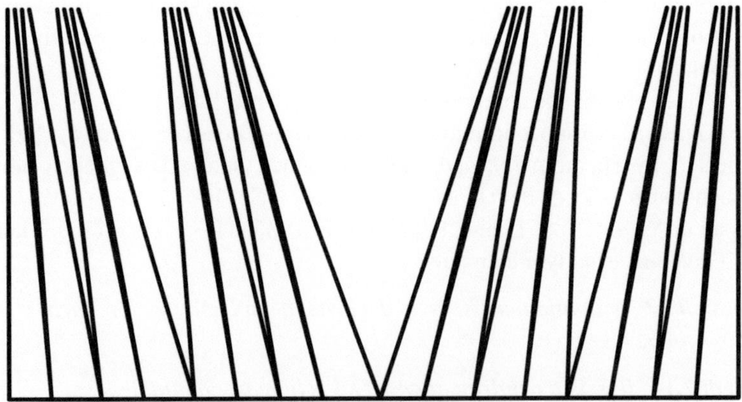

Fig. 7.2 A dendroid with uncountably many ramification points

Fig. 7.3 The Gehman dendrite

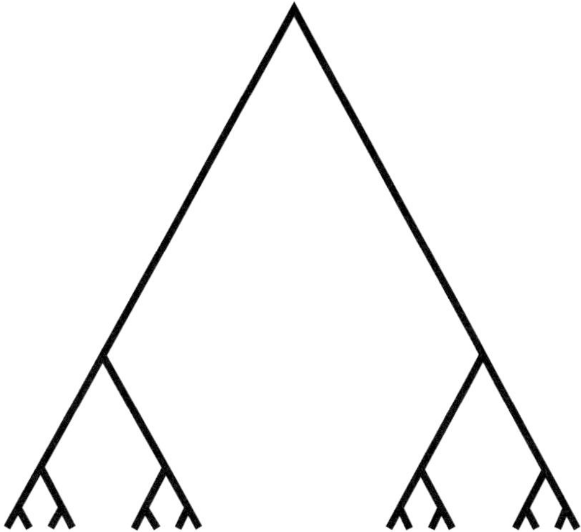

Theorem 7.16 *Let X be a dendrite, $\varepsilon > 0$ and let S be a tree contained in X. Then there exist a tree $T \subset X$ and a retraction $r : X \to T$ such that $S \subset T$ and* $\text{diameter}(r^{-1}(t)) < \varepsilon$ *for all $t \in T$.*

Proof Let d be a metric for X. Let $\mathcal{C} = \{A_1, \ldots, A_n\}$ be a cover for X by subcontinua of X such that for every $i \in \{1, \ldots, n\}$, $\text{diameter}(A_i) < \frac{\varepsilon}{2}$. Given $i \in \{1, \ldots, n\}$, fix a point $p_i \in A_i$. Choose a point $p_0 \in S$. By Exercise 7.42, $T = S \cup p_0p_1 \cup \cdots \cup p_0p_n$ is a tree.

Given $p \in X$, by Exercise 7.29, there exists a unique point $r(p) \in T$ such that $pr(p) \cap T = \{r(p)\}$ and the function $r : X \to T$ is continuous. Clearly, r is a retraction.

Let $t \in T$ and $p, q \in X$ be such that $\{p, q\} \in r^{-1}(t)$. Then there exists an $i \in \{1, \ldots, n\}$ such that $p \in A_i$. Since $pp_i \subset A_i$, we have that $\text{diameter}(pp_i) < \frac{\varepsilon}{2}$. Since $pp_i \cup T$ is a continuum containing p and $r(p)$, we have that $pr(p) \subset pp_i \cup T$. Since $(pr(p) \setminus \{r(p)\}) \cap T = \emptyset$, we have that $pr(p) \setminus \{r(p)\} \subset pp_i$. This implies that $pr(p) \subset pp_i$. Hence $d(p, t) = d(p, r(p)) < \frac{\varepsilon}{2}$. Similarly, $d(q, t) < \frac{\varepsilon}{2}$. Therefore $d(p, q) < \varepsilon$. □

Example 7.17 (The Gehman Dendrite) The Gehman dendrite is the simplest example of a dendrite with uncountably many end-points. The first four steps of the construction of this example are pictured in Fig. 7.3. Let A_1 be the union of the convex segments $(\frac{1}{2}, 2)(0, 0)$ and $(\frac{1}{2}, 2)(1, 0)$ in the Euclidean plane. Let A_2 be the union of A_1 and the two segments $(\frac{1}{6}, \frac{2}{3})(0, \frac{1}{3})$ and $(\frac{5}{6}, \frac{2}{3})(0, \frac{2}{3})$. Let A_3 be the union of A_2 and the four segments $(\frac{1}{18}, \frac{2}{9})(0, \frac{1}{9})$, $(\frac{5}{18}, \frac{2}{9})(0, \frac{2}{9})$, $(\frac{13}{18}, \frac{2}{9})(0, \frac{7}{9})$ and $(\frac{17}{18}, \frac{2}{9})(0, \frac{8}{9})$. Continuing in this way, the sets $A_4, A_5, \ldots$ are defined and the Gehman dendrite is defined as $G = \text{cl}_{\mathbb{R}^2}(A_1 \cup A_2 \cup \cdots)$. Note that the set of end-points of G is the set $C \times \{0\}$, where C is the Cantor ternary set in $[0, 1]$.

7.4 Semi-combs

Semi-brooms and semi-combs were defined by A. Illanes and V. Martínez-de-la-Vega in [67], and it was shown that a dendroid is a dendrite if and only if it contains neither a semi-broom nor a semi-comb. We observed that both definitions can be combined into one. So we defined a new concept, also called a semi-comb, and we obtained a simpler result, which is proved in this section.

Definition 7.18 A subcontinuum Y of a dendroid X is a *semi-comb* if there exist:

(a) an arc $A \subset Y$,
(b) two points $p \neq q$ in A,
(c) a sequence of points $\{p_n\}_{n=1}^{\infty}$ in $Y \setminus A$, and
(d) a sequence of points $\{q_n\}_{n=1}^{\infty}$ in A such that:

(I) $Y = A \cup \mathrm{cl}_X(\bigcup\{p_n q_n : n \in \mathbb{N}\})$,
(II) $\lim_{n\to\infty} p_n = p$, $\lim_{n\to\infty} q_n = q$,
(III) the sets $p_1q_1 \setminus \{q_1\}$, $p_2q_2 \setminus \{q_2\}, \ldots$ are pairwise disjoint, and
(IV) $p_nq_n \cap A = \{q_n\}$ for every $n \in \mathbb{N}$.

Theorem 7.19 *Let X be a dendroid. Then X is a dendrite if and only if X does not contain a semi-comb.*

Proof (Necessity) Suppose that X is a dendrite and it contains a semi-comb. Let $A \subset Y$, $p, q \in A$, and let $\{p_n\}_{n=1}^{\infty}$ and $\{q_n\}_{n=1}^{\infty}$ be sequences in X as in the definition of semi-comb. Since X is locally connected, there exists an open connected subset U of X such that $p \in U$ and $q \notin \mathrm{cl}_X(U)$. Let $N \in \mathbb{N}$ be such that $\{p_N, p_{N+1}\} \subset U$ and there exists a subarc A_1 of A such that $\{q_N, q_{N+1}\} \subset A_1 \subset X \setminus \mathrm{cl}_X(U)$. Set $B = A_1 \cup p_Nq_N \cup p_{N+1}q_{N+1}$. Then B is a subcontinuum of X, and

$$B \cap \mathrm{cl}_X(U) = (p_Nq_N \cap \mathrm{cl}_X(U)) \cup (p_{N+1}q_{N+1} \cap \mathrm{cl}_X(U)).$$

Since the sets $p_Nq_N \cap \mathrm{cl}_X(U)$ and $p_{N+1}q_{N+1} \cap \mathrm{cl}_X(U)$ are disjoint closed nonempty sets, we obtain a contradiction with the fact that X is a dendroid. Therefore X does not contain a semi-comb.

(Sufficiency) Suppose that X is not a dendrite. Suppose also that X does not contain a semi-comb. Since X is not locally connected, we can take $U, V \subset X$, $p, q \in \mathrm{cl}_X(V) \subset U$, $\{p_n\}_{n=1}^{\infty}$, $\{q_n\}_{n=1}^{\infty}$, $\{A_n\}_{n=1}^{\infty}$ and $\{C_n\}_{n=1}^{\infty}$ as in Theorem 3.5. Then $p \neq q$; $\lim_{n\to\infty} p_n = p$; $\lim_{n\to\infty} q_n = q$; $C_1, C_2, \ldots$ are pairwise distinct components of U; for each $n \in \mathbb{N}$, A_n is a subcontinuum of X and $\{p_n, q_n\} \subset A_n \subset C_n \cap \mathrm{cl}_X(V)$.

For each arc L in X and each point $x \in X$, let $P(x, L)$ be the unique point y in L such that $xy \cap L = \{y\}$. The following property is easy to prove.

Claim 1. If K and L are arcs in X such that $K \cap L \neq \emptyset$ and $x \in K$, then $P(x, L) \in K$.

Claim 2. If L is an arc in X, then the sets $\{n \in \mathbb{N} : p_n \in L\}$ and $\{n \in \mathbb{N} : q_n \in L\}$ are finite.

We prove Claim 2. Suppose the contrary. We assume that $\{n \in \mathbb{N} : p_n \in L\}$ is infinite, the case when $\{n \in \mathbb{N} : q_n \in L\}$ is infinite can be treated with similar arguments. Since $\lim_{n\to\infty} p_n = p$, we have $p \in L$. Let W be an open connected set in L such that $p \in W \subset U \cap L$. Since $\{n \in \mathbb{N} : p_n \in L\}$ is infinite, there exist $m, k \in \mathbb{N}$ such that $m \neq n$ and $p_m, p_k \in W$. Since W is a connected subset of U, $W \subset C_m \cap C_k$. This is a contradiction since C_m and C_k are distinct components of U. We have proved Claim 2.

Claim 3. If L is an arc in X, then the set $J = \{n \in \mathbb{N} : P(p_n, L) \neq P(q_n, L)\}$ is finite.

In order to prove Claim 3, suppose to the contrary that J is infinite. By Claim 2, there exists an $N \in \mathbb{N}$ such that, for every $n \geq N$, $p_n \notin L$ and $q_n \notin L$.

Given $n \in J$, with $n \geq N$, let $r_n = P(p_n, L)$ and $s_n = P(q_n, L)$. Then $r_n \neq s_n$. If $L \cap p_nq_n = \emptyset$, there is only one arc connecting L and p_nq_n. This implies that $r_n = s_n$, a contradiction. Hence $L \cap p_nq_n \neq \emptyset$. By Claim 1, $r_n \in p_nq_n \subset A_n \subset C_n \cap \mathrm{cl}_X(V) \subset U$.

Since J is infinite, there exists a convergent subsequence $\{r_{n_k}\}_{k=1}^{\infty}$ of $\{r_n\}_{n=1}^{\infty}$ such that $\lim_{n\to\infty} r_{n_k} = r$, for some $r \in \mathrm{cl}_X(V)$ and for each $k \in \mathbb{N}$, $n_k \in J$ and $n_k \geq N$. Let W be an open connected subset of L such that $r \in W \subset U \cap L$. Thus there exists a $K \in \mathbb{N}$ such that for each $k \geq K$, $r_{n_k} \in W$. Let C' be the component of U containing r. Since W is a connected subset of U, for each $k \geq K$, $r_{n_k} \in C'$. Thus $C' = C_{n_K} = C_{n_{K+1}} = \cdots$. This contradicts the choice of the sets C_n and ends the proof of Claim 3.

By Claims 2 and 3, given an arc L in X, there exists an $M_L \in \mathbb{N}$ such that for each $n \geq M_L$, $p_n \notin L$, $q_n \notin L$ and $P(p_n, L) = P(q_n, L)$. We suppose that M_L is minimal with these properties.

Claim 4. No arc L in X has the following properties: $p \in L$ and the set $R = \{P(p_n, L) \in L : n \in \mathbb{N}\}$ is an infinite subset of L having an accumulation point in $L \setminus \{p\}$.

In order to prove Claim 4, suppose to the contrary that there is an arc L with the described properties. For each $n \in \mathbb{N}$, let $r_n = P(p_n, L)$. Suppose that $r \in L \setminus \{p\}$ is an accumulation point of R. Then there exists a sequence of positive integers $M_L < n_1 < n_2 < \cdots$ such that $\lim_{k\to\infty} r_{n_k} = r$ and $r_{n_k} \neq r_{n_j}$, if $k \neq j$.

Define $Y = L \cup \mathrm{cl}_X(\bigcup\{p_{n_k}r_{n_k} : k \in \mathbb{N}\})$. Given $k, j \in \mathbb{N}$ with $k \neq j$, $(p_{n_k}r_{n_k} \cup p_{n_j}r_{n_j}) \cap L = \{r_{n_k}, r_{n_j}\}$ is not connected. Since X is a dendroid, we have that $p_{n_k}r_{n_k} \cup p_{n_j}r_{n_j}$ is not a subcontinuum of X, so $p_{n_k}r_{n_k} \cap p_{n_j}r_{n_j} = \emptyset$. Then property (III) in the definition of semi-comb is satisfied. Hence it is possible to see that Y is a semi-comb when we take the arc L and the sequences $\{p_{n_k}\}_{k=1}^{\infty}$ and $\{r_{n_k}\}_{k=1}^{\infty}$.

The proof of the following claim is similar to the proof of Claim 4.

Claim 5. No arc L in X has the following properties: $q \in L$; the set $R = \{P(q_n, L) \in L : n \in \mathbb{N}\}$ is an infinite subset of L having an accumulation point in $L \setminus \{q\}$.

Claim 6. For every arc L in X, the sets $\{P(p_n, L) : n \in \mathbb{N}\}$ and $\{P(q_n, L) : n \in \mathbb{N}\}$ are finite.

In order to prove Claim 6, suppose to the contrary that there exists an arc L in X such that $\{P(p_n, L) : n \in \mathbb{N}\}$ is infinite (by Claim 3, it is equivalent to suppose that $\{P(q_n, L) : n \in \mathbb{N}\}$ is infinite). For each $n \in \mathbb{N}$, let $r_n = P(p_n, L)$. Let r be an accumulation point of $\{r_n : n \in \mathbb{N}\}$. Suppose that $L = ab$, where $a, b \in X$.

If $p \notin L$, let $L_1 = aP(p, L) \cup P(p, L)p$ and $L_2 = bP(p, L) \cup P(p, L)p$. Then the set $T = L_1 \cup L_2$ is either an arc (when $P(p, L) \in \{a, b\}$) or a simple triod with the point $P(p, L)$ as its top.

Since r is an accumulation point of $\{r_n : n \in \mathbb{N}\}$, we have that r is also an accumulation point of the set $R = \{r_n \in L : r_n \neq P(p, L)\}$.

Since $R \subset L \subset aP(p, L) \cup pP(p, L) \cup bP(p, L)$, we have that r is an accumulation point of either $R \cap aP(p, L)$ or $R \cap pP(p, L)$ or $R \cap bP(p, L)$.

If there exists a point $r_n \in R \cap pP(p, L)$, then $r_n \in L \cap pP(p, L) = \{P(p, L)\}$, so $r_n = P(p, L)$, which contradicts the definition of R. Therefore r is not an accumulation point of $R \cap pP(p, L)$. Thus we may assume that r is an accumulation point of $R \cap aP(p, L) \subset L$.

We want to prove that $p \in L$. Suppose to the contrary that $p \notin L$. Let $L_1 = aP(p, L) \cup pP(p, L)$. Given $r_n \in R \cap aP(p, L)$, in the previous paragraph, we saw that $r_n \notin pP(p, L)$. Then $r_n \neq P(p, L)$. Thus $r_n \in R \cap aP(p, L) \setminus \{P(p, L)\}$. By definition, $p_n r_n \cap L = \{r_n\}$. Since $aP(p, L) \subset L$, we have that $p_n r_n \cap aP(p, L) = \{r_n\}$. Thus $p_n r_n \cap L_1$ is the union of the closed disjoint sets $\{r_n\}$ and $p_n r_n \cap pP(p, L)$. Since X is a dendroid, this union is a connected set. This implies that $p_n r_n \cap pP(p, L) = \emptyset$, so $p_n r_n \cap L_1 = \{r_n\}$. Hence $P(p_n, L_1) = r_n$.

Since $r \neq p$ and $p \in L_1$, we obtain a contradiction with Claim 4. Since this contradiction comes from the assumption that $p \notin L$, we conclude that $p \in L$. Similarly, $q \in L$. Since $r \neq p$ or $r \neq q$, we obtain a contradiction with Claim 4 or with Claim 5. This finishes the proof of Claim 6.

Claim 7. It is possible to construct:

(a) a sequence of infinite subsets $J_1, J_2, \ldots$ of $\mathbb{N}$, and
(b) a sequence of points $z_0, z_1, z_2, \ldots$ of X such that:

- (A) $z_0 \in \{p, q\}$, $z_1 \in pq \setminus \{z_0\}$,
- (B) if $n_k = \min(J_k)$, then $M_{z_0 z_k} \leq n_k < n_{k+1}$ for every $k \in \mathbb{N}$,
- (C) if $z_0 = q$, then for each $k \geq 2$ and each $n \in J_k$, $z_k = P(q_n, z_0 q_{n_{k-1}})$,
- (D) if $z_0 = p$, then for each $k \geq 2$ and each $n \in J_k$, $z_k = P(p_n, z_0 p_{n_{k-1}})$,
- (E) $z_0 z_{k-1} \subset z_0 z_k$ for every $k \in \mathbb{N}$, and
- (F) $J_1 \supset J_2 \supset J_3 \cdots$.

We construct the sequences $\{J_k\}_{k=1}^{\infty}$ and $\{z_k\}_{k=1}^{\infty}$ by induction.

By Claim 6, $\{P(q_n, pq) : n \in \mathbb{N} \text{ and } M_{pq} \leq n\}$ is finite. Then there exists a $z_1 \in pq$ such that the set $J_1 = \{n \in \mathbb{N} : M_{pq} \leq n \text{ and } P(q_n, pq) = z_1\}$ is infinite. Fix a point $z_0 \in \{p, q\} \setminus \{z_1\}$. Let $n_1 = \min(J_1)$.

Given $n \in J_1$, we have that $n \geq M_{pq}$, since $z_0z_1 \subset pq$, we have that $p_n, q_n \notin pq$ and $p_n, q_n \notin z_0z_1$. Moreover, since $P(p_n, pq) = z_1 = P(q_n, pq)$, we have that $P(p_n, z_0z_1) = z_1 = P(q_n, z_0z_1)$. Hence $M_{z_0z_1} \leq n$ and $q_nz_1 \cap pq = \{z_1\}$. Therefore $z_1 \in z_0q_n$. In particular, $z_1 \in z_0q_{n_1}$ and $M_{z_0z_1} \leq n_1$.

Now, suppose that $J_1, \ldots, J_k$ and $z_1, \ldots, z_k$ have been constructed satisfying properties (A)–(F) and $z_0 = q$ (the construction for the case $z_0 = p$ is similar). Set $n_k = \min(J_k)$.

Set $L' = z_0q_{n_k}$. In the case $k = 1$, we have seen that $z_1 \in z_0q_{n_1} = L'$. In the case $k \geq 2$, since $n_k \in J_k$, $M_{z_0z_k} \leq n_k$, so $q_{n_k} \notin z_0z_k$, and by induction hypothesis, $z_k = P(q_{n_k}, z_0q_{n_{k-1}})$, then $q_{n_k}z_k \cap z_0q_{n_{k-1}} = \{z_k\}$. This implies that $z_k \in z_0q_{n_k} = L'$.

By Claim 6, the set $\{P(q_n, L') \in L' : n \in J_k\}$ is finite. Since J_k is infinite, there exists a $z_{k+1} \in L'$ such that the set $J = \{n \in J_k : n > \max\{n_k, M_{L'}\} \text{ and } z_{k+1} = P(q_n, L')\}$ is infinite. Let $J_{k+1} = \{n \in J : M_{z_0z_{k+1}} < n\}$ and $n_{k+1} = \min(J_{k+1})$.

Fix $n \in J_{k+1}$. Then $n \in J_k$. In the case $k \geq 2$, by (C), $z_k = P(q_n, z_0q_{n_{k-1}})$. This implies that $z_k \in z_0q_n$ and we have seen that $z_k \in L'$. In the case $k = 1$, we have seen that $z_1 \in z_0q_n$ and $z_1 \in L'$. In both cases, $z_k \in z_0q_n \cap L'$. Since $n \in J$, $q_nz_{k+1} \cap L' = \{z_{k+1}\}$. Thus $z_0q_n \cap L' = z_0q_n \cap z_0q_{n_k} = z_0z_{k+1}$. Therefore $z_k \in z_0z_{k+1}$ and $z_0z_k \subset z_0z_{k+1}$.

This completes the inductive construction and the proof of Claim 7.

Thus the sequence $\{z_0z_k\}_{k=1}^{\infty}$ is an increasing sequence of arcs in the dendroid X. By Theorem 7.2, there exists a point $z \in X$ such that $z_0z_k \subset z_0z$ for all $k \in \mathbb{N}$ and $\lim_{k\to\infty} z_k = z$. Set $L = z_0z$. Let $K \in \mathbb{N}$ be such that for each $k \geq K$, $n_k > M_L$.

Claim 8. For each $k \geq K$, $p_{n_k} \notin L$, $q_{n_k} \notin L$, $P(q_{n_k}, L) = P(p_{n_k}, L)$ and if $y_k = P(p_{n_k}, L)$, then $y_k \in z_{k+1}z$. Moreover, the set $\{y_k \in L : k \geq K\}$ is finite.

In order to prove the first part of Claim 8, by the definition of M_L, we only need to prove that given $k \geq K$, $y_k \in z_{k+1}z$. Suppose, for example, that $z_0 = q$. Since $z_{k+1} = P(q_{n_{k+1}}, z_0q_{n_k})$, we have that $z_{k+1} \in z_0q_{n_k}$. Then $z_0z_{k+1} \subset z_0z \cap z_0q_{n_k}$. This implies that $y_k = P(q_{n_k}, z_0z) \in z_{k+1}z$.

The fact that $\{y_k \in L : k \geq K\}$ is finite is immediate from Claim 6. This finishes the proof of Claim 8.

Claim 9. Suppose that $z_0 = p$ and $k, m \geq K$. Let y_k and y_m be defined as in Claim 8. Suppose also that $k \neq m$ and $y_k = y_m$. Then $p_{n_k}y_k \cap p_{n_m}y_m = \{y_k\}$.

In order to prove Claim 9, suppose that $k < m$. Since $n_m \in J_m \subset J_{k+1}$, $z_{k+1} = P(p_{n_m}, z_0p_{n_k})$. This implies that $z_{k+1} \in p_{n_m}p_{n_k}$. Since $z_{k+1} \in L$, the definitions of y_m and y_k imply that $y_m \in p_{n_m}z_{k+1}$ and $y_k \in z_{k+1}p_{n_k}$. Since $y_k = y_m$, we obtain that $y_k = z_{k+1} = y_m$. Hence $p_{n_k}y_k \cap p_{n_m}y_m = p_{n_k}z_{k+1} \cap p_{n_m}z_{k+1} = \{z_{k+1}\} = \{y_k\}$.

The proof of the following claim is similar to the proof of Claim 9.

Claim 10. If $z_0 = q$ and $k, m \geq K$ are such that $k \neq m$ and $y_k = y_m$, then $q_{n_k} y_k \cap q_{n_m} y_m = \{y_k\}$.

By Claim 6, there exists a point $y \in L$ and there exists a sequence of positive integers $K < k_1 < k_2 < \cdots$ such that $y_{k_j} = y$ for every $j \in \mathbb{N}$.

We will obtain a final contradiction by showing that X contains a semi-broom. We assume that $z_0 = q$, the case $z_0 = p$ is similar.

We consider the arc L, the points $q = z_0$ and $y \in z_1 z \subset L \setminus \{z_0\}$ (Claim 8). We also consider the sequences $\{q_{n_{k_j}}\}_{j=1}^{\infty}$ in $X \setminus L$ and the constant sequence $\{y_{k_j}\}_{j=1}^{\infty}$ in L, the continuum $Y = L \cup \mathrm{cl}_X(\bigcup\{q_{n_{k_j}} y : j \in \mathbb{N}\})$. Clearly, Y is a semi-broom in X.

This contradicts the assumption that X does not contain a semi-comb and completes the proof of the sufficiency. □

7.5 Exercises

Exercise 7.20 Show that every non-degenerate subcontinuum of a dendroid is a dendroid.

Exercise 7.21 Prove that if X is a dendroid, then for each pair of distinct points of X there exists a unique arc in X joining them.

Exercise 7.22 Find a uniquely arcwise connected continuum which is not a dendroid.

Exercise 7.23 Prove that a continuum X is hereditarily unicoherent if and only if for each pair of points p, q in X there exists a unique subcontinuum A containing p and q which is contained in every subcontinuum of X that contains p and q.

Exercise 7.24 Let X be a continuum. Prove that X is a dendroid if and only if every subcontinuum of X is uniquely arcwise connected.

Exercise 7.25 Show that the dendroid in Fig. 7.4 is not contractible (this dendroid is the union of two harmonic fans joined by a point).

Exercise 7.26 Prove that the topological cone over a compact metric space Z is a dendroid if and only if Z is totally disconnected.

Exercise 7.27 Show that the dendroid in Fig. 7.5 is contractible but it contains a non-contractible subcontinuum.

Exercise 7.28 The dendroid X in Fig. 7.6 is constructed in the Euclidean space $\mathbb{R}^3$ by taking a simple triod T in the plane $z = 0$. The dendroid X also contains two sequences of subcontinua $\{A_n\}_{n=1}^{\infty}$ and $\{B_n\}_{n=1}^{\infty}$ as in Fig. 7.6, the sequence of simple triods A_n approaches T and the same happens with the sequence of arcs B_n. The subcontinua T, A_n and B_n contain exactly one common point V. This

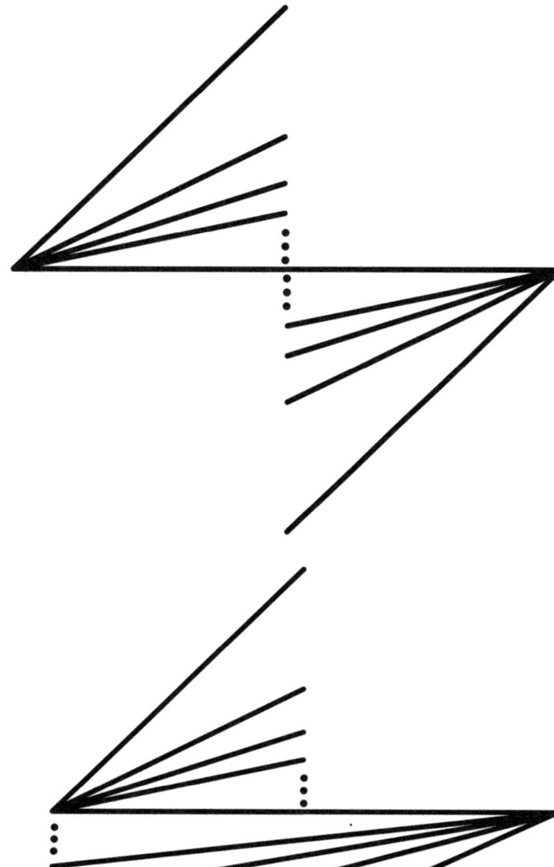

Fig. 7.4 A non-contractible dendroid

Fig. 7.5 Contractibility is not hereditary for dendroids

dendroid is very important for the theory of hyperspaces [24, Theorem 5.78] and [114, Example 4.2]. Prove that X is contractible.

Exercise 7.29 Let X be a dendroid. Given a subcontinuum A of X and $p \in X$, show that there exists a unique point $a \in A$ such that $pa \cap A = \{a\}$. Prove that if $p \in X \setminus A$, then for each $x \in A$, $a \in px$. Define $f_A : X \to A$ by $f_A(p) = a$. Prove that X is a dendrite if and only if for each subcontinuum A of X, f_A is continuous.

Exercise 7.30 Let X be a dendroid. Prove that X is a dendrite if and only if for every pair of points $p, q \in X$, with $p \neq q$, there exist respective neighborhoods U

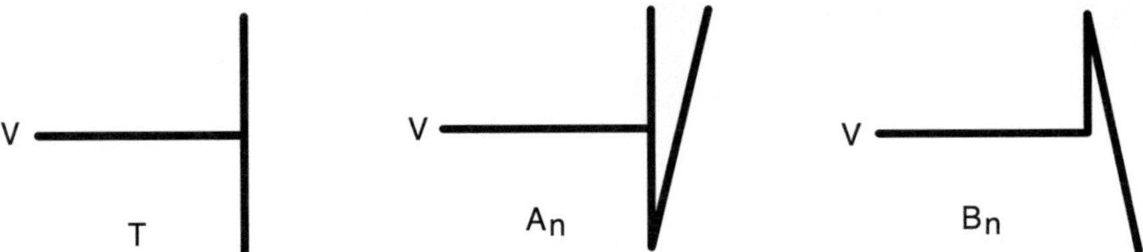

Fig. 7.6 The scissors dendroid

and V of p and q in X such that for every $u \in U$ and every $v \in V$, $uv \cap pq \neq \emptyset$. (Hint: use Theorem 3.5.)

Exercise 7.31 Let X be a dendroid and $p \in X$. Show that

$$X = \bigcup\{pe \subset X : e \in E(X)\}.$$

Exercise 7.32 Show that the monotone image of a dendrite is a dendrite.

Exercise 7.33 Let X be a dendroid. Prove that if $\{A_n\}_{n=1}^{\infty}$ is a sequence of arcs in X such that $A_1 \subset A_2 \subset \cdots$, then $\mathrm{cl}_X(\bigcup\{A_n : n \in \mathbb{N}\})$ is an arc. Show that this property characterizes dendrites among Peano continua.

Exercise 7.34 Prove that the intersection of two connected subsets of a dendrite is also connected. Prove that this property characterizes dendrites among dendroids. (Hint: use Theorem 7.19.) Also find a dendroid X and two distinct points p and q in X such that no point $x \in pq \setminus \{p, q\}$ separates p and q in X.

Exercise 7.35 Let X be a continuum and $p \in X$. Prove that if $o(p, X) = 1$, then p is an end-point of X. Also find a dendroid X without points of order 1.

Exercise 7.36 Prove that if A is a non-degenerate subcontinuum of a dendroid and if A is a finite union of arcs, then A is a tree.

Exercise 7.37 Find a dendroid with exactly one cut point, one with exactly two cut points, one with exactly three cut points, etc.

Exercise 7.38 Prove that every subcontinuum of a dendrite is a dendrite.

Exercise 7.39 Find a fan X with vertex v such that $X \setminus \{v\}$ is connected.

Exercise 7.40 Let X be a continuum and $p \in X$. Prove that $c(p, X) \leq o(p, X)$ and find a dendroid X such that for each $p \in X$, $c(p, X) < o(p, X)$.

Exercise 7.41 Find an uncountable family of pairwise non-homeomorphic dendrites.

Exercise 7.42 Prove that the union of two intersecting trees in a dendroid is also a tree.

Chapter 8
The Cantor Set

P.S. Alexandroff and F. Hausdorff ([1] and [46]) proved that any compact metric space X is a continuous image of the Cantor set. One popular proof of this theorem uses that X is the union of a finite number of sets with arbitrarily small diameters to construct an appropriate sequence of coverings of closed subsets of X, and then uses the fact that each point p of X is determined by the intersection of the elements of the coverings containing p to define the suitable mapping. In this chapter we present a nicer and more elegant proof which, although is contained in some classical books (e.g. [90, Corollary 3a, p. 23] (1968) and [91, Thm. 4, Ch. XVI, §8, p. 214] (1972)), seems to be not very well known. In fact, in 1976 I. Rosenholtz [117] published this proof with the title "Another proof that any compact metric space is the continuous image of the Cantor set".

The pieces we need for the proof of this theorem are: (a) The Cantor C set is homeomorphic to $\{0, 2\} \times \{0, 2\} \times \cdots$ ($\{0, 2\}$ is considered with the discrete topology); (b) C is homeomorphic to $C \times C \times \cdots$; (c) the interval $[0, 1]$ is a continuous image of C; (d) the Hilbert cube is a continuous image of C; (d) any compact metric space is embeddable in the Hilbert cube; and (e) each closed subset of C is a retract of C. As we will see in this chapter, these preliminary results are relatively easy to show. In fact, we leave most of them as exercises.

8.1 The Cantor Set as a Product

First, we recall the usual construction of the Cantor set C as the intersection of a decreasing sequence of nonempty compact subsets $C_1, C_2, \ldots$ in the interval $[0, 1]$ (Fig. 8.1).

The set C_1 is defined as the unit interval $[0, 1]$.

The set C_2 is obtained by removing the middle open interval $(\frac{1}{3}, \frac{2}{3})$ from $[0, 1]$. So $C_2 = [0, \frac{1}{3}] \cup [\frac{2}{3}, 1]$.

A. Illanes, *Continuum Theory*, Universitext,
https://doi.org/10.1007/978-3-031-91011-1_8

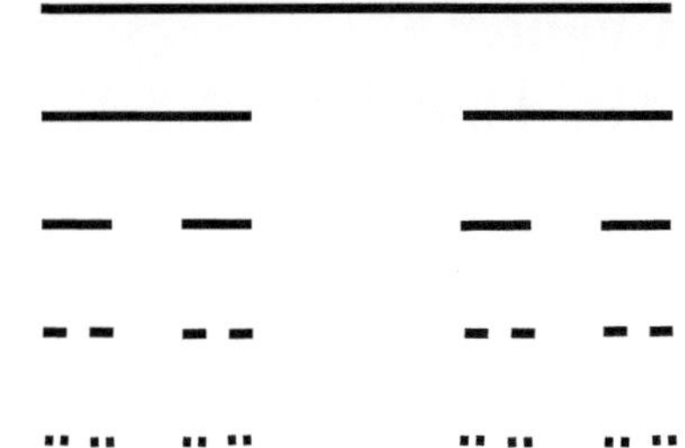

Fig. 8.1 The Cantor set

The set C_3 is obtained by removing the middle open interval from each of the components of C_2. That is, C_3 is C_2 minus the intervals $(\frac{1}{9}, \frac{2}{9})$ and $(\frac{7}{9}, \frac{8}{9})$. So, $C_3 = [0, \frac{1}{9}] \cup [\frac{2}{9}, \frac{3}{9}] \cup [\frac{6}{9}, \frac{7}{9}] \cup [\frac{8}{9}, \frac{9}{9}]$.

In general, C_{n+1} is obtained by removing the middle open interval from each of the components of C_n.

The Cantor set is then defined by

$$C = \bigcap \{C_n : n \in \mathbb{N}\}.$$

For working with the Cantor set in a more formal way, we need an analytic description of C.

In order to obtain this description, we recall that when we write a number t in the interval $[0, 1]$ in base 3, we write:

$$t = \frac{a_1}{3^1} + \frac{a_2}{3^2} + \frac{a_3}{3^3} + \cdots .$$

Where each a_i takes one of the values 0, 1 or 2.

The following notation is used:

$$t = (.a_1a_2a_3 \ldots)_3.$$

The elements of the Cantor set are exactly those numbers in $[0, 1]$ whose coordinates in the base 3 representation are numbers in $\{0, 2\}$ (Exercise 8.5). Set

$$\mathcal{C} = \{0, 1\} \times \{0, 1\} \times \cdots .$$

Consider $\mathcal{C}$ with the product topology, where each set $\{0, 1\}$ is considered with the discrete topology.

Consider the function $\varphi : \mathcal{C} \to C$ given by

$$\varphi((a_1, a_2, a_3, \ldots)) = (.(2a_1)(2a_2)(2a_3) \ldots)_3 = \frac{2a_1}{3^1} + \frac{2a_2}{3^2} + \frac{2a_3}{3^3} + \cdots .$$

In Exercise 8.6, it is asked to prove that φ is a homeomorphism. For the rest of this section, we will identify the Cantor set with the space C.

We will use the well-known fact that a countable union of countable sets is countable. Therefore there exists a one-to-one surjective function $\gamma = (\alpha, \beta) : \mathbb{N} \to \mathbb{N} \times \mathbb{N}$, where α and β are the coordinate functions of γ. We will use the function γ in the following result.

Theorem 8.1 *C is homeomorphic to $C \times C \times \cdots$.*

Proof Given an element $p \in C \times C \times \cdots$, p is a sequence of sequences of numbers in $\{0, 1\}$, so we write $p = (p_1, p_2, p_3, \ldots)$ and, for each $n \in \mathbb{N}$, set

$$p_n = (p((1, n)), p((2, n)), p((3, n)), \ldots),$$

where each $p((m, n))$ belongs to $\{0, 1\}$.

Define $\psi : C \times C \times \cdots \to C$ by

$$\psi(p) = (p(\gamma(1)), p(\gamma(2)), p(\gamma(3)), \ldots).$$

Given $k \in \mathbb{N}$, $\gamma(k) = (\alpha(k), \beta(k))$. Then $p(\gamma(k)) = p((\alpha(k), \beta(k))) \in \{0, 1\}$, so $\psi(p)$ is a sequence of elements in $\{0, 1\}$ and hence $\psi(p) \in C$. Moreover, $p(\gamma(k))$ is the coordinate $\alpha(k)$ of the point $p_{\beta(k)}$. To obtain $p(\gamma(k))$, we first choose the point $p_{\beta(k)}$ and then we choose the $\alpha(k)$-coordinate of p_{β_k}. That is, to obtain the k-coordinate of $\psi(p)$, first we project the point p to obtain its $\beta(k)$-coordinate (the point $p_{\beta(k)}$) in C and then we project $p_{\beta(k)}$ to obtain its $\alpha(k)$-coordinate. Thus, we obtain $p(\gamma(k)) = p((\alpha(k), \beta(k)))$ by composing two continuous functions. Thus each of the coordinate functions of ψ is continuous. Therefore ψ is continuous.

In order to check that ψ is surjective, take $a = (a_1, a_2, \ldots) \in C$. We need to construct a point $p = (p_1, p_2, p_3, \ldots)$ such that $\psi(p) = a$, so for each $n \in \mathbb{N}$ we need to define the sequence $p_n = (p((1, n)), p((2, n)), p((3, n)), \ldots)$. In order to do this, we need to define $p(m, n) \in \{0, 1\}$ for each (m, n). So, for each $(m, n) \in \mathbb{N} \times \mathbb{N}$ define $p((m, n)) = a_{\gamma^{-1}((m,n))}$. Then we define

$$p = (p_1, p_2, p_3, \ldots), \text{ where } p_n = (p((1, n)), p((2, n)), p((3, n)), \ldots).$$

We check that $\psi(p) = a$. Given $k \in \mathbb{N}$, we write $\gamma(k) = (m, n)$. Then

$$p(\gamma(k)) = p((m, n)) = a_{\gamma^{-1}((m,n))} = a_k.$$

Thus $\psi(p) = a$. Therefore ψ is surjective.

In order to show that ψ is one-to-one, take $p, q \in C \times C \times \cdots$ such that $\psi(p) = \psi(q)$. Then

$$(p(\gamma(1)), p(\gamma(2)), p(\gamma(3)), \ldots) = (q(\gamma(1)), q(\gamma(2)), q(\gamma(3)), \ldots).$$

Given $m, n \in \mathbb{N}$, there exists a unique $k \in \mathbb{N}$ such that $\gamma(k) = (m, n)$. Then $p((m,n)) = p(\gamma(k)) = q(\gamma(k)) = q((m,n))$. Since this equality holds for each $m \in \mathbb{N}$, we have that $p_n = q_n$ for every $n \in \mathbb{N}$. Hence $p = q$.

Since $C \times C \times \cdots$ is compact and C is a Hausdorff space, we conclude that ψ is a homeomorphism. □

8.2 Images of the Cantor Set

Theorem 8.2 *Every non-empty compact metric space is a continuous image of the Cantor set.*

Proof Let X be a compact metric space. By Theorem 1.1, we may assume that X is contained in the Hilbert cube $\mathbf{Q}$. By Exercise 8.13, there exists a surjective mapping $g : C \to \mathbf{Q}$.

Let $A = g^{-1}(X)$. Then A is a closed and nonempty subset of C. By Exercise 8.15, there exists a retraction $r : C \to A$.

Define $f : C \to X$ by $f = g \circ r$.

Since $r(C) \subset A$, we have that $f(C) \subset X$. Thus f is well defined and continuous.

Given $x \in X$, since g is surjective, there exists a $p \in C$ such that $x = g(p)$. Then $p \in A$ and $f(p) = g(r(p)) = g(p) = x$. We have shown that f is surjective. □

8.3 A Characterization

Theorem 8.3 *Let X be a nonempty compact metric space. Then X is homeomorphic to the Cantor set if and only if X is 0-dimensional and X does not contain isolated points.*

Proof The necessity is left as an exercise (Exercise 8.21).

To prove the sufficiency, suppose that X is 0-dimensional and it does not contain isolated points.

Given $m \in \mathbb{N}$, define

$$J(m) = \{0, 1\}^m = \{(a_1, \ldots, a_m) : a_i \in \{0, 1\} \text{ for every } i \in \{1, \ldots, n\}\}.$$

We will construct, inductively:

(a) a sequence of positive integers $m_1, m_2, \ldots$, and
(b) a sequence of coverings $\mathcal{A}_n = \{A^{(n)}(s) : s \in J(m_1) \times \cdots \times J(m_n)\}$ of nonempty, pairwise disjoint, open and closed subsets of X,

such that:

(1) for every $n \in \mathbb{N}$ and $s \in J(m_1) \times \cdots \times J(m_n)$, diameter$(A^{(n)}(s)) < \frac{1}{n}$, and
(2) for each $s = (r_1, \ldots, r_n) \in J(m_1) \times \cdots \times J(m_n)$,

$$A^{(n)}(r_1, \ldots, r_n) = \bigcup\{A^{(n+1)}(r_1, \ldots, r_n, r_{n+1}) : r_{n+1} \in J(m_{n+1})\}.$$

For $n = 1$, take a covering $\mathcal{A}_1$ as in Exercise 8.20 (applied to $\varepsilon = 1$) and with as many elements as we wish. So we can take such a covering with 2^{m_1} elements, with $1 \leq m_1$. Then we can write $\mathcal{A}_1 = \{A^{(1)}(s) : s \in J(m_1)\}$. Thus $\mathcal{A}_1$ has the required properties.

Now, suppose that we have constructed positive integers $m_1, \ldots, m_n$ and coverings $\mathcal{A}_1, \mathcal{A}_2, \ldots, \mathcal{A}_n$ satisfying the required properties.

Note that each of the spaces $A^{(n)}(s)$, where $s \in J(m_1)\times\cdots\times J(m_n)$, is compact, 0-dimensional, metric and without isolated points.

By Exercise 8.20, each space $A^{(n)}(s)$ can be divided into as many sets as we wish, so we can choose coverings for all spaces $A^{(n)}(s)$ in such a way that each cover has $2^{m_{n+1}}$ elements, for some positive integer m_{n+1}. Then for each $s \in J(m_1) \times \cdots \times J(m_n)$ there exists a covering $\{A^{(n+1)}(s, r_{n+1}) : r_{n+1} \in J(m_{n+1})\}$ of $A^{(n)}(s)$ by nonempty open and closed subsets which are pairwise disjoint, and each of the elements of the covering having diameter less than $\frac{1}{n+1}$.

Grouping all these covers into one family, we obtain a cover for X:

$$\begin{aligned}\mathcal{A}_{n+1} &= \{A^{(n+1)}(s, r_{n+1}) : s \in J(m_1) \times \cdots \times J(m_n), \quad r_{n+1} \in J(m_{n+1})\} \\ &= \{A^{(n+1)}(s) : s \in J(m_1) \times \cdots \times J(m_{n+1})\}.\end{aligned}$$

Clearly, the family $\mathcal{A}_{n+1} = \{A^{(n+1)}(s) : s \in J(m_1) \times \cdots \times J(m_{n+1})\}$ is a covering of X by nonempty, pairwise disjoint, open and closed subsets of X, with each element having diameter less than $\frac{1}{n+1}$.

This completes the inductive construction of the sequences $\{m_n\}_{n=1}^{\infty}$ and $\{\mathcal{A}_n\}_{n=1}^{\infty}$.

For each $n \in \mathbb{N}$, set $k_n = m_1 + \cdots + m_n$. Let $k_0 = 0$. Note that $k_1 = m_1$ and for each $n \in \mathbb{N}$, $n \leq k_n$.

Define $\psi : C \to X$ as follows.

Given $a = (a_1, a_2, \ldots) \in C$, we consider the following sequence of finite sequences:

$$r_1 = (a_1, \ldots, a_{k_1}),\ r_2 = (a_{k_1+1}, \ldots, a_{k_2}),\ r_3 = (a_{k_2+1}, \ldots, a_{k_3}), \ldots$$

Note that for each $n \in \mathbb{N}$, $r_n \in J(m_n)$.

Consider the family of sets: $A^{(1)}(r_1)$, $A^{(2)}(r_1, r_2)$, $A^{(3)}(r_1, r_2, r_3), \ldots$

By Property (2), we have that

$$A^{(1)}(r_1) \supset A^{(2)}(r_1, r_2) \supset A^{(3)}(r_1, r_2, r_3) \supset \cdots.$$

By Property (1), for each $n \in \mathbb{N}$, diameter$(A^{(n)}(r_1, \ldots, r_n)) < \frac{1}{n}$.

Thus the set $\bigcap\{A^{(n)}(r_1, \ldots, r_n) : n \in \mathbb{N}\}$ is a one-point set which we will denote by $\psi(a)$. That is, $\psi : C \to X$ is defined by

$$\{\psi(a)\} = \bigcap\{A^{(n)}(r_1, \ldots, r_n) : n \in \mathbb{N}\}.$$

Then by the definition, $\psi(a) \in A^{(n)}(r_1, \ldots, r_n)$ for every $n \in \mathbb{N}$.

We check that ψ is continuous. Let $a = (a_1, a_2, \ldots) \in C$ and $\varepsilon > 0$. Take $n \in \mathbb{N}$ such that $\frac{1}{n} < \varepsilon$. Consider $r_1, r_2, \ldots$ as before.

Let $U = \{a_1\} \times \cdots \times \{a_{k_n}\} \times \{0, 1\} \times \{0, 1\} \times \cdots$. Then U is open in C and contains the point a. Given $b \in U$, we have that b is of the form.

$$b = (a_1, a_2, \ldots, a_{k_n}, b_{k_n+1}, b_{k_n+2}, \ldots).$$

If we construct the respective sequence $\{r'_n\}_{n=1}^{\infty}$ for b, we have that $r'_1 = r_1, \ldots, r'_n = r_n$. According to the definition of ψ, $\psi(b) \in A^{(n)}(r'_1, \ldots, r'_n) = A^{(n)}(r_1, \ldots, r_n)$. By (1), this set has diameter less than $\frac{1}{n}$. Thus $\psi(b) \in B(\psi(a), \varepsilon)$. This finishes the proof that ψ is continuous.

In order to show that ψ is one-to-one, take $a = (a_1, a_2, \ldots)$, and $b = (b_1, b_2, \ldots) \in C$ such that $a \neq b$. Let $n \in \mathbb{N}$ be such that $a_n \neq b_n$. Let $s = (a_1, \ldots, a_{k_n})$ and $s' = (b_1, \ldots, b_{k_n})$. Since $n \leq k_n$, we have that $s \neq s'$. Since $A^{(n)}(s)$, $A^{(n)}(s')$ are elements of the covering $\mathcal{A}_n$, they are disjoint. Since $\psi(a) \in A^{(n)}(s)$ and $\psi(b) \in A^{(n)}(s')$, we conclude that $\psi(a) \neq \psi(a')$. Therefore, ψ is one-to-one.

Finally, we see that ψ is surjective.

Let $p \in X$. Since $\mathcal{A}_1$ covers X, there exists $r_1 = (a_1, \ldots, a_{k_1}) \in J(m_1)$ such that $p \in A^{(1)}(r_1)$. By (2), $\{A^{(2)}(r_1, s) : s \in J(m_2)\}$ is a cover of $A^{(1)}(r_1)$, we have that there exists $r_2 \in J(m_2)$ such that $p \in A^{(2)}(r_1, r_2)$. Proceeding in this way, for each $n \in \mathbb{N}$ is possible to find $r_n \in J(m_n)$ such that $p \in A^{(n)}(r_1, \ldots, r_n)$. Then $p \in \bigcap\{A^{(n)}(r_1, \ldots, r_n) : n \in \mathbb{N}\}$.

For each $n \in \mathbb{N}$, set $r_n = (a_{k_{n-1}+1}, \ldots, a_{k_n})$. Define $a = (a_1, a_2, \ldots) \in C$. Clearly, $p = \psi(a)$. Therefore ψ is surjective.

We have shown that C is homeomorphic to X. By Exercise 8.6, C is homeomorphic to C. Therefore C is homeomorphic to X. □

Theorem 8.4 *Let X be a nonempty compact 0-dimensional metric space. Then X can be embedded in the Cantor set.*

Proof By Theorem 1.1, we may assume that X is contained in the Hilbert cube $\mathbf{Q}$. Let d be a metric for $\mathbf{Q}$. By Theorem 8.3, we may suppose that X contains isolated points. Let $R = \{p \in X : p \text{ is an isolated point of } X\}$.

Given $p \in R$, we can fix a number $\varepsilon_p > 0$ such that $\{p\} = X \cap B(p, 3\varepsilon_p)$. Choose a Cantor set C_p contained in $B(p, \varepsilon_p)$. This can be done by fixing an arc α

in $B(p, \varepsilon_p)$ such that $p \in \alpha$ and then constructing C_p in α, with $p \in C_p$. Define

$$Z = X \cup (\bigcup \{C_p : p \in R\}).$$

We prove that Z is compact, 0-dimensional and Z contains no isolated points. In this way we will have a Cantor set Z containing X.

To prove that Z is compact, it is enough to show that Z is closed in $\mathbf{Q}$. Take a sequence $\{z_n\}_{n=1}^{\infty}$ in Z converging to a point $z \in \mathbf{Q}$. We prove that $z \in Z$. If $z_n \in X$ for infinitely many numbers n, then $z \in X$ and we are done. Thus, taking a subsequence if necessary, we may assume that no point z_n belongs to X. Similarly, we may assume that each set of the form C_p contains at most one element of the sequence. Hence there exists a sequence of points $\{p_n\}_{n=1}^{\infty}$ in X such that for each $n \in \mathbb{N}$, $z_n \in C_{p_n}$; and $p_n \neq p_m$, if $n \neq m$. We are going to see that $\lim_{n\to\infty} p_n = z$. Let $\delta > 0$. Then there exists an $n \in \mathbb{N}$ such that for each $n \geq N$, $d(z_n, z) < \delta$. Given $n, m \geq N$, suppose for example that $\varepsilon_{p_m} \leq \varepsilon_{p_n}$. By the choice of ε_{p_n}, C_{p_n} and C_{p_m}, we have that

$$3\varepsilon_{p_n} \leq d(p_n, p_m) \leq d(p_n, z_n) + d(z_n, z_m) + d(z_m, p_m) < 2\varepsilon_{p_n} + 2\delta.$$

Thus $\varepsilon_{q_m} \leq \varepsilon_{p_n} < 2\delta$. Therefore $d(p_n, z) \leq d(p_n, z_n) + d(z_n, z) < 3\delta$. We have shown that for every $n \geq N$, $d(p_n, z) < 3\delta$. This ends the proof that $\lim_{n\to\infty} p_n = z$. Hence $z \in X$. Therefore, Z is compact.

Let $p, q \in R$ be such that $p \neq q$. We check that $C_q \cap B(p, \varepsilon_p) = \emptyset$. Suppose that $\varepsilon_p \leq \varepsilon_q$ (the other case is similar). If there exists a point $x \in C_q \cap B(p, \varepsilon_p) \subset B(q, \varepsilon_q) \cap B(p, \varepsilon_p)$, then $d(p, q) \leq 2\varepsilon_q$ and $p \in B(q, 2\varepsilon_q) \cap X = \{q\}$, so $p = q$, a contradiction. We have shown that $C_q \cap B(p, \varepsilon_p) = \emptyset$.

Given $p \in R$, since $X \cap B(p, 3\varepsilon_p) = \{p\}$, we infer that $Z \cap B(p, \varepsilon_p) = C_p$. Since C_p is compact, we conclude that C_p is open and closed in Z.

In order to show that Z is 0-dimensional, by Exercise 3.6, we only need to prove that the only connected subsets of Z are the one-point sets. Let $E \subset Z$ be connected. We show that E is a one-point set. Since X is 0-dimensional, in the case that $E \subset X$, we are done. Suppose then that there exists a $p \in R$ such that $E \cap C_p \neq \emptyset$. By the connectedness of E and the previous paragraph, we have that $E \subset C_p$. Since C_p is a Cantor set, we conclude that E is a one-point set. Therefore Z is 0-dimensional.

Finally, we prove that Z does not have isolated points. Let $z \in Z$. If $z \in X \setminus R$, then z is an accumulation point of $X \subset Z$ and we are done. If z belongs to some C_p, then since C_p is a Cantor set, z is an accumulation point of $C_p \subset Z$ and we are done. Therefore Z does not have isolated points.

By Theorem 8.3, Z is a Cantor set. □

8.4 Two Important Mappings

We finish this chapter by defining two mappings on C that have important dynamic properties.

The *shift mapping* is the function $\sigma : C \to C$ defined by

$$\sigma(a_1, a_2, a_3, \ldots) = (a_2, a_3, \ldots)$$

The *adding machine* is the function $\omega : C \to C$ defined by

$$\omega(1, 1, \ldots) = (0, 0, \ldots), \text{ and}$$
$$\omega(a_1, a_2, \ldots) = (0, \ldots, 0, 1, a_{m+1}, a_{m+2}, \ldots),$$

where $m = \min\{i \in \mathbb{N} : a_i = 0\}$, if $(a_1, a_2, \ldots) \neq (1, 1, \ldots)$.

Another way to describe ω is the following. If $(a_1, a_2, \ldots) \neq (1, 1, \ldots)$, then $\omega(a_1, a_2, \ldots)$ changes the first ones to zeros, the first zero changes to one, and the rest of the numbers do not change.

The reason of calling ω the "adding machine" is the following.

We think of the sequence $a = (a_1, a_2, \ldots)$ backwards, without parenthesis or commas. That is, we think of the sequence as follows.

$$\ldots a_3 a_2 a_1.$$

To obtain $\omega(a)$, add the number 1 by the usual algorithm for the sum (modulo 2), that is, add 1 to a_1 with "carry over" to the left. This means that if $a_1 = 0$, then the result is to change a_1 to 1, and if $a_1 = 1$, then add 1 to a_1 to obtain the number 10, so change a_1 to 0 and carry over 1. This number 1 is added to a_2 in the same way. Continue this procedure until the first m for which a_m is equal to 0, which is changed to 1. The rest of the coordinates $a_{m+1}, a_{m+2}, \ldots$ do not change.

8.5 Exercises

Exercise 8.5 Show that the elements of the Cantor set are exactly those numbers in $[0, 1]$ of the form $t = (.a_1 a_2 a_3 \ldots)_3$, where $a_i \in \{0, 2\}$ for all $i \in \mathbb{N}$.

Exercise 8.6 Let

$$C = \{0, 1\} \times \{0, 1\} \times \cdots,$$

where C is taken with the product topology and $\{0, 1\}$ is considered with the discrete topology.

Let $\varphi : C \to C$ be the function given by

$$\varphi((a_1, a_2, a_3, \ldots)) = (.(2a_1)(2a_2)(2a_3)\ldots)_3 = \frac{2a_1}{3^1} + \frac{2a_2}{3^2} + \frac{2a_3}{3^3} + \cdots.$$

Prove that φ is a homeomorphism.

Exercise 8.7 Prove that the Cantor set is homogeneous (see Definition 15.1).

Exercise 8.8 Prove that the Cantor set is uncountable.

Exercise 8.9 Prove that the Cantor set can be embedded in every continuum.

Exercise 8.10 Find a subset of the real line homeomorphic to the Cantor set and not containing any rational numbers.

Exercise 8.11 Prove that C is homeomorphic to $C \times C$.

Exercise 8.12 Let $f : C \to [0, 1]$ be given by

$$f((a_1, a_2, a_3, \ldots)) = \sum_{n=1}^{\infty} \frac{a_n}{2^n}.$$

Prove that f is a surjective continuous function. (Hint: after showing that f is continuous, by the compactness of C, in order to show that f is surjective, it is enough to show that the image of f contains the dense subset of $[0, 1]$ given by $\{\frac{k}{2^m} : k, m \in \mathbb{N} \text{ and } 1 \leq k \leq 2^m\}$.)

Exercise 8.13 Using Theorem 8.1, prove that the Hilbert cube is a continuous image of the Cantor set.

Exercise 8.14 Let $d_0 : C \times C \to [0, 1]$ be given by

$$d_0(a, b) = \sum_{n=1}^{\infty} \frac{|a_n - b_n|}{4^n}.$$

Prove that d_0 is a metric for C that generates the product topology on C, and satisfies the following property:

(a) there are no three distinct points a, b and c in C such that $d_0(a, b) = d_0(a, c)$.

Exercise 8.15 Let X be a compact metric space with a metric d satisfying property (a) in Exercise 8.14. Given a nonempty closed subset A of X, define $r : X \to A$ by: $r(p)$ is the point of A closest to p. Prove that r is a well-defined retraction from X onto A. As a consequence we obtain that the Cantor set can be retracted onto each of its nonempty closed subsets.

Exercise 8.16 Prove that the Cantor set is topologically characterized as the only non-degenerate compact metric space that can be retracted onto each of its nonempty closed subsets and which does not contain isolated points.

Exercise 8.17 Find a metric compact space Y, non-homeomorphic to the Cantor set, such that every nonempty compact metric space is a continuous image of Y.

Exercise 8.18 Suppose that Y is a compact metric space with the property that every nonempty compact metric space is a continuous image of Y. Prove that the Cantor set can be embedded in Y. (Hint: at some point use Exercise 1.46.)

Exercise 8.19 Let G be the Gehman dendrite discussed in Example 7.17. Let E be the set of end-points of G. Then E is homeomorphic to the Cantor set. Prove that for each compact metric space Z there exists a locally connected compactification of $G \setminus E$ with remainder Z.

Exercise 8.20 Let X be a compact, 0-dimensional, metric space without isolated points. Without using Theorem 8.3, prove that X satisfies the following property.

For each $\varepsilon > 0$, there exists an $N \in \mathbb{N}$ such that for each $n \geq N$, there exist nonempty pairwise disjoint open and closed subsets $A_1, \ldots, A_n$ such that $X = A_1 \cup \cdots \cup A_n$ and for each $i \in \{1, \ldots, n\}$, diameter$(A_i) < \varepsilon$.

Exercise 8.21 Prove that C does not contain isolated points.

Exercise 8.22 Let $\{A_n\}_{n=1}^{\infty}$ be a sequence of discrete non-degenerate finite topological spaces. Prove that $\prod\{A_n : n \in \mathbb{N}\}$ is homeomorphic to the Cantor set.

Exercise 8.23 Let σ be the shift mapping. A *periodic point for* σ is a point $p \in C$ for which there exists an $n \in \mathbb{N}$ such that $\sigma^n(p) = p$. Prove that the sets: $\{p \in C : p$ is a periodic point of $\sigma\}$ and $\{p \in C : \{\sigma^n(p) : n \in \mathbb{N}\}$ is dense in $C\}$ are dense in C.

Exercise 8.24 Let ω be the adding machine. Prove that ω does not contain fixed points (see Definition 13.1), and for each $p \in C$, $\{\omega^n(p) : n \in \mathbb{N}\}$ is dense in C.

Exercise 8.25 Let X be the cone over the Cantor set. Prove that there exists a mapping from X onto X with only one fixed point.

Exercise 8.26 Let G be the Gehman dendrite. Prove that there exists a mapping from G onto G with only one fixed point.

Exercise 8.27 Find two arcs in the plane whose intersection is the Cantor set.

Chapter 9
Hyperspaces of Continua

A hyperspace of a continuum X is a family of subsets of X satisfying a specific property, the most commonly studied are the following ($n \in \mathbb{N}$).

$$
\begin{aligned}
2^X &= \{A \subset X : A \text{ is a nonempty closed subset of } X\},\\
C_n(X) &= \{A \in 2^X : A \text{ has at most } n \text{ components}\},\\
F_n(X) &= \{A \in 2^X : A \text{ has at most } n \text{ points}\}.
\end{aligned}
$$

Although hyperspaces can be defined for more general topological spaces, we will see in this chapter that the hyperspaces of continua have very strong properties. This makes them a fertile field for research. A very important promoter of their study was Sam B. Nadler, Jr. His 1978 book "Hyperspaces of Sets" not only taught us the most important knowledge up to this year but offered us a large list of significant open problems that stimulated much of the research on this area.

In the middle of the 1990s, Sam gave me the honor of inviting me to collaborate in the writing of a new book on hyperspaces. His ideas for the book were the following:

(a) to present the fundamentals of hyperspaces in a pedagogically appropriate way,
(b) to survey the research in hyperspaces that had occurred since 1978, and
(c) to present the proofs of what he considered to be the most important results in this period of time, namely: (i) if X is a continuum such that $\dim[X] = 2$, then $\dim[C(X)] = \infty$ (M. Levin and Y. Sternfeld [92, Theorem 2.1] or [69, Theorem 73.9]); (ii) for a continuum X, $C(X)$ contains an n-cell if and only if X contains an n-od (A.I. [51, Theorem 1.9] or [69, Theorem 70.1], see Exercise 9.74 for the sufficiency); and (iii) for a continuum X, $C(X)$ is the product of two non-degenerate finite-dimensional continua if and only if X is either an arc or a simple closed curve (A.I. [53, Theorem 3.8] or [69, Theorem 79.2]).

A. Illanes, *Continuum Theory*, Universitext,
https://doi.org/10.1007/978-3-031-91011-1_9

In this chapter we develop the basic results of the theory of hyperspaces. These results are necessary (and in some cases sufficient) to understand most of the research on this topic. Of course, if the reader wants to go further in the study of hyperspaces, both books [112] and [69] offer many opportunities to increase their knowledge in this area.

For the interested readers, there are several sources on the history of hyperspaces we can mention: (a) S.B. Nadler, Jr.'s "Very brief history of hyperspace theory" in [112, pp. xii–xiv]; (b) section 12 ("Hyperspaces") by J.J. Charatonik in [22, pp. 750–753]; and (c) §5 ("Hyperspaces of continua, absolute retracts, and infinite-dimensional manifolds") and §6 ("Whitney maps") by V.V. Fedorchuk and A.A. Odintsov in [39, pp. 2337–2344].

Now, we describe the contents of this chapter.

We start by defining the Hausdorff metric on the hyperspace 2^X (introduced by F. Hausdorff in 1914 [45]).

In Sect. 9.2 we give two proofs showing that the hyperspace 2^X is compact. Both proofs are different to the one given by Nadler in [112, Theorem (0.8)]. As a consequence we obtain that $C_n(X)$ is also compact. As an application (Theorem 9.11) we prove that a continuum is hereditarily locally connected if and only if it does not contain convergence subcontinua.

It is easy to prove that the hyperspaces $F_n(X)$ are continua and, as a consequence it follows that 2^X is connected (see Exercise 9.29), so 2^X is a continuum. It is more complicated to prove the connectedness of $C_n(X)$. To do this, it is necessary to introduce order arcs. These arcs allow us to show that in fact 2^X and $C_n(X)$ are always arcwise connected (even if X does not contain any arc!). For this and other reasons specialists usually say that the hyperspaces 2^X and $C_n(X)$ have better properties that the continuum X.

Order arcs are introduced in Sect. 9.4. In order to work with them we first introduce Whitney mappings in Sect. 9.3. These mappings are very useful in the theory of hyperspaces, since they provide a way to define the size of elements of 2^X.

Finally, we include a section on Whitney levels. Nadler's original proof of the connectedness of Whitney levels is a consequence of the fact that $C(X)$ is unicoherent, which in turn was proved using inverse limits. In Theorem 9.21, we offer an elementary proof of the connectedness of Whitney levels.

The long list of exercises at the end of the chapter shows that the basic elements presented in this chapter are enough to construct a large portion of the theory of hyperspaces.

9.1 The Hausdorff Metric

Throughout this chapter X and Y will denote continua, X with metric d. Given a nonempty subset Z of X and $p \in X$, let

$$\text{dist}(p, Z) = \inf\{d(p, y) : y \in Z\}.$$

In the theory of hyperspaces:

- The hyperspace $C_1(X)$ is denoted by $C(X)$ and is called the *hyperspace of subcontinua* of X.
- The hyperspace $F_1(X)$ coincides with the set of one-point sets of X.

Note that $F_1(X) \subset F_2(X) \subset F_3(X) \subset \cdots$ and $C(X) \subset C_2(X) \subset C_3(X) \subset \cdots$.

The space $F_n(X)$ is called the n-th symmetric product.

Definition 9.1 The *Hausdorff metric* H for 2^X is defined in the following way:

$$H(A, B) = \inf\{\varepsilon > 0 : A \subset N(B, \varepsilon) \text{ and } B \subset N(A, \varepsilon)\}.$$

Proposition 9.2 *The function H defined in Definition 9.1 is a metric for* 2^X.

Proof Given $A, B \in 2^X$, define

$$E(A, B) = \{\varepsilon > 0 : A \subset N(B, \varepsilon) \text{ and } B \subset N(A, \varepsilon)\}.$$

Then $H(A, B) = \inf E(A, B)$.

Note that $A \subset N(B, \text{diameter}(X) + 1)$, so that $E(A, B)$ is a nonempty set of positive numbers. This shows that H is well defined and $H(A, B) \geq 0$. Since A and B play symmetric roles in the definition of H, we have that $H(A, B) = H(B, A)$.

Since $A \subset N(A, \varepsilon)$ for every $\varepsilon > 0$, we obtain that $H(A, A) = 0$.

Suppose that $H(A, B) = 0$. We prove that $A = B$. Since the roles of A and B are symmetric, it is enough to show that $A \subset B$.

Let $a \in A$ and $\varepsilon > 0$. Since $\inf E(A, B) < \varepsilon$, there exists a $\delta > 0$ such that $0 < \delta < \varepsilon$, $A \subset N(B, \delta)$ and $B \subset N(A, \delta)$. Then there exists a $b \in B$ such that $d(a, b) < \delta$. Thus $b \in B \cap B(a, \varepsilon)$. We have shown that $A \subset \text{cl}_X(B) = B$. Hence $A \subset B$. Therefore $A = B$.

Finally, we show the triangle inequality.

Let $A, B, D \in 2^X$. We want to show that $\inf E(A, D) \leq \inf E(A, B) + \inf E(B, D)$.

By known properties of the infimum of a set, $\inf E(A, B) + \inf E(B, D) = \inf\{\delta + \lambda : \delta \in E(A, B) \text{ and } \lambda \in E(B, D)\}$. Now we only need to prove that $\inf E(A, D) \leq \delta + \lambda$ for every $\delta \in E(A, B)$ and $\lambda \in E(B, D)$. Let $\delta \in E(A, B)$ and $\lambda \in E(B, D)$. By definition, $A \subset N(B, \delta)$, $B \subset N(A, \delta)$, $B \subset N(D, \lambda)$ and $D \subset N(B, \lambda)$. Given $x \in A$, there exists a $y \in B$ such that $d(x, y) < \delta$. Since $y \in N(D, \lambda)$, there exists a $z \in D$ such that $d(y, z) < \lambda$. Thus $d(x, z) < \delta + \lambda$. Hence

$x \in N(D, \delta + \lambda)$. This shows that $A \subset N(D, \delta + \lambda)$. Similarly, $D \subset N(A, \delta + \lambda)$. Therefore $\delta + \lambda \in E(A, D)$.

We have shown that $\inf E(A, D)$ is a lower bound of the set $\inf\{\delta + \lambda : \delta \in (A, B) \text{ and } \lambda \in (B, D)\}$, thus $\inf E(A, D) \leq \inf E(A, B) + \inf E(B, D)$. This proves that $H(A, D) \leq H(A, B) + H(B, D)$, and finishes the proof that H is a metric. □

Given a finite family of subsets $U_1, \ldots, U_n$ of X, define

$$\langle U_1, \ldots, U_n \rangle = \{A \in 2^X : A \subset \bigcup\{U_i : i \in \{1, \ldots, n\}\} \text{ and } A \cap U_i \neq \emptyset \text{ for every } i \in \{1, \ldots, n\}\}.$$

Also define

$$\mathcal{B} = \{\langle U_1, \ldots, U_n \rangle : n \in \mathbb{N} \text{ and } U_1, \ldots, U_n \text{ are open subsets of } X\}.$$

Theorem 9.3 *The family $\mathcal{B}$ is base for a topology τ in 2^X, called the* Vietoris topology. *The topology τ coincides with the topology induced on 2^X by the Hausdorff metric and does not depend on the metric of X.*

Proof By Exercise 9.32, $\mathcal{B}$ is closed under finite intersections. Moreover, since $2^X = \langle X \rangle$, we conclude that $\mathcal{B}$ is base for a topology τ_V on 2^X.

Let τ_H be the topology on 2^X induced by the Hausdorff metric (for some metric d in X). By Exercise 9.31, the elements of $\mathcal{B}$ are open in τ_H. Thus $\tau_V \subset \tau_H$.

In order to show the other inclusion, take $A \in 2^X$ and $\varepsilon > 0$. By the compactness of A, there exist $n \in \mathbb{N}$ and $a_1, \ldots, a_n \in A$ such that $A \subset B(a_1, \frac{\varepsilon}{3}) \cup \cdots \cup B(a_n, \frac{\varepsilon}{3})$.

For each $i \in \{1, \ldots, n\}$, let $U_i = B(a_i, \frac{\varepsilon}{3})$. Then $\mathcal{U} = \langle U_1, \ldots, U_n \rangle \in \mathcal{B}$.

Clearly, $A \subset U_1 \cup \cdots \cup U_n$ and for every $i \in \{1, \ldots, n\}$, $a_i \in A \cap U_i$. Thus $A \in \langle U_1, \ldots, U_n \rangle$.

Given $D \in \langle U_1, \ldots, U_n \rangle$, we have that $D \subset B(a_1, \frac{\varepsilon}{3}) \cup \cdots \cup B(a_n, \frac{\varepsilon}{3}) \subset N(A, \frac{\varepsilon}{3})$.

Given $a \in A$, there exists an $i \in \{1, \ldots, n\}$ such that $a \in U_i$. Since $D \cap U_i \neq \emptyset$ and diameter$(U_i) < \frac{2\varepsilon}{3}$, we have that $a \in N(D, \frac{2\varepsilon}{3})$. This shows that $A \subset N(D, \frac{2\varepsilon}{3})$. We have shown that $H(A, D) \leq \frac{2\varepsilon}{3}$. This proves that $\langle U_1, \ldots, U_n \rangle$ is contained in the ε-ball of 2^X, centered at A.

We have shown that $\tau_H \subset \tau_V$. Therefore $\tau_H = \tau_V$.

Finally, note that the definition of τ_V only depends on the topology of X and this topology does not depend on the given metric for X. Therefore, the topology for 2^X does not depend on the metric for X. □

9.2 Compactness

There are several ways to prove the compactness of 2^X. Here, we present two very different ones. The first is based on Alexander's Lemma, it uses Zorn's Lemma and can be applied for more general spaces (not necessarily metric spaces). The second does not use Zorn's Lemma, is constructive, is valid for metric spaces and uses the characterization of compactness given in Exercise 1.26: a metric space is compact if and only if each of its sequences has a convergent subsequence.

Theorem 9.4 (Alexander's Lemma) *Let Z be a topological space and let $\mathcal{S}$ be a subbase for the topology of Z. Suppose that every cover of Z by elements of $\mathcal{S}$ has a finite subcover. Then Z is compact.*

Proof Suppose that Z is not compact. Then there exists an open cover $\mathcal{U}$ by open subsets of Z having no finite subcover.

We will use Zorn's Lemma to show that there exists a maximal open cover of Z containing $\mathcal{U}$ and having no finite subcover.

Define $\mathcal{W} = \{\mathcal{V} : \mathcal{V}$ is an open cover of Z such that $\mathcal{U} \subset \mathcal{V}$ and $\mathcal{V}$ does not have finite subcovers$\}$. As usual, the family $\mathcal{W}$ is considered with the order given by inclusion.

Let C be a nonempty totally ordered subset of $\mathcal{W}$. Let $\mathcal{V}_0 = \bigcup\{\mathcal{V} : \mathcal{V} \in C\}$. Note that $\mathcal{V}_0$ is a family of open subsets of Z such that $\mathcal{U} \subset \mathcal{V}_0$. Thus $\mathcal{V}_0$ is a cover of Z.

If $\mathcal{V}_0$ has a finite subcover $\{V_1, \ldots, V_n\}$, for each $i \in \{1, \ldots, n\}$, there exists a $\mathcal{V}_i \in C$ such that $V_i \in \mathcal{V}_i$. Since C is totally ordered, we may suppose that $\mathcal{V}_1 \subset \mathcal{V}_2 \subset \cdots \subset \mathcal{V}_n$. Then $\{V_1, \ldots, V_n\}$ is a finite subcover of $\mathcal{V}_n$, which contradicts the fact that $\mathcal{V}_n$ belongs to $\mathcal{W}$. Therefore $\mathcal{V}_0$ does not have a finite subcover. We have shown that $\mathcal{V}_0$ belongs to $\mathcal{W}$, so that $\mathcal{V}_0$ is an upper bound of C in $\mathcal{W}$.

Applying Zorn's Lemma, we obtain that $\mathcal{W}$ has a maximal element $\mathcal{V}_M$.

Given an open subset V of Z such that $V \notin \mathcal{V}_M$, the maximality of $\mathcal{V}_M$ implies that $\mathcal{V}_M \cup \{V\}$ does not belong to $\mathcal{W}$. Note that $\mathcal{V}_M \cup \{V\}$ is an open cover of Z and contains $\mathcal{U}$. Since this family does not belong to $\mathcal{W}$, we conclude that $\mathcal{V}_M \cup \{V\}$ has a finite subcover. This implies that $Z \setminus V$ can be covered by a finite number of elements of $\mathcal{V}_M$.

Now, we see that $\mathcal{V}_M \cap \mathcal{S}$ is a cover of Z. Let $p \in Z$. Then there exists a $U \in \mathcal{U}$ such that $p \in U$. Thus there exist $n \in \mathbb{N}$ and $S_1, \ldots, S_n \in \mathcal{S}$ such that $p \in S_1 \cap \cdots \cap S_n \subset U$.

In the case that no set of the form S_i belongs to $\mathcal{V}_M$, by the previous paragraph, we have that for each $i \in \{1, \ldots, n\}$, $Z \setminus S_i$ can be covered by a finite number of elements of $\mathcal{V}_M$. This implies that the union $(Z \setminus S_1) \cup \cdots \cup (Z \setminus S_n) = Z \setminus (S_1 \cap \cdots \cap S_n)$ can also be covered by a finite number of $\mathcal{V}_M$. Adding U to this finite number of elements of $\mathcal{V}_M$, we obtain that Z can be covered by a finite number of elements of $\mathcal{V}_M$. This contradicts the fact that $\mathcal{V}_M$ belongs to $\mathcal{W}$. Therefore there exists an $i \in \{1, \ldots, n\}$ such that $S_i \in \mathcal{V}_M$. Since $p \in S_i$ and $S_i \in \mathcal{V}_M \cap \mathcal{S}$, we conclude that $\mathcal{V}_M \cap \mathcal{S}$ is a cover of Z.

By hypothesis, $\mathcal{V}_M \cap \mathcal{S}$ has a finite subcover. This again contradicts the fact that $\mathcal{V}_M \in \mathcal{W}$.

Since this contradiction comes from supposing that Z is not compact, we conclude that Z is indeed compact. □

Theorem 9.5 *For every continuum X, 2^X is compact.*

Proof *(First Proof)* Given an open subset U of X, we consider the sets

$$C(U) = \{A \in 2^X : A \subset U\}, \text{ and}$$

$$\mathcal{D}(U) = \{A \in 2^X : A \cap U \neq \emptyset\}.$$

By Exercise 9.31, $C(U)$ and $\mathcal{D}(U)$ are open in 2^X.

Note that for any finite family $\{U_1, \ldots, U_n\}$, the set $\langle U_1, \ldots, U_n\rangle$ is equal to $\mathcal{D}(U_1) \cap \cdots \cap \mathcal{D}(U_n) \cap C(U_1 \cup \cdots \cup U_n)$. Thus, by Theorem 9.3, the family

$$\mathcal{S} = \{C(U) : U \text{ is open in } X\} \cup \{\mathcal{D}(U) : U \text{ is open in } X\}$$

is a subbase for the topology of 2^X.

By Theorem 9.4 , to prove the compactness of 2^X, we only need to take a cover $\mathcal{U}$ of 2^X such that $\mathcal{U} \subset \mathcal{S}$ and to show that $\mathcal{U}$ has a finite subcover.

We can put $\mathcal{U} = \{C(U) : U \in J\} \cup \{\mathcal{D}(U) : U \in L\}$.

We consider two cases.

Case 1. $X = \bigcup\{U : U \in L\}$.

In this case, since X is compact, there exist $n \in \mathbb{N}$ and $U_1, \ldots, U_n \in L$ such that $X = U_1 \cup \cdots \cup U_n$. Given $A \in 2^X$, we choose $p \in A$. Then there exists an $i \in \{1, \ldots, n\}$ such that $p \in U_i$. Thus $A \cap U_i \neq \emptyset$ and $A \in \mathcal{D}(U_i)$. This proves that $\{\mathcal{D}(U_1), \ldots, \mathcal{D}(U_n)\}$ is a cover of 2^X extracted from $\mathcal{U}$. Therefore $\mathcal{U}$ has a finite subcover.

Case 2. $A = X \setminus \bigcup\{U : U \in L\}$ is nonempty.

In this case, $A \in 2^X$. Since A does not intersect any element $U \in L$, we have that $A \notin \mathcal{D}(U)$ for any $U \in L$. Since $\mathcal{U}$ is a cover of 2^X, there exists a $V \in J$ such that $A \in C(V)$. That is, $A \subset V$.

Thus $X \setminus V$ is a compact subset of X contained in $\bigcup\{U : U \in L\}$. Then there exist $n \in \mathbb{N}$ and $U_1, \ldots, U_n \in L$ such that $X \setminus V \subset U_1 \cup \cdots \cup U_n$.

Given $B \in 2^X$ we have that either B is contained in V or intersects some set U_i. Therefore $2^X = C(V) \cup \mathcal{D}(U_1) \cup \cdots \cup \mathcal{D}(U_n)$.

This completes the proof that $\mathcal{U}$ has a finite subcover. Therefore 2^X is compact.

Second Proof.

According to Exercise 1.26 we need to prove that each sequence in 2^X has a convergent subsequence. Let $\{A_n\}_{n=1}^{\infty}$ be a sequence in 2^X. We will need the following claim.

Claim Let $\varepsilon > 0$, $B \in 2^X$ and $J \subset \mathbb{N}$ be such that J is infinite and $A_n \subset B$ for each $n \in J$. Then there exist $B^* \in 2^X$ and $J^* \subset J$ such that J^* is infinite and for each $n \in J^*$, $A_n \subset B^* \subset B$ and $H(A_n, B^*) \leq 3\varepsilon$.

We prove this claim. By the compactness of B, there exist $m \in \mathbb{N}$ and $p_1, \ldots, p_m \in B$ such that $B \subset B(p_1, \varepsilon) \cup \cdots \cup B(p_m, \varepsilon)$.

Given $n \in J$, let $L_n = \{i \in \{1, \ldots, m\} : A_n \cap B(p_i, \varepsilon) \neq \emptyset\}$. Then we have a function $n \to L_n$ from J into the power set of $\{1, \ldots, m\}$. Since this power set is finite and J is infinite, there exists an $L \subset \{1, \ldots, m\}$ such that the set $J^* = \{n \in J : L_n = L\}$ is infinite.

Define

$$B^* = B \cap \mathrm{cl}_X(\bigcup\{B(p_i, \varepsilon) : i \in L\}).$$

Take $n \in J^*$. Then $L = \{i \in \{1, \ldots, m\} : A_n \cap B(p_i, \varepsilon) \neq \emptyset\}$. Given $i \in L$, $A_n \cap B(p_i, \varepsilon) \neq \emptyset$, so there exists an $a \in A_n$ such that $B(p_i, \varepsilon) \subset B(a, 2\varepsilon)$. Thus $\mathrm{cl}_X(B(p_i, \varepsilon)) \subset B(a, 3\varepsilon) \subset N(A_n, 3\varepsilon)$. Therefore $B^* \subset N(A_n, 3\varepsilon)$. On the other hand, given $p \in A_n$, since $n \in J$, $A_n \subset B$, then there exists an $i \in \{1, \ldots, m\}$ such that $p \in B(p_i, \varepsilon)$. Thus $i \in L_n = L$. Hence $p \in B^*$. This proves that $A_n \subset B^*$, in particular, $B^* \neq \emptyset$.

Since $A_n \subset B^*$ and $B^* \subset N(A_n, 3\varepsilon)$, we conclude that $H(A_n, B^*) \leq 3\varepsilon$. This completes the proof of the claim.

We apply the claim to $\varepsilon = \frac{1}{4}$, $B = X$ and $J = \mathbb{N}$. Then there exist $B_1 \in 2^X$ and $J_1 \subset \mathbb{N}$ such that J_1 is infinite and for each $n \in J_1$, $A_n \subset B_1$ and $H(A_n, B_1) < 1$.

Applying the claim again, there exist $B_2 \in 2^X$ and $J_2 \subset J_1$ such that J_2 is infinite and for each $n \in J_2$, $A_n \subset B_2 \subset B_1$ and $H(A_n, B_2) < \frac{1}{2}$.

Proceeding in this way it is possible to construct sequences $\{J_m\}_{m=1}^{\infty}$ of infinite subsets of $\mathbb{N}$ and $\{B_m\}_{m=1}^{\infty}$ of elements of 2^X such that:

$$J_1 \supset J_2 \supset \cdots,$$

$$B_1 \supset B_2 \supset \cdots$$

and for each $m \in \mathbb{N}$ and $n \in J_m$,

$$A_n \subset B_m \text{ and } H(A_n, B_m) < \frac{1}{2^{m-1}}.$$

Define $B_0 = \bigcap\{B_m : m \in \mathbb{N}\}$. By Exercise 9.40, $B_0 = \lim_{m\to\infty} B_m$.

Since each J_m is infinite, it is possible to find a sequence $n_1 < n_2 < \cdots$ such that for each $m \in \mathbb{N}$, $n_m \in J_m$.

Given $m \in \mathbb{N}$, $H(A_{n_m}, B_m) < \frac{1}{2^{m-1}}$. It follows that $B_0 = \lim_{m\to\infty} A_{n_m}$. We have constructed a convergent subsequence of the sequence $\{A_n\}_{n=1}^{\infty}$. Therefore 2^X is compact. □

Theorem 9.6 *For every continuum X, $C(X)$ is compact.*

Proof By Theorem 9.5, we only need to show that $C(X)$ is closed in 2^X. In order to do this, take a sequence of elements $\{A_n\}_{n=1}^{\infty}$ in $C(X)$ converging to an element A in 2^X. We need to show that $A \in C(X)$. Suppose to the contrary that A is not connected. Let K and L be nonempty disjoint compact subsets of X such that $A = K \cup L$. Let U and V be disjoint open subsets of X such that $K \subset U$ and $L \subset V$. Then $A \cap U \neq \emptyset \neq A \cap V$ and $A \subset U \cup V$. Thus $A \in \langle U, V\rangle$ and $\langle U, V\rangle$ is open in 2^X (by Theorem 9.3). Thus there exists an $n \in \mathbb{N}$ such that $A_n \in \langle U, V\rangle$. This implies that $A_n \cap U \neq \emptyset \neq A_n \cap V$ and $A_n \subset U \cup V$. This contradicts the connectedness of A_n and completes the proof of the theorem. □

The following theorem give us a better geometric way to visualize the upper semi-continuous decompositions defined in Sect. 1.5 of Chap. 1. This result states that a decomposition is semi-continuous if and only if a limit of elements of the decomposition intersects at most one element of the decomposition.

Theorem 9.7 *Let X be a continuum and $\mathcal{D}$ a decomposition of X. Then $X/\mathcal{D}$ is a continuum if and only if for every sequence $\{A_n\}_{n=1}^{\infty}$ of elements of $\mathcal{D}$ converging to an element $A \in 2^X$, there exists a $D \in \mathcal{D}$ such that $A \subset D$.*

Proof (Necessity) Let $\{A_n\}_{n=1}^{\infty}$ be a sequence of elements of $\mathcal{D}$ such that $\lim_{n\to\infty} A_n = A$.

We need to show that A is contained in some element of $\mathcal{D}$. Suppose the contrary. Then there exist two distinct elements D_1 and $D_2 \in \mathcal{D}$ such that $A \cap D_1 \neq \emptyset \neq A \cap D_2$. By Theorem 1.14 , there exist two disjoint open and $\mathcal{D}$-saturated subsets U and V of X such that $D_1 \subset U$ and $D_2 \subset V$. Then $A \in \langle X, U, V\rangle$ and since this set is open in 2^X, there exists an $n \in \mathbb{N}$ such that $A_n \in \langle X, U, V\rangle$. This implies that we can choose points $p \in A_n \cap U$ and $q \in A_n \cap V$. Since $A_n \in \mathcal{D}$, $\{\pi(p)\} = \pi(A_n) = \{\pi(q)\}$. Thus $q \in \pi^{-1}(\pi(U)) = U$. This contradicts the fact that U and V are disjoint and finishes the proof of the necessity.

(Sufficiency) First, we show that for each $B \in 2^X$, the set $E = \bigcup\{D \in \mathcal{D} : D \cap B \neq \emptyset\}$ is closed in X and the sets E and $X \setminus E$ are $\mathcal{D}$-saturated.

In order to do this, take $p \in \mathrm{cl}_X(E)$. Then there exists a sequence of points $\{p_n\}_{n=1}^{\infty}$ in E such that $\lim_{n\to\infty} p_n = p$. For each $n \in \mathbb{N}$, there exists an $A_n \in \mathcal{D}$ such that $p_n \in A_n$ and $A_n \cap B \neq \emptyset$, so we can choose a point $q_n \in A_n \cap B$. Taking subsequences, if necessary, by Theorem 9.5, we may suppose that $\lim_{n\to\infty} q_n = q$, for some $q \in B$ and that $\lim_{n\to\infty} A_n = A$ for some $A \in 2^X$. Then $p, q \in A$ (see Exercise 9.34). By hypothesis, there exists a $D \in \mathcal{D}$ such that $A \subset D$. Since $q \in D \cap B$, we conclude that $p \in E$. Therefore E is closed.

Since each element of $\mathcal{D}$ is $\mathcal{D}$-saturated, it follows that E is $\mathcal{D}$-saturated.

Given $x \in X \setminus E$ and $u \in \pi^{-1}(\pi(x))$, if $u \in E$, we have that $x \in \pi^{-1}(\pi(E)) = E$, a contradiction. Thus $u \notin E$. We have shown that $\pi^{-1}(\pi(X \setminus E)) = X \setminus E$. Therefore $X \setminus E$ is $\mathcal{D}$-saturated.

We are ready to show that $X/\mathcal{D}$ is a Hausdorff space. Let A and B be two distinct elements in $\mathcal{D}$. Let S and T be disjoint open subsets of X such that $A \subset S$ and $B \subset T$. Set $E = \bigcup\{D \in \mathcal{D} : D \cap (X \setminus S) \neq \emptyset\}$, $F = \bigcup\{D \in \mathcal{D} : D \cap X \setminus T \neq \emptyset\}$,

$U = X \setminus E$ and $V = X \setminus F$. By the previous steps in this proof, U and V are open and $\mathcal{D}$-saturated.

Since $X \setminus S \subset E$, we have that $U \subset S$. Similarly, $V \subset T$. Hence U and V are disjoint and open.

Since $A \in \mathcal{D}$ and $A \cap (X \setminus S) = \emptyset$, we have that $A \cap E = \emptyset$, then $A \subset U$. Similarly, $B \subset V$. According to Theorem 1.14, we conclude that $X/\mathcal{D}$ is a Hausdorff space. □

Definition 9.8 Let X be a continuum. A decomposition $\mathcal{D}$ of X is a *continuous decomposition* provided that $\mathcal{D}$ is a closed subset of 2^X, equivalently, $\mathcal{D}$ is compact.

In 1985, W. Lewis [93] constructed a very interesting family of continua that admit continuous decompositions. He proved the following theorem.

Theorem 9.9 (W. Lewis, [93]) *For every one-dimensional continuum M, there is a one-dimensional continuum M^* such that M^* has a continuous decomposition $\mathcal{D}$ into pseudo-arcs with decomposition space $M^*/\mathcal{D}$ homeomorphic to M. Every homeomorphism of M lifts to a homeomorphism of M^*, with a free motion within the decomposition elements. Moreover each element of $\mathcal{D}$ is terminal in M^*.*

Lewis' theorem extended previously known constructions. In 1959, RH Bing and F.B. Jones [13] constructed the circle of pseudo-arcs. In Lewis' notation, the circle of pseudo-arcs is M^* for $M = S^1$. In this case, Bing and Jones proved that M^* is embeddable in the plane and it is one of only three plane homogeneous continua (the other two being the pseudo-arc and the simple closed curve) [47], see more comments on this topic at the beginning of Chap. 17.

If $M = [0, 1]$, M^* is known as the arc of pseudo-arcs. This continuum is useful in the theory of hyperspaces since the structure of $C(M^*)$ is relatively simple. This is because each subcontinuum of M^* is either contained in one element of the decomposition or it is an arc of pseudo-arcs.

The arc of pseudo-arcs is also an interesting example of the irreducible continua treated in Chap. 11.

Definition 9.10 Given a continuum X, a non-degenerate subcontinuum A of X is a *convergence continuum* if there exists a sequence of pairwise disjoint subcontinua $\{A_n\}_{n=1}^{\infty}$ of X such that $\lim_{n\to\infty} A_n = A$ and for each $n \in \mathbb{N}$, $A_n \cap A = \emptyset$. A continuum is *hereditarily locally connected* if all its subcontinua are locally connected.

Theorem 9.11 *A continuum X is hereditarily locally connected if and only if it does not contain convergence subcontinua.*

Proof (Necessity) Suppose that X is hereditarily locally connected, but it contains a convergence subcontinuum A. Let $\{A_n\}_{n=1}^{\infty}$ be a sequence of pairwise disjoint subcontinua such that $\lim_{n\to\infty} A_n = A$ and for each $n \in \mathbb{N}$, $A_n \cap A = \emptyset$. We choose two distinct points $p, q \in A$. Since X is locally connected, there exists a subcontinuum M of X such that $p \in \mathrm{int}_X(M)$ and $q \notin M$.

Let $N \in \mathbb{N}$ be such that $A_n \cap M \neq \emptyset$ for all $n \geq N$.

Set $Y = A \cup M \cup (\bigcup\{A_n : n \geq N\})$.

Clearly, Y is a connected subset of X. Since $\lim_{n\to\infty} A_n = A$, the set $\{A\}\cup\{A_n : n \geq N\}$ is closed in 2^X. By Exercise 9.37 (a), its union is a closed subset of X. Thus Y is closed in X. Therefore Y is a subcontinuum of X.

Since X is hereditarily locally connected, there exists a subcontinuum K of Y such that $q \in \text{int}_Y(K)$ and $K \cap M = \emptyset$. Let $n_0 \geq N$ be such that $K \cap A_{n_0} \neq \emptyset$.

Set $Z = A\cup(\bigcup\{A_n : n \geq n_0+1\})$. Arguing as we did for Y, it is possible to show that A is closed in X. Note that $K \subset \bigcup\{A_n : n \leq n_0\})\cup Z$, these two sets are disjoint and closed and K intersects both of them. This contradicts the connectedness of K and ends the proof that X does not contain convergence continua.

(Sufficiency) Suppose that X is not hereditarily locally connected. Let Y be a non-locally connected subcontinuum of X. Then Y is non-degenerate. By Theorem 3.5, there exist a sequence of pairwise disjoint subcontinua $\{A_n\}_{n=1}^{\infty}$ of Y; two distinct points p and q of X; two sequences $\{p_n\}_{n=1}^{\infty}$, $\{q_n\}_{n=1}^{\infty}$ in Y and a closed subset D of Y such that $\lim_{n\to\infty} p_n = p \in D$, $\lim_{n\to\infty} q_n = q \in D$ and for every $n \in \mathbb{N}$, $\{p_n, q_n\} \subset A_n \subset D \setminus C_0$, where C_0 is the component of D containing p. Since $C(Y)$ is compact (Theorem 9.6), we may suppose, without loss of generality, that $\lim_{n\to\infty} A_n = A$ for some $A \in C(Y)$. By Exercise 9.34 (a), $\{p, q\} \subset A \subset D$, thus $A \subset C_0$. Thus for every $n \in \mathbb{N}$, $A_n \cap A = \emptyset$. Therefore A is a convergence subcontinuum of X. □

9.3 Whitney Mappings

Definition 9.12 Let X be a continuum and let $\mathcal{H}(X)$ be one of the hyperspaces 2^X or $C_n(X)$. A mapping $\mu : \mathcal{H}(X) \to [0, 1]$ is a *Whitney mapping* for $\mathcal{H}(X)$ if the following hold:

(a) for every $p \in X$, $\mu(\{p\}) = 0$,
(b) if $A, B \in \mathcal{H}(X)$ and $A \subsetneq B$, then $\mu(A) < \mu(B)$, and
(c) $\mu(X) = 1$.

Theorem 9.13 *For every continuum X, there exist Whitney mappings for* 2^X.

Proof Let d be a metric for X such that 1 is an upper bound of d. By Exercise 1.19, we can take a countable dense subset $D = \{p_1, p_2, \ldots\}$ of X.

For each $n \in \mathbb{N}$, define $f_n : X \to [0, 1]$ by

$$f_n(p) = d(p_n, p).$$

Since for every $p, q \in X$, $|d(p_n, p) - d(p_n, q)| \leq d(p, q)$, we have that f_n is continuous.

Define ω_n, ω and μ as in Exercises 9.45 and 9.46. We show that μ is a Whitney mapping.

Take $A, B \in 2^X$ such that $A \subsetneq B$. By Exercise 9.46, we only need to show that there exists an $n \in \mathbb{N}$ such that $\omega_n(A) < \omega_n(B)$.

Choose $b_0 \in B \setminus A$ and $\varepsilon > 0$ such that $B(b_0, 2\varepsilon) \cap A = \emptyset$.

By the density of D, there exists an $n \in \mathbb{N}$ such that $d(b_0, p_n) < \varepsilon$.

If there exists an $a \in A$ such that $d(a, p_n) \leq \varepsilon$, then $d(b_0, a) \leq d(b_0, p_n) + d(p_n, a) < 2\varepsilon$. This contradicts the choice of ε. We have shown that for each $a \in A$, $d(a, p_n) > \varepsilon$. Thus $\min(f_n(B)) = \min\{d(p_n, b) : b \in B\} \leq d(p_n, b_0) < \varepsilon < \min\{d(p_n, a) : a \in A\} = \min(f_n(A))$. Therefore

$$\min(f_n(B)) < \min(f_n(A)).$$

Since $A \subset B$, $\max(f_n(A)) \leq \max(f_n(B))$.

Therefore

$$\begin{aligned}
\omega_n(A) &= \text{diameter}(f_n(A)) \\
&= \max(f_n(A)) - \min(f_n(A)) \\
&< \max(f_n(B)) - \min(f_n(B)) \\
&= \text{diameter}(f_n(B)) = \omega_n(B).
\end{aligned}$$

Hence $\omega_n(A) < \omega_n(B)$.

This completes the proof that μ is a Whitney mapping. □

9.4 Order Arcs and Connectedness

Definition 9.14 Let X be a continuum. Let $A, B \in 2^X$ such that $A \subsetneq B$. An *order arc from A to B* is a mapping $\alpha : [0, 1] \to 2^X$ such that $\alpha(0) = A$, $\alpha(1) = B$ and if $0 \leq s < t \leq 1$, then $\alpha(s) \subsetneq \alpha(t)$.

Lemma 9.15 *Let A, B be subcontinua of a continuum X such that $A \subsetneq B$. Let $\mu : 2^X \to [0, 1]$ be a Whitney mapping and $t \in (\mu(A), \mu(B))$. Then there exists a $C \in C(X)$ such that $A \subset C \subset B$ and $\mu(C) = t$.*

Proof Let

$$\mathcal{A} = \mu^{-1}([t, 1]) \cap \{D \in C(X) : A \subset D \subset B\}.$$

By Exercise 9.31 (b), $\mathcal{A}$ is a closed subset of $C(X)$, so $\mathcal{A}$ is compact and $B \in \mathcal{A}$. Then μ has a minimum in $\mathcal{A}$. Thus there exists an $E \in \mathcal{A}$ such that $\mu(E) \leq \mu(D)$ for every $D \in \mathcal{A}$.

Now, set

$$\mathcal{B} = \mu^{-1}([0, t]) \cap \{D \in C(X) : A \subset D \subset E\}.$$

Again, by Exercise 9.31 (b), we have that $\mathcal{B}$ is compact, and is nonempty since $A \in \mathcal{B}$. Thus μ has a maximum in $\mathcal{B}$. Then there exists an $F \in \mathcal{B}$ such that $\mu(D) \leq \mu(F)$ for all $D \in \mathcal{B}$.

If $\mu(E) = t$ or $\mu(F) = t$, we are done. Therefore we may assume that $\mu(F) < t < \mu(E)$. Since $F \in \mathcal{B}$, $F \subset E$, so $F \subsetneq E$.

By Exercise 3.13, there exists a $G \in C(X)$ such that $A \subset F \subsetneq G \subsetneq E \subset B$.

If $t \leq \mu(G)$, then $G \in \mathcal{A}$ and $\mu(G) < \mu(E)$, which contradicts the choice of E. On the other hand, if $\mu(G) \leq t$, then $G \in \mathcal{B}$ and $\mu(F) < \mu(G)$, which contradicts the choice of F.

We have shown that either $\mu(E) = t$ or $\mu(F) = t$, so the lemma is proved. □

The following theorem gives a complete characterization of elements $A, B \in 2^X$ such that there exists an order arc from A to B. A proof of this theorem appears at the beginning of Chapter I of [112]. That proof uses Zorn's Lemma and can be used in the more general setting of non-necessarily metric compact connected Hausdorff spaces. Here we offer a more constructive proof.

Theorem 9.16 *Let X be a continuum and let $A, B \in 2^X$ be such that $A \subsetneq B$. Then there exists an order arc from A to B if and only if each component of B intersects A.*

Proof (Necessity) Suppose that there exists an order arc $\alpha : [0, 1] \to 2^X$ from A to B. Suppose also that there exists a component C of B such that $C \cap A = \emptyset$. Then A and C are closed subsets of B and no connected subset of B intersects both sets A and C. By Theorem 3.3, there exist disjoint compact subsets K and L of B such that $A \subset K$, $C \subset L$ and $B = K \cup L$.

Define

$$\mathcal{A} = \{D \in 2^X : D \subset K\} \text{ and } \mathcal{B} = \{D \in 2^X : D \cap L \neq \emptyset\}.$$

By Exercise 9.31 (b), $\mathcal{A}$ and $\mathcal{B}$ are closed in 2^X and clearly they are disjoint.

Given $t \in [0, 1]$, since $\alpha(t) \subset \alpha(1) = B$, we have that either $\alpha(t) \subset K$ or $\alpha(t) \cap L \neq \emptyset$. This shows that $\alpha([0, 1]) \subset \mathcal{A} \cup \mathcal{B}$. Since $\alpha(0) = A \subset K$, we have that $A \in \alpha([0, 1]) \cap \mathcal{A} \neq \emptyset$. Since $\alpha(1) = B$ and $C \subset B \cap L$, we have that $\alpha(1) \in \mathcal{B}$. Thus $\alpha([0, 1]) \cap \mathcal{B} \neq \emptyset$. This contradicts the connectedness of $\alpha([0, 1])$ and finishes the proof of the necessity.

(Sufficiency) Suppose that each component of B intersects A. Fix a Whitney mapping $\mu : 2^X \to [0, 1]$.

Consider the set $\mathcal{R} = \{C \in C(X) : C \subset B\}$. By Exercise 9.31 (b), $\mathcal{R}$ is closed in 2^X.

Given $a \in A$ and $t \in [0, 1]$, define

$$\beta(a, t) = \bigcup\{C \in \mathcal{R} : a \in C \text{ and } \mu(C) \leq t\}.$$

Also define

$$\alpha(t) = \bigcup\{\beta(a, t) : a \in A\}.$$

Given $a \in A$, $\{a\}$ is the unique element in $\mathcal{R}$ containing a at which the value of μ is 0. Thus $\beta(a, 0) = \{a\}$. Therefore $\alpha(0) = A$.

Given $a \in A$, let C_a be the component of B containing a. Note that $\beta(a, t)$ is union of connected subsets of B containing a. Thus $\beta(a, t) \subset C_a$. Since $\mu(C_a) \leq 1$, we have that $C_a \subset \beta(a, 1)$. Hence $\beta(a, 1) = C_a$. Therefore $\alpha(1) = \bigcup\{C_a : a \in A\}$. Since every component of B intersects A, we have that each component of B is of the form C_a for some $a \in A$. Therefore $\alpha(1) = B$.

Given $0 \leq s \leq t \leq 1$, for every $a \in A$ we have that $\beta(a, s) \subset \beta(a, t)$. This implies that $\alpha(s) \subset \alpha(t)$.

We prove that for every $t \in [0, 1]$, $\alpha(t)$ is a closed subset of X. Take a point $p \in X$ such that $p = \lim_{n\to\infty} p_n$, where $\{p_n\}_{n=1}^{\infty}$ is a sequence of points of $\alpha(t)$.

Given $n \in \mathbb{N}$, there exist $C_n \in \mathcal{R}$ and $a_n \in A$ such that $\mu(C_n) \leq t$ and $a_n, p_n \in C_n$.

Since A and $\mathcal{R}$ are compact, we may assume that the sequence $\{C_n\}_{n=1}^{\infty}$ converges to an element C of $\mathcal{R}$ and the sequence $\{a_n\}_{n=1}^{\infty}$ converges to an element $a \in A$.

By Exercise 9.34, $a, p \in C$. By the continuity of μ, $\mu(C) \leq t$. Thus $p \in \beta(a, t) \subset \alpha(t)$. Therefore $\alpha(t)$ is closed.

Now, we show that the function $\alpha : [0, 1] \to 2^X$ is continuous.

Take a monotone sequence $\{t_n\}_{n=1}^{\infty}$ in $[0, 1]$ converging to a number t. By Exercise 9.59, we only have to prove that $\lim_{n\to\infty} \alpha(t_n) = \alpha(t)$. We consider two cases.

Case 1. The sequence $\{t_n\}_{n=1}^{\infty}$ is increasing.

In this case $t_1 \leq t_2 \leq \cdots$ and $t_n \leq t$ for all $n \in \mathbb{N}$. So $\alpha(t_1) \subset \alpha(t_2) \subset \cdots$ and $\alpha(t_n) \subset \alpha(t)$ for all $n \in \mathbb{N}$.

Set $E = \text{cl}_X(\bigcup\{\alpha(t_n) : n \in \mathbb{N}\})$.

By Exercise 9.40, $\lim_{n\to\infty} \alpha(t_n) = E$. Then $E \subset \alpha(t)$.

In order to prove that $E = \alpha(t)$. Suppose to the contrary that there exists a point $p \in \alpha(t) \setminus E$. Thus there exist $a \in A$ and $C \in \mathcal{R}$ such that $a, p \in C$ and $\mu(C) \leq t$. If $\mu(C) < t$, then there exists an $m \in \mathbb{N}$ such that $\mu(C) < t_m$. Thus $p \in \beta(a, t_m) \subset \alpha(t_m) \subset E$, which is a contradiction. Hence $\mu(C) = t$.

Given $n \in \mathbb{N}$, since $\{a\} \subset C$ and $\mu(\{a\}) = 0 \leq t_n \leq t$, by Lemma 9.15 there exists a $C_n \in C(X)$ such that $a \in C_n \subset C$ and $\mu(C_n) = t_n$. So $C_n \subset \beta(a, t_n) \subset \alpha(t_n) \subset E$. Then $\lim_{n\to\infty} \mu(C_n) = \mu(C)$. The second part of Exercise 9.57 implies that $\lim_{n\to\infty} C_n = C$. By Exercise 9.42, there exists a sequence $\{p_n\}_{n=1}^{\infty}$ in X such that $\lim_{n\to\infty} p_n = p$ and $p_n \in C_n \subset E$ for every $n \in \mathbb{N}$. Then $p \in E$, a contradiction. Therefore $E = \alpha(t)$.

Case 2. The sequence $\{t_n\}_{n=1}^{\infty}$ is decreasing.

In this case $t_1 \geq t_2 \geq \cdots$ and for each $n \in \mathbb{N}$, $t_n \geq t$. Then $\alpha(t_1) \supset \alpha(t_2) \supset \cdots$ and $\alpha(t) \subset \alpha(t_n)$ for all $n \in \mathbb{N}$.

Set $E = \bigcap\{\alpha(t_n) : n \in \mathbb{N}\}$. By Exercise 9.40, $\lim_{n\to\infty} \alpha(t_n) = E$.

Given $p \in E$ and $n \in \mathbb{N}$, $p \in \alpha(t_n)$, so there exist $a_n \in A$ and $C_n \in \mathcal{R}$ such that $a_n, p \in C_n$ and $\mu(C_n) \leq t_n$. Since $\mathcal{R}$ and A are compact, taking subsequences if necessary, we may suppose that $\lim_{n\to\infty} C_n = C$ and $\lim_{n\to\infty} a_n = a$ for some $C \in \mathcal{R}$ and $a \in A$. By Exercise 9.34, $a, p \in C$, the continuity of μ implies that $\mu(C) \leq t$. Thus $p \in \beta(a, t) \subset \alpha(t)$. We have shown that $E \subset \alpha(t)$.

In order to prove the other inclusion, take $p \in \alpha(t)$. Then there exist $a \in A$ and $C \in \mathcal{R}$ such that $a, p \in C$ and $\mu(C) \leq t$. Given $n \in \mathbb{N}$, $\mu(C) \leq t \leq t_n$. Then $p \in \beta(a, t_n) \subset \alpha(t_n)$. Thus $p \in E$. We have shown that $E = \alpha(t)$.

This completes the proof that $\lim_{n\to\infty} \alpha(t_n) = \alpha(t)$. Therefore α is continuous.

We finally reparametrize the mapping α to obtain an order arc from A to B. Let $u = \mu(A)$ and $v = \mu(B)$. Given $t \in [0, 1]$, we have that $A = \alpha(0) \subset \alpha(t) \subset \alpha(1) = B$. Then $u = \mu(A) \leq \mu(\alpha(t)) \leq \mu(B) = v$. Thus $\mu(\alpha(t)) \in [u, v]$.

Let $\mathcal{C} = \alpha([0, 1])$. Then $\mathcal{C}$ is a subcontinuum of 2^X such that $\mu(\mathcal{C}) = [u, v]$ and for every $A_1, B_1 \in \mathcal{C}$, either $A_1 \subset B_1$ or $B_1 \subset A_1$.

Given $0 \leq s \leq t \leq 1$ such that $\alpha(s) \neq \alpha(t)$, we have that $\alpha(s) \subsetneq \alpha(t)$. Thus $\mu(\alpha(s)) < \mu(\alpha(t))$. This shows that $\mu|_{\mathcal{C}} : \mathcal{C} \to [u, v]$ is one-to-one and surjective mapping, so $\mu|_{\mathcal{C}}$ is a homeomorphism. Since $A \subsetneq B$, we have that $u < v$.

Let $\varphi : [0, 1] \to [u, v]$ be a strictly increasing surjective mapping. Define $\gamma : [0, 1] \to \mathcal{C}$ by $\gamma = (\mu|_{\mathcal{C}})^{-1} \circ \varphi$. Since $\mu(A) = u = \varphi(0)$, we have that $\gamma(0) = A$. Similarly, $\gamma(1) = B$. Given $0 \leq s < t \leq 1$, we have $\varphi(s) < \varphi(t)$. let $A_1, B_1 \in \mathcal{C}$ be such that $\mu(A_1) = \varphi(s)$ and $\mu(B_1) = \varphi(t)$. Then either $A_1 \subset B_1$ or $B_1 \subset A_1$. If $B_1 \subset A_1$, then $\varphi(t) = \mu(B_1) \leq \mu(A_1) = \varphi(s)$, a contradiction. Thus $\gamma(s) = A_1 \subsetneq B_1 = \gamma(t)$. Therefore γ is an order arc from A to B. □

Corollary 9.17 *For every continuum X, the hyperspace 2^X is arcwise connected.*

Proof Given an element $A \in 2^X \setminus \{X\}$, by Theorem 9.16, there exists an order arc from A to X. □

Corollary 9.18 *Let X be a continuum and $A, B \in C(X)$ satisfying $A \subsetneq B$. Then there exists an order arc $\alpha : [0, 1] \to C(X)$ from A to B. In consequence, the hyperspace $C(X)$ is arcwise connected.*

Proof By Theorem 9.16, there exists an order arc $\alpha : [0, 1] \to 2^X$ from A to B. Given $s \in (0, 1]$, the mapping $\gamma : [0, 1] \to 2^X$ given by $\gamma(t) = \alpha(ts)$ is an order arc from A to $\alpha(s)$. By Theorem 9.16, each component of $\alpha(s)$ intersects A. Since A is connected, we infer that $\alpha(s)$ is connected. We have shown that the image of α is contained in $C(X)$. Therefore $\operatorname{Im}\alpha$ is an arc in $C(X)$ joining A to B. □

9.5 Whitney Levels

Definition 9.19 Let X be a continuum. A *Whitney level* for $C(X)$ is a subset of $C(X)$ of the form $\mu^{-1}(t)$, where μ is a Whitney mapping and $t \in [0, 1]$.

Lemma 9.20 *Let X be a continuum and $\mathcal{A}$ be a Whitney level for $C(X)$. Let $A, B \in \mathcal{A}$ and C be a component of $A \cap B$. Then there exists a path $\alpha : [0, 1] \to \mathcal{A}$, joining A to B, such that for every $t \in [0, 1]$, $C \subset \alpha(t) \subset A \cup B$.*

Proof We may suppose that $A \neq B$. Let $\mu : C(X) \to [0, 1]$ be a Whitney mapping and $t \in [0, 1]$ be such that $\mathcal{A} = \mu^{-1}(t)$.

Since $\mu(A) = \mu(B)$ and $A \neq B$, we have that $A \not\subset B$ and $B \not\subset A$. This implies that $C \subsetneq A$ and $C \subsetneq B$. By Theorem 9.18, there exist order arcs $\alpha, \beta : [0, 1] \to C(X)$, from C to A and from C to B, respectively.

Given $s \in [0, 1]$, we consider the mapping $f : [0, 1] \to C(X)$ given by

$$f(r) = \alpha(s) \cup \beta(r).$$

By Exercise 9.37, f is continuous.

Note that $f(0) = \alpha(s) \cup C = \alpha(s) \subset A$, and $f(1) = \alpha(s) \cup B \supset B$. Thus

$$\mu(f(0)) \leq \mu(A) = t = \mu(B) \leq \mu(f(1)).$$

So, by the Intermediate Value Theorem, applied to the mapping $\mu \circ f$, there exists a number $r(s) \in [0, 1]$ such that $\mu(f(r(s))) = t$.

Define $\gamma : [0, 1] \to \mathcal{A}$ by

$$\gamma(s) = \alpha(s) \cup \beta(r(s)).$$

Since $t = \mu(f(r(s))) = \alpha(s) \cup \beta(r(s))$, we have that $\gamma(s) \in \mathcal{A}$.

We check that $\gamma(s)$ does not depend of the choice of the number $r(s)$. In order to do this, take $u \in [0, 1]$ such that $\mu(\alpha(s) \cup \beta(u)) = t$. We are going to show that $\alpha(s) \cup \beta(u) = \alpha(s) \cup \beta(r(s))$. Since β is an order arc, we have that either $\beta(u) \subset \beta(r(s))$ or $\beta(r(s)) \subset \beta(u)$. This implies that one of the sets $\alpha(s) \cup \beta(u)$ or $\alpha(s) \cup \beta(r(s))$ is contained in the other, and the inclusion is not proper since μ takes the same value in both sets. Thus both sets are the same. Therefore γ does not depend on the number $r(s)$.

We are ready to prove the continuity of γ. Let $\{s_n\}_{n=1}^{\infty}$ be a sequence in $[0, 1]$ converging to a number s. Suppose that $\lim_{n\to\infty} \gamma(s_n) = E$ for some $E \in \mathcal{A}$. By Exercise 1.50, in order to complete the proof of the continuity of γ, we only need to check that $E = \gamma(s)$. By the compactness of $[0, 1]$, we can obtain a subsequence $\{r(s_{n_k})\}_{k=1}^{\infty}$ of the sequence $\{r(s_n)\}_{n=1}^{\infty}$ converging to a number $u \in [0, 1]$. Then $\lim_{k\to\infty} \alpha(s_{n_k}) = \alpha(s)$, $\lim_{k\to\infty} \beta(r(s_{n_k})) = \beta(u)$, so, $E = \lim_{k\to\infty} \gamma(s_{n_k}) = \lim_{k\to\infty} \alpha(s_{n_k}) \cup \beta(r(s_{n_k})) = \alpha(s) \cup \beta(u)$. Since $\mu(\alpha(s_{n_k}) \cup \beta(r(s_{n_k}))) = t$ for all $k \in \mathbb{N}$, we have that $\mu(\alpha(s) \cup \beta(u)) = t$. By the previous paragraph, we conclude that $E = \alpha(s) \cup \beta(r(s)) = \gamma(s)$. Therefore $E = \gamma(s)$ and γ is continuous.

Observe that $\gamma(0) = \alpha(0) \cup \beta(r(0)) = C \cup \beta(r(0)) = \beta(r(0)) \subset B$. Since μ takes the value t in both sets, we have that $\gamma(0) = B$. On the other hand, $\gamma(1) = \alpha(1) \cup \beta(r(1)) = A \cup \beta(r(1)) \supset A$. Then $\gamma(1) = A$. Finally, note that for each $s \in [0, 1]$, $C \subset \gamma(s)$. □

Theorem 9.21 *Whitney levels are connected.*

Proof Let X be a continuum and $\mathcal{A}$ be a Whitney level for $C(X)$. Suppose that $\mathcal{A}$ is not connected. Since $\mathcal{A}$ is compact, there exist two nonempty, disjoint compact subsets $\mathcal{K}$ and $\mathcal{L}$ of $\mathcal{A}$ such that $\mathcal{A} = \mathcal{K} \cup \mathcal{L}$.

Set $K_0 = \bigcup\{K : K \in \mathcal{K}\}$ and $L_0 = \bigcup\{L : L \in \mathcal{L}\}$. Since $\mathcal{K}$ is nonempty and its elements are nonempty, we have that K_0 is nonempty. Similarly, L_0 is nonempty. By Exercise 9.37, each of the sets K_0 and L_0 belongs to 2^X, so they are close in X.

Given $p \in X$, by Exercise 9.82, there exists an $A \in \mathcal{A} = \mathcal{K} \cup \mathcal{L}$ such that $p \in A$. Depending on whether $A \in \mathcal{K}$ or $A \in \mathcal{L}$, we have that $p \in K_0$ or $p \in L_0$. Therefore $X = K_0 \cup L_0$.

We check that K_0 and L_0 are disjoint. Suppose to the contrary that there exists a point $p \in K_0 \cap L_0$. Then there exist $K \in \mathcal{K}$ and $L \in \mathcal{L}$ such that $p \in K \cap L$. By Lemma 9.20, there exists a mapping $\gamma : [0, 1] \to \mathcal{A}$ such that $\gamma(0) = K$ and $\gamma(1) = L$. Then $\gamma([0, 1])$ is a connected subset of $\mathcal{A}$ intersecting $\mathcal{K}$ and $\mathcal{L}$. This contradiction proves that $K_0 \cap L_0 = \emptyset$.

We have shown that K_0 and L_0 form a disconnection of X. This contradiction proves that $\mathcal{A}$ is connected. □

Definition 9.22 A continuum X is *arcwise smooth at a point* $p \in X$ if there exists a mapping $\alpha : X \to C(X)$ such that:

(a) $\alpha(p) = \{p\}$,
(b) for every $x \in X \setminus \{p\}$, $\alpha(x)$ is an arc joining p and x, and
(c) if $y \in \alpha(x) \setminus \{p\}$, then $\alpha(y)$ is the subarc of $\alpha(x)$ joining p and y.

The continuum X is *arcwise smooth* if there exists a $p \in X$ such that X is arcwise smooth at p.

Theorem 9.23 *If a continuum X is arcwise smooth, then X is contractible.*

Proof Let $p \in X$ be such that X is arcwise smooth at p. Let α be as in Definition 9.22. Let $\mu : C(X) \to [0, 1]$ be a Whitney mapping.

Define $G : X \times [0, 1] \to X$ by:

$$G(x, t) \text{ is the unique point in } \alpha(x) \text{ such that } \mu(\alpha(G(x, t))) = t\mu(\alpha(x)).$$

If $\alpha(x)$ is a one-point set, we have that $\alpha(x) = \{p\}$, for each $t \in [0, 1]$, $G(x, t) = p$ and since $t\mu(\alpha(x)) = 0$, the equality $\mu(\alpha(G(x, t))) = 0 = t\mu(\alpha(x))$ is satisfied.

If $\alpha(x)$ is an arc, the end points of $\alpha(x)$ are p and x. Since $\mu(\{p\}) = 0 \leq t\mu(\alpha(x)) \leq \mu(\alpha(x))$, there exists a unique point $q \in \alpha(x)$ such that the subarc pq of $\alpha(x)$ joining p to q satisfies $\mu(pq) = t\mu(\alpha(x))$. By property (c) in Definition 9.22, we have that $pq = \alpha(q)$, so $q = G(x, t)$. Therefore G is well defined.

We prove that G is continuous. Take a sequence $\{(x_n, t_n)\}_{n=1}^{\infty}$ in $X \times [0, 1]$ and a point $(x, t) \in X \times [0, 1]$ such that $\lim_{n\to\infty}(x_n, t_n) = (x, t)$. This implies that $\lim_{n\to\infty} x_n = x$, $\lim_{n\to\infty} t_n = t$, $\lim_{n\to\infty} \alpha(x_n) = \alpha(x)$ and $\lim_{n\to\infty} t_n\mu(\alpha(x_n)) = t\mu(\alpha(x))$.

Suppose that $\lim_{n\to\infty} G(x_n, t_n) = q$. According to Exercise 1.50, we only have to show that $q = G(x, t)$.

By Exercise 9.42, $q \in \alpha(x)$. By the continuity of α, we have that

$$t\mu(\alpha(x)) = \lim_{n\to\infty} t_n\mu(\alpha(x_n)) = \lim_{n\to\infty} \mu(\alpha(G(x_n, t_n))) = \mu(\alpha(q)).$$

By the definition of G, we conclude that $q = G(x, t)$. Therefore G is continuous.

Note that for each $x \in X$, $G(x, 0)$ is the unique point in $\alpha(x)$ such that $\mu(\alpha(G(x, 0))) = 0$. Thus $\alpha(G(x, 0))$ is a one-point set. By Definition 9.22 (b), we conclude that $G(x, 0) = \{p\}$. On the other hand, $G(x, 1)$ is the unique point in $\alpha(x)$ such that $\mu(\alpha(G(x, 1))) = \mu(\alpha(x))$. Since $\alpha(G(x, 1))$ is a subarc of $\alpha(x)$ joining p to $G(x, 1)$, $\alpha(x)$ is an arc joining p and x and μ has the same value on both arcs $\alpha(G(x, 1))$ and $\alpha(x)$, we conclude that $G(x, 1) = x$. Therefore X is contractible. □

9.6 Exercises

Throughout this section X denotes a continuum with metric d and Y denotes a continuum.

Exercise 9.24 Let $A, B \in 2^X$. Show that $H(A, B) < \varepsilon$ if and only if $A \subset N(B, \varepsilon)$ and $B \subset N(A, \varepsilon)$.

Exercise 9.25 Show that $H(A, B)$ can be defined for every pair of nonempty subsets A and B of X, but this extension does not give a metric.

Exercise 9.26 Prove that $F_1(X)$ and X are isometric.

Exercise 9.27 Given $A, B \in 2^X$ and $p \in X$, recall that

$$d(p, A) = \min\{d(p, a) : a \in A\},$$

and define

$$D(A, B) = \max\{\sup\{\text{dist}(a, B) : a \in A\}, \sup\{\text{dist}(b, A) : b \in B\}\}.$$

Prove that $D(A, B) = H(A, B)$.

Exercise 9.28 Prove that the set $F(X) = \bigcup\{F_n(X) : n \in \mathbb{N}\}$ is dense in 2^X.

Exercise 9.29 Given $n \in \mathbb{N}$, let $\varphi : X^n \to F_n(X)$ be the function given by $\varphi((x_1, \ldots, x_n)) = \{x_1, \ldots, x_n\}$. Prove that φ is continuous and surjective. Conclude that $F_n(X)$ is a continuum and 2^X is connected.

Exercise 9.30 Suppose that X is arcwise connected. Prove that for each $n \in \mathbb{N}$, $F_n(X)$ is arcwise connected.

Exercise 9.31 Given a subset A of X, we consider the families

$$\mathcal{C}(A) = \{B \in 2^X : B \subset A\},$$

$$\mathcal{D}(A) = \{B \in 2^X : B \cap A \neq \emptyset\}, \text{ and}$$

$$\mathcal{E}(A) = \{B \in 2^X : A \subset B\}.$$

Prove that:

(a) if A is open, then $\mathcal{C}(A)$ and $\mathcal{D}(A)$ are open in 2^X,
(b) if A is closed, then $\mathcal{C}(A)$, $\mathcal{D}(A)$ and $\mathcal{E}(A)$ are closed in 2^X, and
(c) if A is open, then $\mathcal{E}(A)$ is not necessarily open in 2^X.

Exercise 9.32 Let $U_1, \ldots, U_n$ and $V_1, \ldots, V_m$ be two finite sequences of sets of X. Set $U = U_1 \cup \cdots \cup U_n$ and $V = V_1 \cup \cdots \cup V_m$. Prove that

$$\langle U_1, \ldots, U_n \rangle \cap \langle V_1, \ldots, V_m \rangle = \langle U_1 \cap V, \ldots, U_n \cap V, V_1 \cap U, \ldots, V_m \cap U \rangle .$$

Exercise 9.33 Let $f : X \to Y$ be a mapping. Define $2^f : 2^X \to 2^Y$ by $2^f(A) = f(A)$ (the image of A under f). The mapping f is called an *induced mapping*. Prove that:

(a) 2^f is well defined,
(b) 2^f is continuous,
(c) for each $n \in \mathbb{N}$, $2^f(C_n(X)) \subset C_n(Y)$ and $2^f(F_n(X)) \subset F_n(Y)$,
(d) 2^f is one-to-one if and only if f is one-to-one,
(e) 2^f is surjective if and only if f is surjective,
(f) even if f is surjective, $2^f|_{C(X)} : C(X) \to C(Y)$ is not necessarily surjective.

Exercise 9.34 Let $\{A_n\}_{n=1}^{\infty}$ and $\{B_n\}_{n=1}^{\infty}$ be sequences of elements of 2^X such that $\lim_{n\to\infty} A_n = A$ and $\lim_{n\to\infty} B_n = B$. Prove that:

(a) if for each $n \in \mathbb{N}$, $A_n \subset B_n$, then $A \subset B$,
(b) $\lim_{n\to\infty} A_n \cup B_n = A \cup B$,
(c) if for each $n \in \mathbb{N}$, $A_n \cap B_n \neq \emptyset$, then $A \cap B \neq \emptyset$,
(d) the equality $\lim_{n\to\infty} A_n \cap B_n = A \cap B$ does not hold in general.

Exercise 9.35 Prove that the diameter function diameter : $2^X \to [0, \infty)$ is continuous.

Exercise 9.36 Show that there exists a metric ρ for the interval $[0, 1]$ for which the topology is the usual one and the function $f : [0, \infty) \to 2^{[0,1]}$ given by $f(t) = \{p \in [0, 1] : \rho(0, p) \leq t\}$ is not continuous. (Hint: each arc in the plane induces a metric on $[0, 1]$.)

Exercise 9.37 Consider the function $\cup : 2^{2^X} \to 2^X$ defined by

$$\cup \mathcal{A} = \bigcup \{A : A \in \mathcal{A}\}.$$

Prove that:

(a) $\cup$ is well defined,
(b) $\cup$ is continuous,
(c) $\cup$ is surjective,
(d) if $\mathcal{A}$ is connected, $\cup\mathcal{A}$ is not necessarily connected,
(e) if $\mathcal{A}$ is connected and there exists an $n \in \mathbb{N}$ such that $\mathcal{A} \cap C_n(X) \neq \emptyset$, then $\cup\mathcal{A} \in C_n(X)$. (Hint: use Exercise 1.35.)

Exercise 9.38 Prove that the following are equivalent:

(a) X is locally connected,
(b) 2^X is locally connected,
(c) $C_n(X)$ is locally connected for every $n \in \mathbb{N}$,
(d) $C_n(X)$ is locally connected for some $n \in \mathbb{N}$,
(e) $F_n(X)$ is locally connected for every $n \in \mathbb{N}$, and
(f) $F_n(X)$ is locally connected for some $n \in \mathbb{N}$.

Exercise 9.39 Prove that for every $n \in \mathbb{N}$, $C_n(X)$ is compact. (Hint: use Exercise 1.35.)

Exercise 9.40 Let $\{A_n\}_{n=1}^{\infty}$ be a sequence in 2^X. Prove that:

(a) if $A_1 \supset A_2 \supset \cdots$, then $\lim_{n\to\infty} A_n = \bigcap\{A_n : n \in \mathbb{N}\}$, and
(b) if $A_1 \subset A_2 \subset \cdots$, then $\lim_{n\to\infty} A_n = \mathrm{cl}_X(\bigcup\{A_n : n \in \mathbb{N}\})$.

Exercise 9.41 Let $\{A_n\}_{n=1}^{\infty}$ be a sequence in $C(X)$ such that $\lim_{n\to\infty} A_n = A \in C(X)$. Prove that:

(a) $\lim_{n\to\infty} 2^{A_n} = 2^A$,
(b) the following equality is not necessarily true: $\lim_{n\to\infty} C(A_n) = C(A)$.

Exercise 9.42 Let $\{A_n\}_{n=1}^{\infty}$ be a sequence in 2^X such that $\lim_{n\to\infty} A_n = A$. Prove that $p \in A$ if and only if there exists a sequence $\{p_n\}_{n=1}^{\infty}$ in X such that $\lim_{n\to\infty} p_n = p$ and $p_n \in A_n$ for all $n \in \mathbb{N}$. (Hint: for the necessity, for each $n \in \mathbb{N}$, take $p_n \in A_n$ such that $d(p, p_n) = \min\{d(p, x) : x \in A_n\}$.)

Exercise 9.43 Determine which properties in Definition 9.12 are satisfied by the diameter mapping for 2^X.

Exercise 9.44 Prove that there are metrics for $[0, 1]$, inducing the usual topology, for which the diameter mapping defined on $C([0, 1])$ is a Whitney mapping and there are others for which the diameter is not a Whitney mapping.

Exercise 9.45 Suppose that $\{f_n\}_{n=1}^{\infty}$ is a sequence of mappings from X into $[0, 1]$. For each $n \in \mathbb{N}$, consider the mapping $\omega_n : 2^X \to [0, 1]$ given by

$$\omega_n(A) = \text{diameter}(f_n(A)).$$

Show that for each $n \in \mathbb{N}$, ω_n is continuous, $\omega_n(\{p\}) = 0$ for each $p \in X$ and $\omega_n(A) \leq \omega_n(B)$, when $A \subset B$.

Exercise 9.46 Let $\{\omega_n\}_{n=1}^{\infty}$ be a sequence of mappings from 2^X into $[0, 1]$ such that for each $n \in \mathbb{N}$, $\omega_n(\{p\}) = 0$ for each $p \in X$ and $\omega_n(A) \leq \omega_n(B)$ when $A \subset B$.

Define $\omega : 2^X \to [0, 1]$ by

$$\omega(A) = \sum_{n=1}^{\infty} \frac{\omega_n(A)}{2^n}.$$

Prove that ω is continuous, $\omega(\{p\}) = 0$ for each $p \in X$ and $\omega(A) \leq \omega(B)$, when $A \subset B$.

Suppose also that the following property holds:

$$\text{if } A, B \in 2^X \text{ and } A \subsetneq B, \text{ then there exists an } m \in \mathbb{N} \text{ such that } \omega_m(A) < \omega_m(B).$$

Prove that the mapping $\mu : 2^X \to [0, 1]$ given by

$$\mu(A) = \frac{\omega(A)}{\omega(X)}$$

is a Whitney mapping.

Exercise 9.47 This exercise gives another way to define Whitney mappings on the continuum X. Suppose that for every $x, y \in X$, $d(x, y) < 1$.

Given $A \in 2^X$ and $n \in \mathbb{N}$, define:

$$\begin{aligned} \Lambda_n(A) = \{0\} \cup \{\varepsilon > 0 : &\text{ there exist non-necessarily pairwise distinct points} \\ &a_1, \ldots, a_{n+1} \in A \text{ such that if } i \neq j, \text{ then} \\ &B(a_i, \varepsilon) \cap B(a_j, \varepsilon) = \emptyset\}. \end{aligned}$$

Define $\omega_n(A) = \sup \Lambda_n(A)$.

Prove that:

(a) for every $n \in \mathbb{N}$ and $A \in 2^X$, 1 is an upper bound of $\Gamma_n(A)$. So ω_n is well defined,
(b) if $A \in 2^X$ has at least $n + 1$ distinct points, then $\omega_n(A) > 0$,
(c) if $A \in 2^X$ contains exactly n points, then $0 = \omega_n(A) = \omega_{n+1}(A) = \cdots$,
(d) if $A \in 2^X$, then the sequence $\{\omega_n(A)\}_{n=1}^{\infty}$ converges to 0,
(e) if $A, B \in 2^X$ and $A \subsetneq B$, then for each $n \in \mathbb{N}$, $\omega_n(A) \leq \omega_n(B)$. Moreover, there exists an $m \in \mathbb{N}$ such that $\omega_m(A) < \omega_m(B)$,
(f) the mapping μ defined as in Exercise 9.46 is a Whitney mapping.

Exercise 9.48 For each $n \in \mathbb{N}$, define $\omega_n : 2^X \to [0, 1]$ by

$$\begin{aligned} \omega_n(A) = \inf\{\varepsilon > 0 : &\text{ there exist points } p_1, \ldots, p_n \in X \text{ such that} \\ &A \subset N(\{p_1, \ldots, p_n\}, \varepsilon). \end{aligned}$$

Prove that the function μ defined as in Exercise 9.46 is a Whitney mapping.

Exercise 9.49 Let Z be the topologist's curve. Let $\pi_1, \pi_2 : Z \to [-1, 1]$ be the natural projections. Show that the mapping $\mu : C(Z) \to [0, 1]$ given by $\mu(A) = \frac{1}{3}(\text{diameter}(\pi_1(A)) + \text{diameter}(\pi_2(A)))$ is a Whitney mapping for $C(Z)$.

Exercise 9.50 Prove that the finite product, the minimum, the maximum and any convex combination of Whitney mappings are Whitney mappings.

Exercise 9.51 Let G be a finite graph. Find a simple way of defining a Whitney mapping for $C(G)$.

Exercise 9.52 Let $\mu : 2^X \to [0, 1]$ be a Whitney mapping for 2^X and $E \in 2^X$. Let $\omega : 2^X \to [0, 1]$ be given by $\omega(A) = (\mu(A)\mu(A \cup E))^{\frac{1}{2}}$. Prove that ω is also a Whitney mapping. Determine for which elements $A \in 2^X$ the equality $\omega(A) = \mu(A)$ holds.

Exercise 9.53 Let $A, B \in 2^X$ be such that $A \not\subset B$ and A has more than one point. Prove that there exists a Whitney mapping $\mu : 2^X \to [0, 1]$ such that $\mu(A) > \mu(B)$.

Exercise 9.54 Suppose that X is arcwise connected and the diameter mapping is a Whitney mapping for $C(X)$. Prove that X is an arc.

Exercise 9.55 Prove that for any given metric d defined on the topologist's curve Z, the diameter mapping defined on $C(Z)$ is not a Whitney mapping. (Hint: if the diameter mapping is a Whitney mapping, then for each arc α in Z, with end points p and q, diameter$(\alpha) = d(p, q)$.)

Exercise 9.56 Let Z be a simple closed curve. Prove that for any finite sequence of mappings $\{f_1, f_2, \ldots, f_n\}$ from Z to $[0, \frac{1}{n}]$ the mapping $\mu : C(Z) \to [0, 1]$ given by

$$\mu(A) = \text{diameter}(f_1(A)) + \cdots + \text{diameter}(f_n(A))$$

is not a Whitney mapping.

Exercise 9.57 Let $\mu : 2^X \to [0, 1]$ be a Whitney mapping. Prove that for every $\varepsilon > 0$, there exists a $\delta > 0$ such that if $A \subset B$ and $\mu(B) - \mu(A) < \delta$, then $H(A, B) < \varepsilon$. Conclude that if $\{A_n\}_{n=1}^{\infty}$ is a sequence of elements in 2^X and $A \in 2^X$ are such that $A_n \subset A$ for all $n \in \mathbb{N}$ and $\lim_{n\to\infty} \mu(A_n) = \mu(A)$, then $\lim_{n\to\infty} A_n = A$.

Exercise 9.58 Let $\mu : 2^X \to [0, 1]$ be a Whitney mapping. Prove that for every $\varepsilon > 0$, there exists a $\delta > 0$ such that if $A \in 2^X$ and diameter$(A) < \delta$, then $\mu(A) < \varepsilon$.

Exercise 9.59 Let Z be a metric space and let $f : [0, 1] \to Z$ be a mapping. Then f is continuous if and only if for every monotone sequence $\{t_n\}_{n=1}^{\infty}$ in $[0, 1]$, converging to a number $t \in [0, 1]$, we have that $\lim_{n\to\infty} f(t_n) = f(t)$.

Exercise 9.60 Let $A, B \in 2^X$ be such that $A \subsetneq B$ and there exists an order arc from A to B. Suppose that $A \in C_n(X)$. Prove that $B \in C_n(X)$. Conclude that $C_n(X)$ is arcwise connected.

Exercise 9.61 Prove that 2^X, $C(X)$ and $C_n(X)$ are locally connected at the element X. Show that there exists a continuum X for which X is the only element of local connectedness in these hyperspaces.

Exercise 9.62 Let $p \in X$ be such that $X \setminus \{p\} = U \cup V$, where U and V are nonempty disjoint open subsets of X. Let $A \in C(X)$ be such that $A \cap U \neq \emptyset \neq A \cap V$. Prove that $C(X)$ is connected im kleinen at A, but this conclusion does not follow if we only ask that $p \in A$.

Exercise 9.63 A subcontinuum $\mathcal{A}$ of 2^X is an *ordered arc* if for every $A, B \in \mathcal{A}$, we have that $A \subset B$ or $B \subset A$. Prove that $\mathcal{A}$ is a ordered arc if and only if $\mathcal{A}$ is either a one-point set or the image of an order arc.

Exercise 9.64 Let $\mathfrak{A} = \{\mathcal{A} \in C(2^X) : \mathcal{A}$ is a linear arc in $2^X\}$. Prove that $\mathfrak{A}$ is a subcontinuum of $C(2^X)$.

Exercise 9.65 Given $p \in X$. Show that $C(X) \setminus \{\{p\}\}$ is arcwise connected.

Exercise 9.66 Let $A \in 2^X \setminus C(X)$. Prove that $2^X \setminus \{A\}$ is arcwise connected.

Exercise 9.67 Let $B \in 2^X$. Prove that $\{A \in C(X) : A \cap B \neq \emptyset\}$ is a subcontinuum of $C(X)$.

Exercise 9.68 Suppose that $B \in 2^X$ contains exactly $n \in \mathbb{N}$ components. Estimate the number of components of $\{A \in 2^X : A \subset B\}$ in terms of n.

Exercise 9.69 Let $B \in C(X)$. Suppose that B is decomposable and there exists a $C \in C(X)$ such that $B \cap C \neq \emptyset$, $B \setminus C \neq \emptyset$ and $C \setminus B \neq \emptyset$. Prove that $C(X) \setminus \{B\}$ is arcwise connected.

Exercise 9.70 A proper subcontinuum B of X is *terminal* if it is non-degenerate and has the following property: if $C \in C(X)$ and $C \cap B \neq \emptyset$, then either $C \subset B$ or $B \subset C$. Prove that if B is terminal, then $C(X) \setminus \{B\}$ is not arcwise connected.

Exercise 9.71 Let $A \in C(X)$. Prove that $C(X) \setminus \{A\}$ is connected.

Exercise 9.72 Let $\mu : C(X) \to [0, 1]$ be a Whitney mapping. Prove that for each $t \in [0, 1]$, $\mu^{-1}([0, t])$ and $\mu^{-1}([t, 1])$ are connected.

Exercise 9.73 Prove that 2^X contains Hilbert cubes. (Hint: use Exercise 3.19.)

Exercise 9.74 Let $A, B \in C(X)$ be such that $A \subset B$ and $B \setminus A$ has at least n components. Prove that $C(B)$ contains n-cells. (Hint: use Exercises 1.35 and 4.12.)

Exercise 9.75 Prove that $C_n(X)$ contains n-cells.

Exercise 9.76 Suppose that X is a planar continuum with nonempty interior. Prove that $C(X)$ contains Hilbert cubes.

Exercise 9.77 Prove that X is indecomposable if and only $C(X)\setminus\{X\}$ is not arcwise connected.

Exercise 9.78 Prove that X is hereditarily indecomposable if and only if for each pair of distinct elements of $C(X)$, there exists a unique arc in $C(X)$ joining them.

Exercise 9.79 Suppose that $A, B \in C(X)$ are such that $A \cap B$ has at least n components. Prove that $C(X)$ contains n-cells.

Exercise 9.80 Suppose that $A, B \in C(X)$ are such that $A \cap B$ has infinitely many components, then $C(X)$ contains Hilbert cubes.

Exercise 9.81 Prove that if every non-degenerate element of X is decomposable, then $C_n(X)$ contains $2n$-cells.

Exercise 9.82 Let $\mathcal{A}$ be a Whitney level for $C(X)$. Prove that the union of elements of $\mathcal{A}$ is X.

Exercise 9.83 Let $\mu : C(X) \to [0, 1]$ be a Whitney mapping and $0 \leq t < 1$. Prove that $\mu^{-1}(t)$ is a non-degenerate subcontinuum of $C(X)$.

Exercise 9.84 Show that every Whitney level for $C([0, 1])$, distinct from $\{[0, 1]\}$, is an arc.

Exercise 9.85 Prove that Whitney levels for $C(S^1)$, distinct from $\{S^1\}$, are simple closed curves.

Exercise 9.86 Let $\mathcal{A}$ be a Whitney level for $C(X)$ and $A \in C(X)$. Prove that the set $\{B \in \mathcal{A} : B \cap A \neq \emptyset\}$ is a subcontinuum of $\mathcal{A}$.

Exercise 9.87 Prove that if X is arcwise connected, then every Whitney level for $C(X)$ is arcwise connected.

Exercise 9.88 Prove that if X is locally connected, then every Whitney level for $C(X)$ is locally connected.

Exercise 9.89 Suppose that X is a compactification of the ray $[0, \infty)$. Prove that there exist Whitney levels for $C(X)$ that are arcs.

Exercise 9.90 Prove that if a Whitney level for $C(X)$ contains arcs, then $C(X)$ contains simple closed curves.

Exercise 9.91 Let $\mathcal{A} = \mu^{-1}(t)$ be a Whitney level for $C(X)$, where $\mu : C(X) \to [0, 1]$ is a Whitney mapping and $t > 0$. Prove that if X is a finite union of elements of $\mu^{-1}([0, t))$, then $\mathcal{A}$ is arcwise connected. Also show that this result does not hold if we change finite to countable.

Exercise 9.92 Suppose that every Whitney level for $C(X)$, different from $F_1(X)$, is locally connected. Prove that X is locally connected.

Exercise 9.93 Let $\mu : C(X) \to [0, 1]$ be a Whitney mapping. Prove that the function $f : [0, 1] \to C(C(X))$ given by $f(t) = \mu^{-1}(t)$ is continuous.

Exercise 9.94 Suppose that X is arcwise smooth at the point p. Prove that X is connected im kleinen at p.

Exercise 9.95 Prove that a dendroid is a dendrite if and only if it is arcwise smooth at each of its points.

Exercise 9.96 Let $f : X \to Y$ be a surjective function. Prove that f is continuous if and only if for every sequence $\{y_n\}_{n=1}^{\infty}$ converging to a point $y \in Y$ satisfying that $\{f^{-1}(y_n)\}_{n=1}^{\infty}$ converges to an element in 2^X, we have that $\lim_{n\to\infty} f^{-1}(y_n) \subset f^{-1}(y)$.

Exercise 9.97 Let $p \in X$. Prove that the set $\mathcal{A} = \{A \in C(X) : p \in A\}$ is locally connected, but this result does not hold if we change $C(X)$ to 2^X.

Exercise 9.98 Let $A, Z \in 2^X$. Let $\mathcal{A} = \{B \in C(X) : B$ is a component of Z and $B \cap A \neq \emptyset\}$. Prove that $\bigcup \mathcal{A}$ is closed in X.

Chapter 10
Models of Hyperspaces

In the theory of hyperspaces it is very useful to have a geometric idea of what they look like. Since they are defined as certain classes of subsets of a given space, this task is not easy. For this reason, we try to construct models of them. A *model* of a given hyperspace $\mathcal{K}(X)$ is a topologically equivalent space, where the elements are points instead of subsets of X.

From the geometric point of view, the subject of models of hyperspaces is very attractive. Moreover, models are a very powerful tool which suggest properties and results on hyperspaces. Unfortunately, as we will see, there are only a few hyperspaces that can be modeled.

In this chapter we present a survey of what is known about models of hyperspaces of metric continua. With minor changes, this chapter is a copy of the expository paper published by the author in *Topology Proceedings* in 2013 [64]. We thank the editors of that journal for allowing me to include the material of [64] in this book. Some previous versions on this topic were published in Spanish by the author in [57] and [58]. We privilege geometric ideas; for example, we do not prove the continuity of any function. Some of the models included here are explained in more detail in Chapter II of [69].

10.1 $C([0, 1])$

The simplest continuum is the unit interval [0, 1]. Notice that

$$C([0,1]) = \{[a,b] : 0 \le a \le b \le 1\}.$$

It is easy to check that the function $\varphi : C([0,1]) \to \mathbb{R}^2$ ($\mathbb{R}^2$ is the Euclidean plane) given by $\varphi([a,b]) = (a,b)$ is a homeomorphism between $C([0,1])$ and the triangle $T = \{(a,b) \in \mathbb{R}^2 : 0 \le a \le b \le 1\}$, represented in Fig. 10.1.

A. Illanes, *Continuum Theory*, Universitext,
https://doi.org/10.1007/978-3-031-91011-1_10

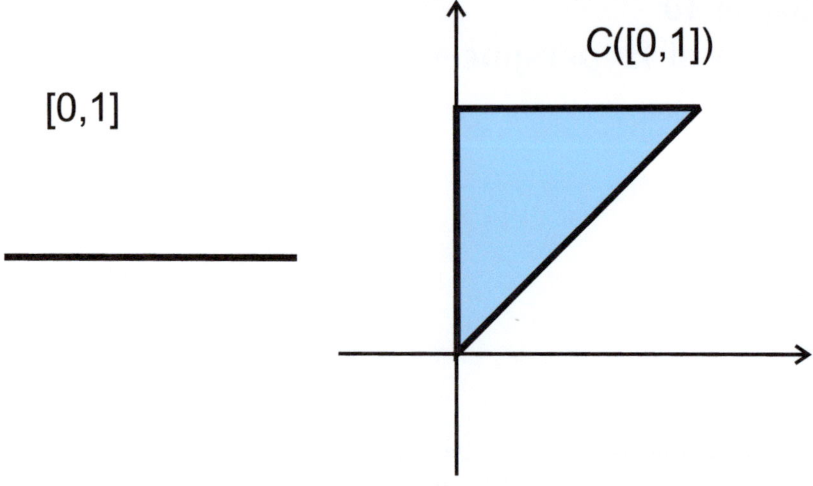

Fig. 10.1 $C([0, 1])$

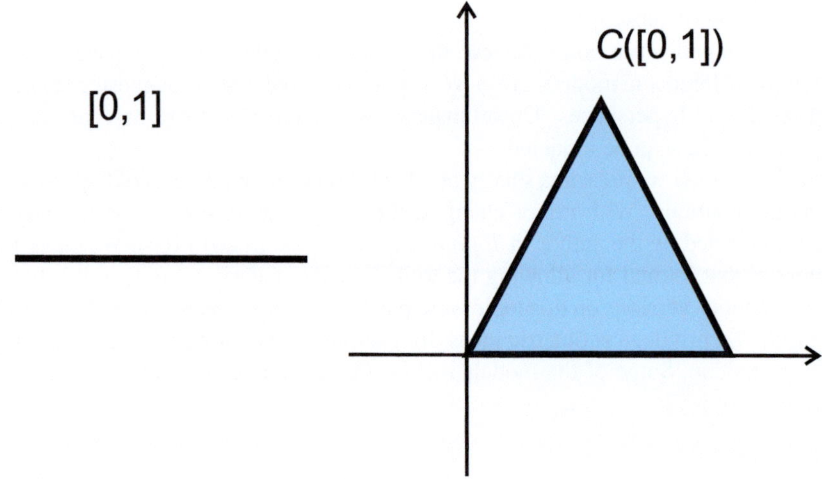

Fig. 10.2 $C([0, 1])$

Thus, we can say that this triangle is a model of $C([0, 1])$. Observe that the set of elements in $C([0, 1])$ that contain 0 (intervals of the form $[0, b]$) is represented by an edge of T. Similarly, the set of elements of $C([0, 1])$ containing 1 is represented by another edge of T. The set of singletons $F_1([0, 1])$ is represented by the third edge of T (the diagonal).

Sometimes it is more useful to represent $C([0, 1])$ by using the mapping $\psi : C([0, 1]) \to \mathbb{R}^2$ given by $\psi([a, b]) = (\frac{a+b}{2}, b - a)$. Notice that the image of ψ is the triangle illustrated in Fig. 10.2.

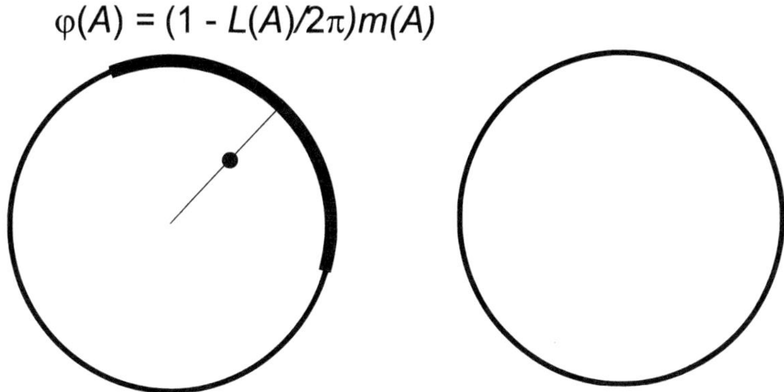

Fig. 10.3 $C(S^1)$

10.2 $C(S^1)$

Recall that S^1 denotes the unit circle in $\mathbb{R}^2$, centered at the origin. For each subarc A of S^1 let $m(A)$ be the middle point of A in S^1 and let $L(A)$ be the length of A. Then define $\varphi : C(S^1) \to \mathbb{R}^2$ by

$$\varphi(A) = \begin{cases} (1 - (L(A)/2\pi))m(A), & \text{if } A \neq S^1, \\ (0, 0), & \text{if } A = S^1. \end{cases}$$

It is easy to check that φ is a homeomorphism between $C(S^1)$ and the unit disc. Thus, a model of $C(S^1)$ is this disc (Fig. 10.3).

Take a point $p \in S^1$. For later use, we need to identify the image under φ of the set $\mathcal{D} = \{A \in C(S^1) : p \in A\}$. By the homogeneity of S^1 we suppose that $p = (0, 1)$. The best way to visualize $\mathcal{D}$ is to recognize its boundary, which is given by the set $\{A \in C(S^1) : A \text{ is a subarc of } S^1 \text{ and } p \text{ is an end point of } A\} \cup \{S^1, \{p\}\}$. Starting at $\{p\}$, we consider arcs having p as their end point and draw the images of them under φ, then we obtain the curve in Fig. 10.4. Now we see that $\mathcal{D}$ has the shape of a heart.

10.3 C(Simple Triod)

Another continuum that can be modeled is the *simple triod* T defined as the union of three arcs L_1, L_2 and L_3, called the *legs* of T, joined by a point v called the *top* of T (Fig. 10.5). The hyperspace $C(T)$ is the union of $C(L_1)$, $C(L_2)$, $C(L_3)$ and $C_v(T) = \{A \in C(T) : v \in A\}$. By the model in Sect. 10.1, each set $C(L_i)$ can be represented as a convex triangle. Given an element A of $C_v(T)$, A is uniquely

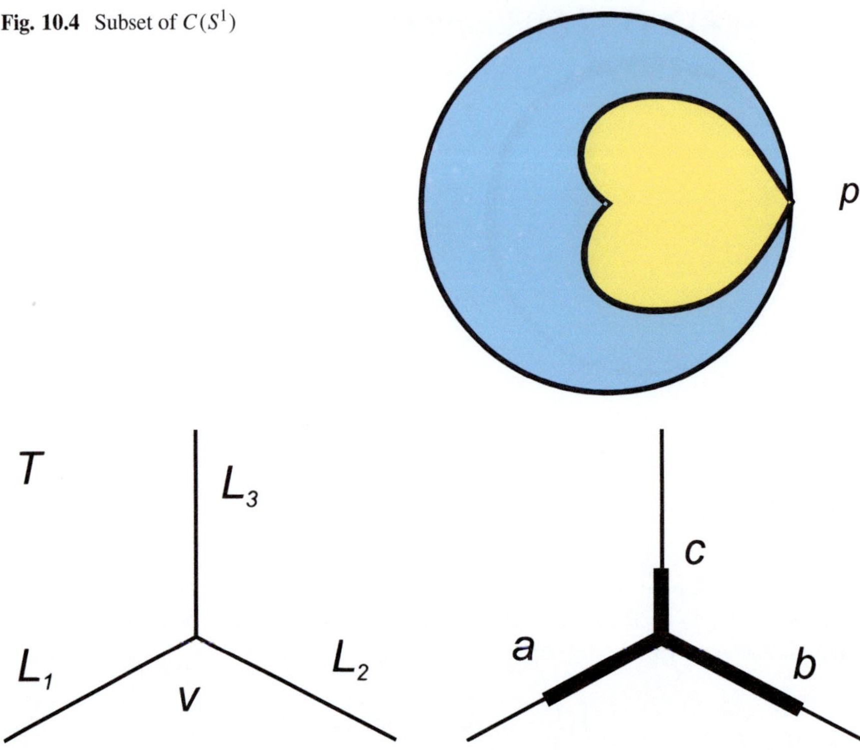

Fig. 10.4 Subset of $C(S^1)$

Fig. 10.5 C(Simple Triod)

determined by the lengths of the intersections of A with the legs of T. So, they can be represented by a vector with three coordinates (a, b, c).

Varying the three lengths a, b and c we obtain a convex cube in $\mathbb{R}^3$. Thus $C(T)$ is the union of a convex cube in $\mathbb{R}^3$ with three convex triangles attached, as pictured in Fig. 10.6.

10.4 C(Noose)

The next continuum we consider is the *noose* N, which is the union of a simple closed curve S and an arc J intersecting S at a point v that is an end point of J. The hyperspace $C(N)$ is the union of $C(S)$, $C(J)$ and $C_v(N) = \{A \in C(N) : v \in A\}$. By the previous examples, the set $C(J)$ can be represented as a convex triangle and $C(S)$ can be represented as a disc. Moreover, the elements A of $C_v(N)$ are uniquely determined by $A \cap S$ and by the length of $A \cap J$ (Fig. 10.7).

For each element $B \in C(S)$ such that $v \in B$, we can enlarge B by using a subarc of J containing v. Thus, for each such B, in the model of $C(N)$ we have to put an

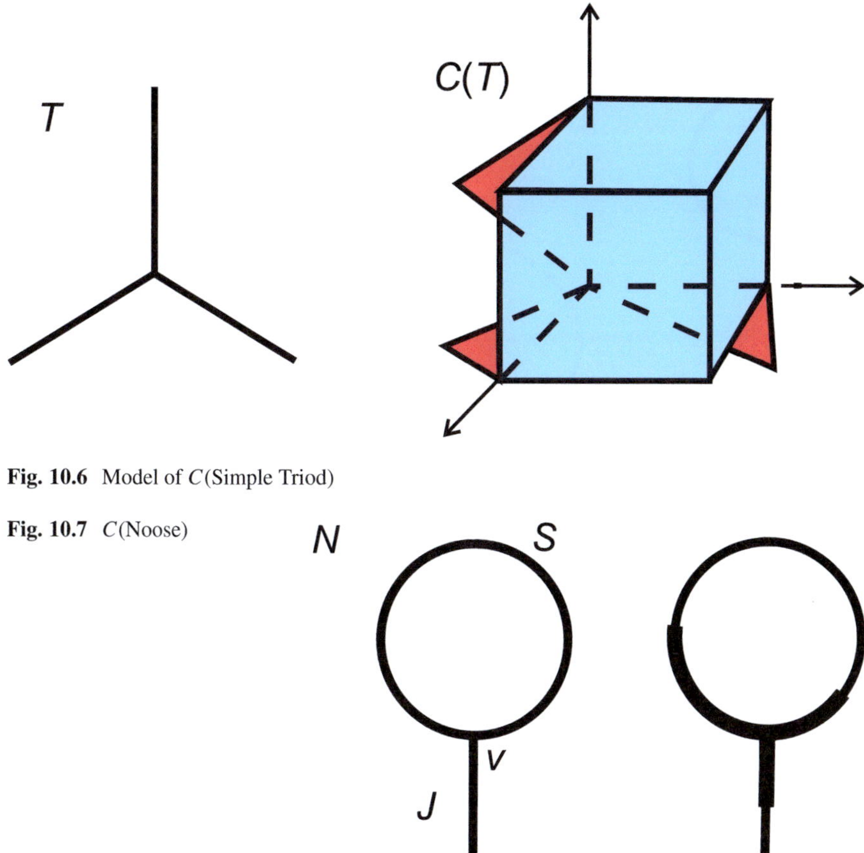

Fig. 10.6 Model of C(Simple Triod)

Fig. 10.7 C(Noose)

arc. As we saw in Sect. 10.3, the set $C_v(S)$ of all such elements B is a two cell with the shape of a heart. Hence, a model of $C_v(S)$ is the cylinder $C_v(S) \times [0, 1]$. To this cylinder we add the disc $C(S)$ and the convex triangle $C(J)$. Now, it is not difficult to see that a model of $C(N)$ is the space represented in Fig. 10.8.

10.5 No More Peano Models of $C(X)$ in $\mathbb{R}^3$

Now consider the continuum $\mathcal{H}$ with the shape of the letter H. Let J be the transversal arc, as in Fig. 10.9.

Since the subcontinua of $\mathcal{H}$ containing J can be enlarged in four independent directions, thus obtaining four lengths a, b, c and d, $C(\mathcal{H})$ contains a 4-cell and $C(\mathcal{H})$ cannot be embedded in $\mathbb{R}^3$.

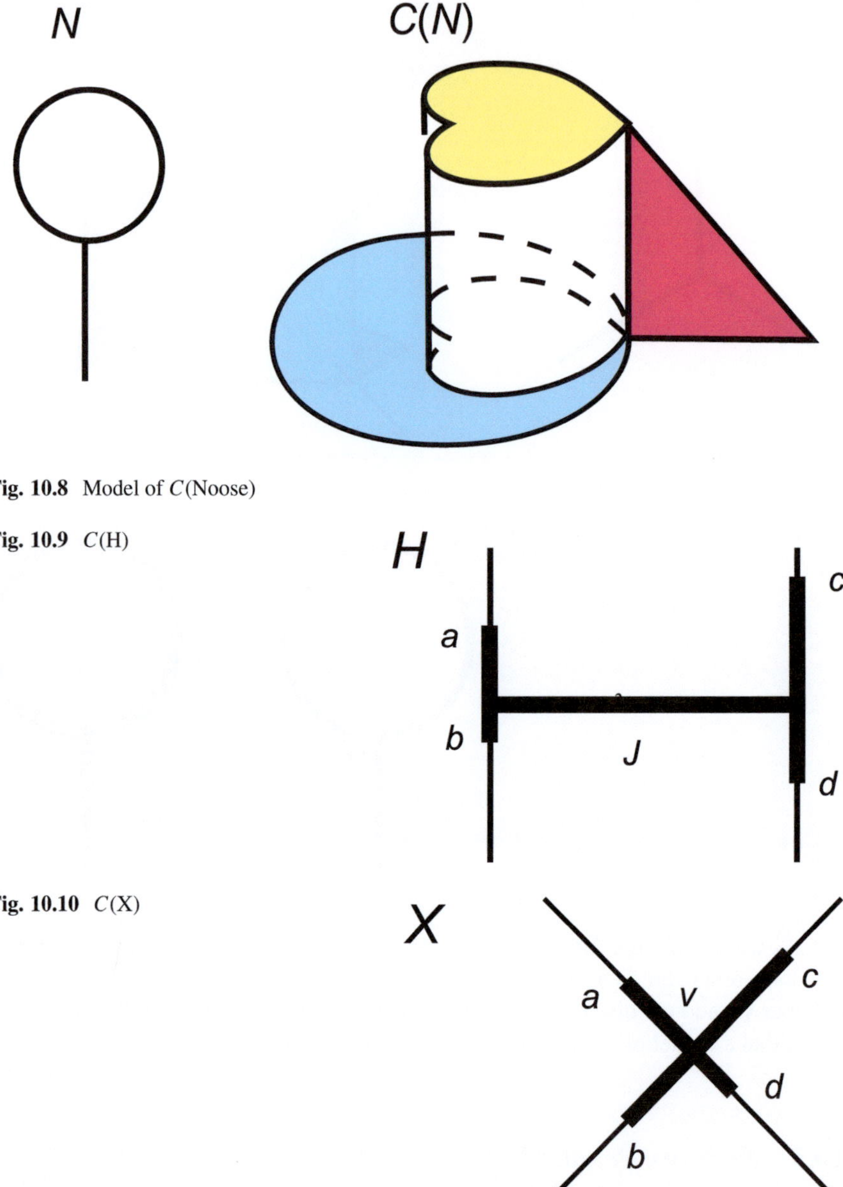

Fig. 10.8 Model of C(Noose)

Fig. 10.9 C(H)

Fig. 10.10 C(X)

If X contains a simple 4-od with top v, similarly as we did with the simple triod, it can be seen that $C_v(X) = \{A \in C(X) : v \in A\}$ contains a 4-cell. Thus $C(X)$ is not embeddable in $\mathbb{R}^3$ (Fig. 10.10).

Let Z be a locally connected continuum such that $C(Z)$ is embeddable in $\mathbb{R}^3$. Then $C(Z)$ is finite-dimensional. Thus, (see the historical remarks in [69, p. 44]) Z

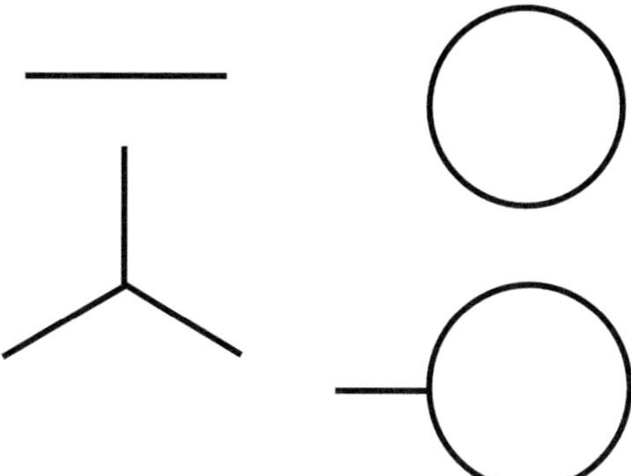

Fig. 10.11 Embedding $C(X)$ in $\mathbb{R}^3$

is a finite graph. By the paragraphs above, Z does not contain two ramification points, nor does it contain a simple 4-od. This implies that Z has at most one ramification point and it is of order at most 3. Therefore, Z is either an arc, a simple closed curve, a simple triod or a noose. Thus, if Z is a locally connected continuum, then $C(Z)$ is embeddable in $\mathbb{R}^3$ if and only if Z is one of the continua described in Sects. 10.1, 10.2, 10.3 or 10.4 (Fig. 10.11).

10.6 More Continua X for Which $C(X)$ is Embeddable in $\mathbb{R}^3$

There are more continua X for which $C(X)$ is embeddable in $\mathbb{R}^3$. For example, S.B. Nadler, Jr. [111] showed that there are exactly eight hereditarily decomposable continua X such that cone(X) is homeomorphic to $C(X)$. These continua are pictured in Fig. 10.12.

Since almost all of them can be embedded in $\mathbb{R}^2$, their hyperspace $C(X)$ can be embedded in $\mathbb{R}^3$. Another example X such that $C(X)$ is embeddable in $\mathbb{R}^3$ is the buckethandle continuum, for which it is also known that $C(X)$ is homeomorphic to its cone. One more example is the continuum X consisting of a simple triod with a ray surrounding it, illustrated in Fig. 10.13.

A model of $C(X)$ is in Fig. 10.14.

This model consists of a solid rocket (homeomorphic to the cube with three triangles of Fig. 10.6) surrounded by an infinite sheet converging to it. This model was useful for showing the existence of a tree-like continuum X such that its hyperspace $C(X)$ does not have the fixed point property [61]. The general problem of characterizing those continua X for which $C(X)$ is embeddable in $\mathbb{R}^3$ seems to be very difficult. In fact, an answer to the following old problem due to J. Krazinkiewicz is not known.

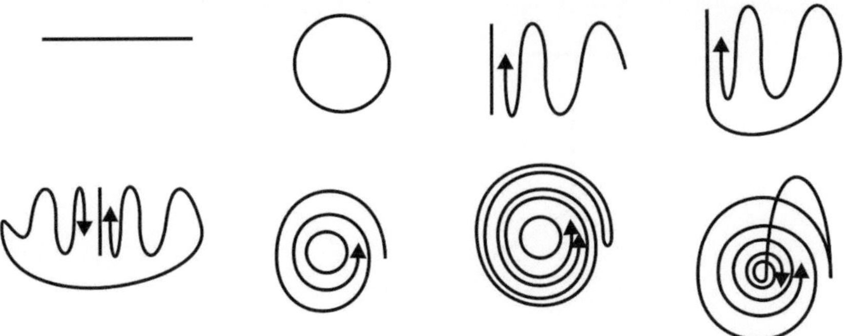

Hereditarily decomposable continua *X* such that *C(X)* is homeomorphic to its cone

Fig. 10.12 Nadler's continua

Fig. 10.13 Triod with spiral

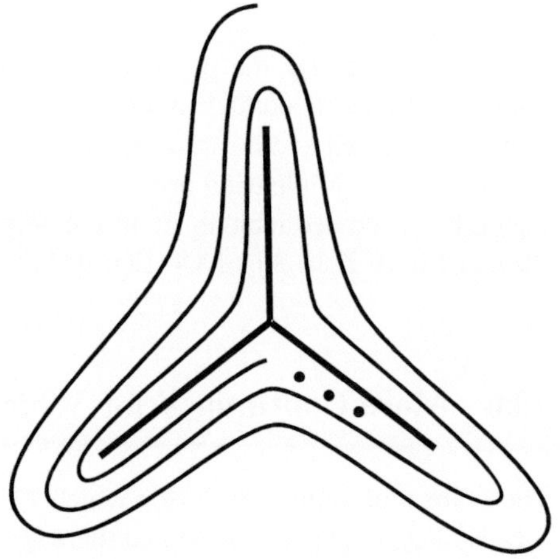

Problem 10.1 ([112, Question 3.5]) *Is it true that if $C(X)$ is embeddable in $\mathbb{R}^3$, then X is embeddable in $\mathbb{R}^2$?*

We can also ask a similar question as Problem 10.1 for $n \geq 3$, that is, we can ask if the fact that $C(X)$ is embeddable in $\mathbb{R}^{n+1}$ implies that X is embeddable in $\mathbb{R}^n$. This question can be easily solved since if $n \geq 3$ and $C(X)$ is embeddable in $\mathbb{R}^{n+1}$, then $C(X)$ is finite-dimensional. This implies that X is 1-dimensional [69, Corollary 73.11], so X is embeddable in $\mathbb{R}^3$.

An *n-od* in a continuum X is a subcontinuum B of X for which there exists a subcontinuum A of B such that $B \setminus A$ has at least n components. It is known [69, Theorem 70.1] that X contains an n-od if and only if $C(X)$ contains an n-cell.

V. Martínez-de-la-Vega and N. Ordoñez have found a characterization of locally connected continua X for which $C(X)$ is embeddable in $\mathbb{R}^4$ (and $\mathbb{R}^5$) [105].

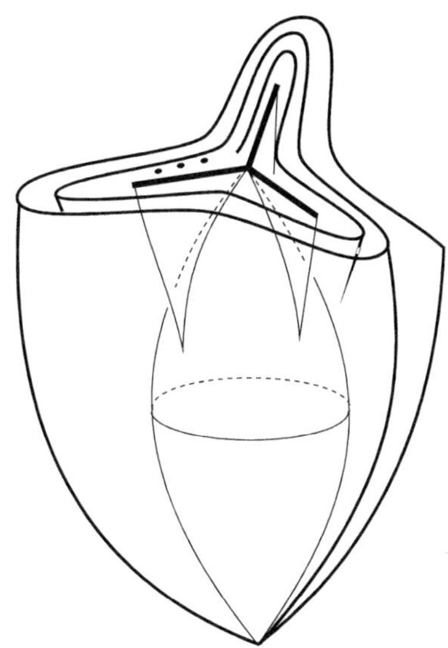

Fig. 10.14 C(Triod with spiral)

10.7 Peano X for Which $C(X)$ is Embeddable in $\mathbb{R}^4$ and $\mathbb{R}^5$

Theorem 10.2 ([105]) *Let X be a locally connected continuum. Then the following are equivalent.*

(a) *$C(X)$ is embeddable in $\mathbb{R}^4$,*
(b) $\dim(C(X)) \leq 4$,
(c) *X contains no 5-ods,*
(d) *$C(X)$ contains no 5-cells,*
(e) *X is one of the continua in Fig. 10.15.*

Theorem 10.3 ([105]) *Let X be a locally connected continuum. Then the following are equivalent.*

(a) *$C(X)$ is embeddable in $\mathbb{R}^5$,*
(b) $\dim(C(X)) \leq 5$,
(c) *X contains no 6-ods,*
(d) *$C(X)$ contains no 6-cells,*
(e) *X is one of the continua in Fig. 10.15 or $X = Z \cup J$, where Z is one of the continua in Fig. 10.15, J is an arc such that $Z \cap J = \{p\}$ (for some $p \in Z$) and p is an end point of J.*

The proofs of Theorems 10.2 and 10.3 depend on the construction of the models of $C(X)$ (in $\mathbb{R}^4$) of the continua in Fig. 10.16.

The construction of the models of the continua in Fig. 10.16 is difficult. In [105], V. Martínez-de-la-Vega and N. Ordoñez gave explicit formulas for embedding their hyperspaces $C(X)$ in $\mathbb{R}^4$. In particular, the formulas for the θ-curve are really

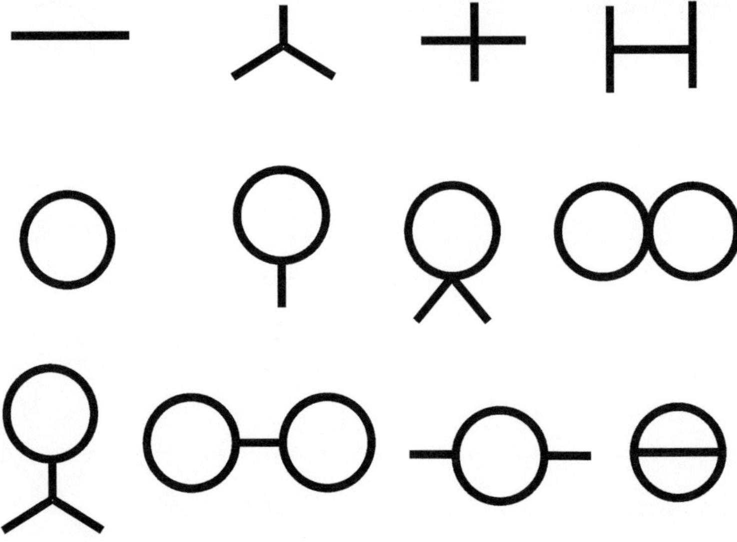

Locally connected continua X such that C(X) is embeddable in R4

Fig. 10.15 Embedding $C(X)$ in $\mathbb{R}^4$

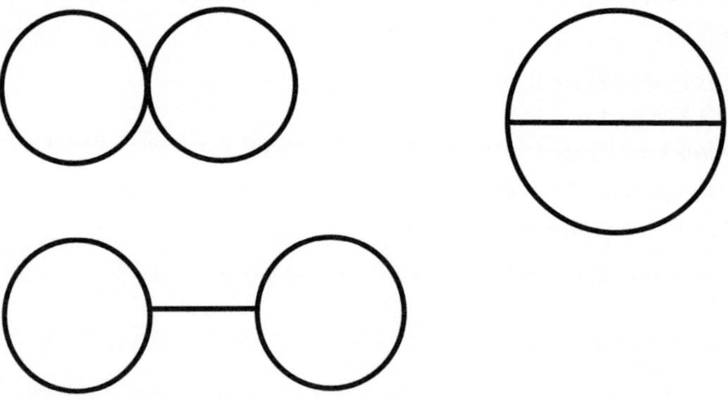

Fig. 10.16 Embedding $C(X)$ in $\mathbb{R}^4$

complex. So this procedure does not seem to be useful for proving a similar result for $\mathbb{R}^n$, for $n \geq 6$. The following question remains open.

Problem 10.4 ([105, Problem 2]) *Given $n \geq 6$ and a continuum X, are the following equivalent?*

(a) *$C(X)$ is embeddable in $\mathbb{R}^n$,*
(b) $\dim(C(X)) \leq n$.

More results and questions on the topic of embedding the hyperspace $C(X)$ in some space $\mathbb{R}^n$ can be found in Chapter III of [112].

10.8 Infinite-Dimensional Models of $C_n(X)$

In the literature, we can find some models of the hyperspace $C(X)$, in the case when $C(X)$ is infinite dimensional. For example, C. Eberhart and S.B. Nadler, Jr. constructed models of smooth fans in [34]. In [69, Example 6.1] a model of the hyperspace $C(F_\omega)$ is constructed. The most important result about infinite-dimensional models is the one given by the following fundamental theorem.

Theorem 10.5 ([30] and [27] for the case $n \geq 2$**)** *Let X be a continuum. Then the following are equivalent.*

(a) $C(X)$ *is homeomorphic to the Hilbert cube,*
(b) *X is locally connected and each arc in X has empty interior,*
(c) $C_n(X)$ *is homeomorphic to the Hilbert cube.*

10.9 $C_n([0, 1])$ for $n \geq 2$

R.M. Schori has shown that $C_2([0, 1])$ is a 4-cell by using the following argument [54, Lemma 2.2]. Let $C_0^1 = \{A \in C_2([0, 1]) : 0, 1 \in A\}$ and $C_1 = \{A \in C_2([0, 1]) : 1 \in A\}$. The typical elements of C_0^1 are of the form $A = [0, a] \cup [b, 1]$, where $0 \leq a \leq b \leq 1$. We can define $\varphi(A) = (a, b)$. Then φ is not a function since $\varphi([0, 1]) = \varphi([0, a] \cup [a, 1]) = (a, a)$ for each $a \in [0, 1]$. The image of φ is the triangle T in Fig. 10.1. If we identify the diagonal Δ of T with a point we obtain the space T/Δ and now φ is a well-defined homeomorphism between C_0^1 and T/Δ. This proves that C_0^1 is a 2-cell. It is easy to show that the function $\psi : C_0^1 \times [0, 1] \rightarrow C_1$ given by $\psi(A, t) = t + (1 - t)A$ is continuous, surjective and its only non-degenerate fiber is the set $C_0^1 \times \{1\}$. Thus, C_1 is homeomorphic to the cone of C_0^1. Hence, C_1 is a 3-cell. Finally, the function $\sigma : C_1 \times [0, 1] \rightarrow C_2([0, 1])$ given by $\sigma(A, t) = tA$ is continuous, surjective and its only non-degenerate fiber is $C_1 \times \{0\}$. Hence, $C_2([0, 1])$ is homeomorphic to the cone over C_1. Therefore, $C_2([0, 1])$ is a 4-cell.

In [55], it has been shown that, if $n \geq 3$, then $\{A \in C_n([0, 1]) : A$ has a neighborhood in $C_n(X)$ that is a $2n$-cell$\} = C_n([0, 1]) \setminus C_{n-1}([0, 1])$. In particular, this implies that $C_n([0, 1])$ is not a $2n$-cell. The author has constructed a model of $C_3([0, 1])$ and he is able to show that $C_3([0, 1])$ can be embedded in $\mathbb{R}^6$. The following problem remains unsolved.

Problem 10.6 *Is* $C_n([0, 1])$ *embeddable in* $\mathbb{R}^{2n}$ *for each (for some)* $n \geq 4$*?*

10.10 $C_n(S^1)$ for $n \geq 2$

In [56] it was shown that $C_2(S^1)$ is the cone over a solid torus. The proof is difficult and it seems that it cannot be generalized for $n \geq 3$. Nothing else is known for $C_n(S^1)$ ($n \geq 3$). The following problem is interesting.

Problem 10.7 *(a) Find a model of $C_3(S^1)$; (b) Is $C_n(S^1)$ the cone over a continuum for some (for all) $n \geq 3$?; (c) Is $C_n(S^1)$ embeddable in $\mathbb{R}^{2n}$ for each (for some) $n \geq 3$?*

In [98, Theorem 3.1] it was shown that if X is a simple m-od, then $C(X)$ is the cone over the set $\{A \in C_n(X) : A$ contains an end point of $X\}$. In [104] V. Martínez-de-la-Vega proved that if G is a finite graph such that $C_n(G)$ is a cone for some $n \geq 2$, then G is either an m-od or a simple closed curve. So, the answer to Problem 10.7 (b) could give a characterization of those finite graphs G for which $C_n(G)$ is a cone.

10.11 Continua for Which $C(X)$ is a Cone

Suppose that X is a continuum such that $C(X)$ is homeomorphic to cone(Z) for some continuum Z. The problem of finding such continua X has been widely studied. Based on the deep results obtained by María de Jesús López [31] in her dissertation, in [66], A.I. and J. de M. López obtained the following very complete characterization.

Theorem 10.8 *Let X be a hereditarily decomposable continuum for which there exists a finite-dimensional continuum Z such that $C(X)$ is homeomorphic to cone(Z). Then X is one of the continua in Fig. 10.12 or X is in one of the classes of continua described as follows:*

(a) a union of a finite number of simple spirals that share their limit circle (a simple spiral is a space homeomorphic to the continuum described in Fig. 13.2),
(b) a union of a finite number of $\sin(\frac{1}{x})$-continua that share their limit arc,
(c) a union of a finite number of $\sin(\frac{1}{x})$-continua that share their limit arc, with the limit arc extended to a larger arc by either of its two end-points,
(d) a simple n-od, for some $n \in \mathbb{N}$.

Conversely, for each of the described continua, its hyperspace $C(X)$ is homeomorphic to a finite-dimensional cone.

In order to obtain a model of each of the continua described in Theorem 10.8 it is enough to know the respective continuum Z. For the continua X in Fig. 10.12, $C(X)$ is homeomorphic to cone(X) [111]. For each of the continua described in (a)–(d), the respective Z is constructed in [66].

10.12 Models of 2^X

In the early 1920s, in Poland, it was conjectured that $2^{[0,1]}$ is a Hilbert cube. It was not until the 1970s that this problem was solved by D.W. Curtis and R.M. Schori, who proved the following fundamental theorem.

Theorem 10.9 ([29] and [30]) *Let X be a continuum. Then the following are equivalent.*

(a) X is locally connected,
(b) 2^X is homeomorphic to the Hilbert cube.

Although models of some very specific examples have been constructed for 2^X, the only significant result about models of 2^X is Theorem 10.9.

10.13 $F_n([0, 1])$

The simplest continuum is the unit interval $[0, 1]$.

Let $\varphi : F_2([0, 1]) \to \mathbb{R}^2$ be given by $\varphi(\{a, b\}) = (\min\{a, b\}, \max\{a, b\})$. Clearly, φ is an embedding whose image is the convex triangle $\{(a, b) \in \mathbb{R}^2 : 0 \leq a \leq b \leq 1\}$. Thus, $F_2([0, 1])$ is a 2-cell (Fig. 10.17).

In order to construct a model of $F_3([0, 1])$ let us consider again the map $\varphi : F_3([0, 1]) \to \mathbb{R}^2$ given by $\varphi(A) = (\min A, \max A)$. Then φ is a continuous function whose image is the triangle T in Fig. 10.7. Given $(a, b) \in T$, the fiber $\varphi^{-1}((a, b))$ is the set $\{\{a, b, c\} : a \leq c \leq b\}$. If $a < b$, the set $\{\{a, b, c\} : a \leq c \leq b\}$ is a simple closed curve since c runs over the interval $[a, b]$ and $\{a, a, b\} = \{a, b, b\}$. If

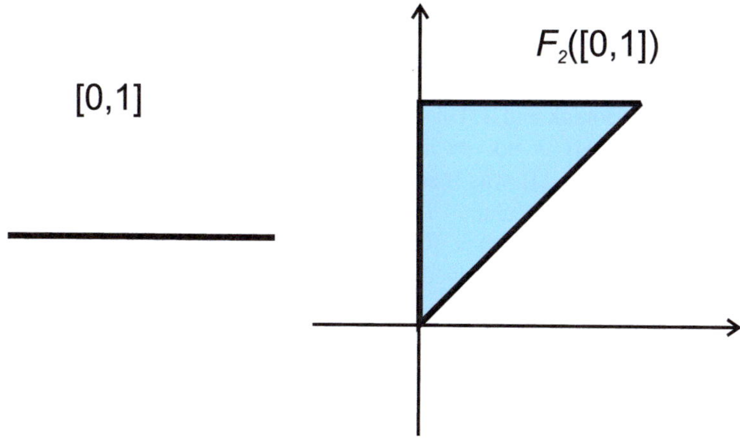

Fig. 10.17 $F_2([0, 1])$

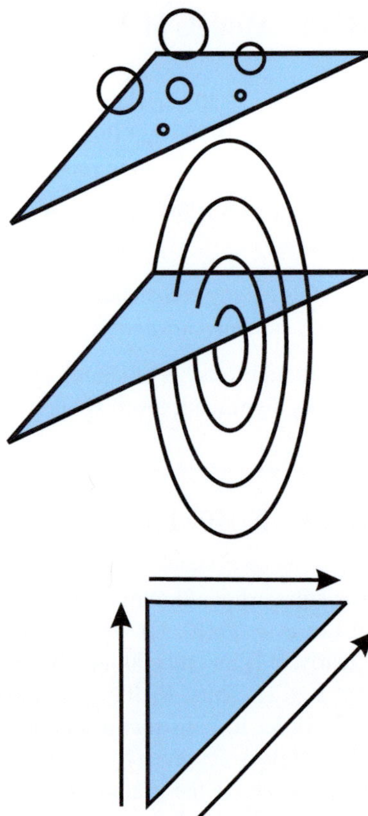

Fig. 10.18 $F_3([0, 1])$

Fig. 10.19 Constructing the dunce hat

$a = b$, $\varphi^{-1}((a, b)) = \{\{a, b, c\} : a \le c \le b\} = \{\{a\}\}$. Thus to obtain a model of $F_3([0, 1])$ we need to put a circle at each point $(a, b) \in T$ such that $a < b$. This can be realized by taking the revolution body that can be obtained by rotating T around its diagonal. Therefore, $F_3([0, 1])$ is a 3-cell (Fig. 10.18).

For $n \ge 4$, K. Borsuk and S. Ulam, in the first paper about symmetric products, proved that $F_n([0, 1])$ is not an n-cell [19, Theorem 7]. A detailed study of the hyperspaces $F_n([0, 1])$ was made by R.N. Andersen, M.M. Marjanović and R.M. Schori in [5]. In particular, in Theorem 2.1 of [5], it was shown that $F_4([0, 1])$ is homeomorphic to cone$(D_2) \times [0, 1]$, where D_2 is the *dunce hat*. Recall that D_2 is the space that can be obtained by identifying the edges of a convex triangle according to the arrows shown in Fig. 10.19.

First, identify two of the arrows to obtain a cone; second, identify the vertex of the cone with a point in its base to obtain the space in Fig. 10.20; finally, identify the two simple bold closed curves in Fig. 10.20.

It is easy to see that D_2 can be constructed in $\mathbb{R}^3$. So, $F_4([0, 1])$ is embeddable in $\mathbb{R}^5$. This answers the following question for the case $n = 4$. This question is open for $n \ge 5$.

Fig. 10.20 Dunce hat

Problem 10.10 ([19, Last Paragraph]) *Is $F_n([0,1])$ embeddable in $\mathbb{R}^{n+1}$ for every $n \geq 5$?*

10.14 $F_n(S^1)$

The symmetric product $F_2(S^1)$ is a Moebius strip. We can see this by using the following argument. Let $\mathcal{NA} = \{\{p,q\} \in F_2(S^1) : p \neq -q\}$. Let $A : \mathcal{NA} \to C(S^1)$, $m : \mathcal{NA} \to S^1$, $L : \mathcal{NA} \to \mathbb{R}$ and $\varphi : \mathcal{NA} \to \mathbb{R}^2$ be given by

$$A(\{p,q\}) = \text{ the shorter arc joining} p \text{ and } q \text{ in } S^1,$$

$$m(\{p,q\}) = \text{ the middle point of} A(\{p,q\}),$$

$$L(\{p,q\}) = \text{ the length of} A(\{p,q\}), \text{ and}$$

$$\varphi(\{p,q\}) = (1 - (\frac{1}{2\pi}L(\{p,q\})))m(\{p,q\}).$$

Notice that φ is a homeomorphism between $\mathcal{NA}$ and the annulus $R = \{z \in \mathbb{R}^2 : \frac{1}{2} < |z| \leq 1\}$. If we want to extend φ to the set $\mathcal{A} = \{\{z,-z\} \in F_2(S^1) : z \in S^1\}$, by continuity and depending on the way we approximate $\{z,-z\}$ by elements $\{p,q\} \in \mathcal{NA}$, we should define $\varphi(\{z,-z\})$ to have two values, namely, $\varphi(\{z,-z\}) = w$ or $-w$, where $2w$ is the point obtained by rotating z by $\frac{\pi}{2}$. To solve this ambiguity, we need to identify points w and $-w$. Notice that the points w are the points of the circle $B = \{u \in \mathbb{R}^2 : |u| = \frac{1}{2}\}$. The quotient space obtained from $R_0 = R \cup B$ by this identification is the Moebius strip M. In Fig. 10.21 we show how this strip can be obtained. We start with the annulus and we cut it by two arrows a and b. Then we make the transformations marked in Fig. 10.21 to get the strip M.

For further use, fix a point $Q \in S^1$. We need to represent the set $D_Q = \{\{Q,z\} : z \in S^1\}$ in M. According to the definition of φ, the image of this set consists of two arcs B_1 and B_2 in the annulus R_0. If we follow the transformations that we have made to obtain the Moebius strip M, we can see that D_Q is homeomorphic to a simple closed curve B that touches the boundary of M in exactly one point. This

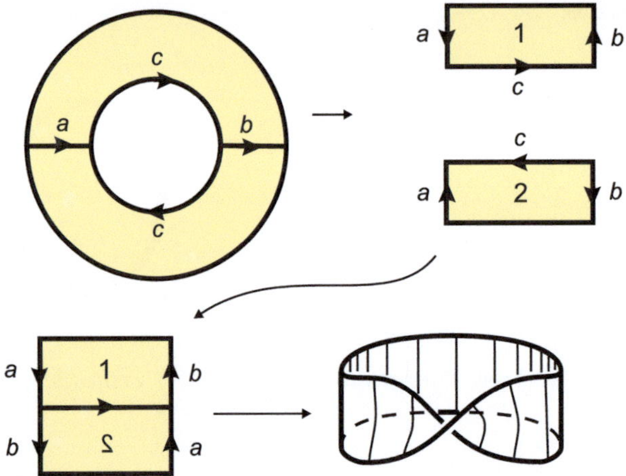

Fig. 10.21 $F_2(S^1)$

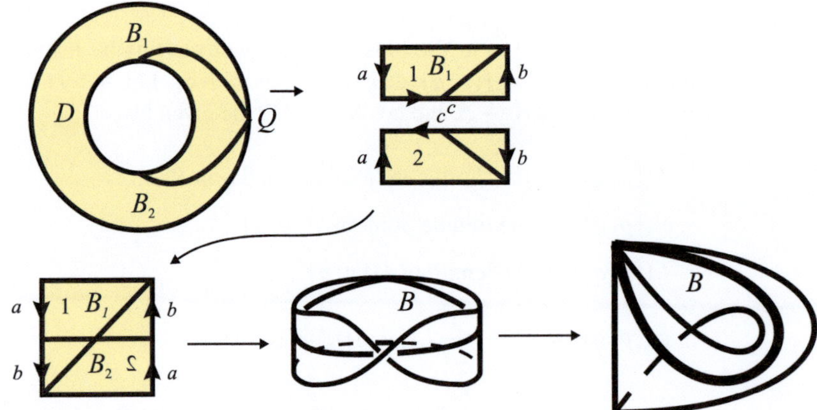

Fig. 10.22 A subset of $F_2(S^1)$

simple closed curve is represented at the end of Fig. 10.22. It is important to remark that in this representation of the Moebius strip the curve B can be pictured entirely without dotted lines.

As we have seen, some models of hyperspaces are easy to construct. There are other more complicated examples for which a specific approach is necessary. To illustrate how difficult it may be to construct a model, let us mention that K. Borsuk made a mistake. He published a paper [17] claiming that $F_3(S^1)$ is homeomorphic to $S^1 \times S^2$, where S^n is the unit sphere in $\mathbb{R}^{n+1}$. Three years later, R. Bott [20] corrected this fact by proving that $F_3(S^1)$ is homeomorphic to S^3. J. Mostovoy pointed out that an interesting illustration of the non-triviality of Bott's theorem is the result attributed to E. Schepin which says that the embedding knot $(F_1(S^1))$ is a Trefoil knot (see Theorem 2 of [110]).

Although no models of $F_n(S^1)$ $(n \geq 4)$ have been constructed, in [122] and [126] some topological properties of these spaces have been studied.

10.15 F_2(Simple Triod)

Consider a simple 3-od Y as illustrated in Fig. 10.23. Let $J_1 = L_1 \cup L_2$, $J_2 = L_2 \cup L_3$ and $J_3 = L_3 \cup L_1$. Note that $F_2(T) = F_2(L_1) \cup F_2(L_2) \cup F_2(L_3)$. Since each J_i is an arc, we know that $F_2(J_i)$ can be viewed as a convex triangle. Thus, to obtain a model of $F_2(Y)$ we need to take the three triangles $F_2(J_1)$, $F_2(J_2)$ and $F_2(J_3)$ and identify the points that represent elements of $F_2(Y)$ appearing in more than one triangle. For example, $F_2(J_1) \cap F_2(J_2) = F_2(L_2)$, which is a subtriangle of both triangles $F_2(J_1)$ and $F_2(J_2)$. In Fig. 10.23, we picture the triangles $F_2(J_1)$, $F_2(J_2)$ and $F_2(J_3)$ with the parts that need to be identified. The resulting space is a convex triangle with three wings attached to it.

In [64, pages 55 and 56], the author wrote "E. Castañeda-Alvarado has recently found a model of $F_3(Y)$ [123]. He showed that $F_3(Y)$ is the cone over a torus with four disks attached to it, one as an "equator" and the three other ones as "meridians"

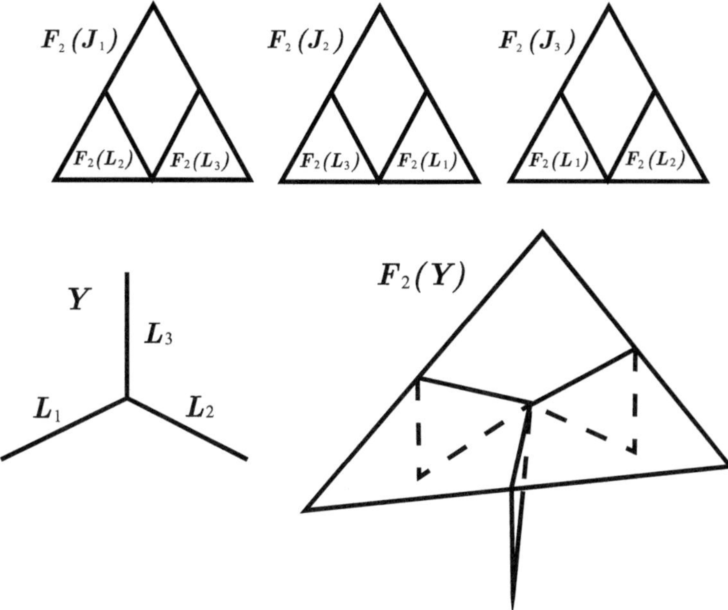

Fig. 10.23 F_2(Simple triod)

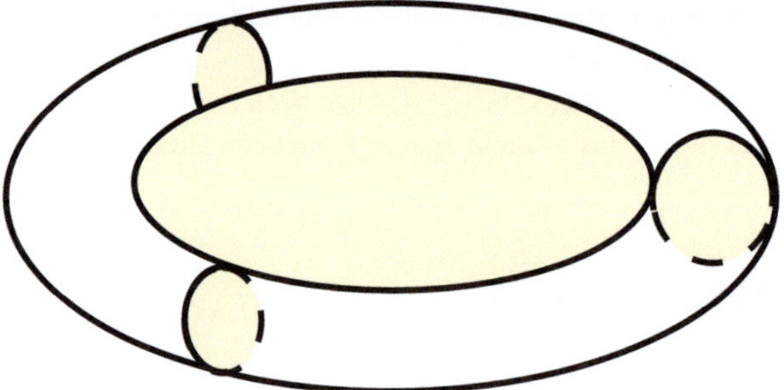

Fig. 10.24 F_3(Simple triod)

(Fig. 10.24)". F. Corona-Vázquez, R.A. Quiñones-Estrella, J. Sánchez-Martínez and H. Villanueva showed that the Castañeda–Alvarado model is wrong and to construct the correct model one must replace the torus by a Klein bottle, with the four disks attached in a similar way [26, section 5].

10.16 F_2(Simple 4-od)

Let X be a simple 4-od with top v, as in Fig. 10.25, where L is one of the legs. Let T be the simple triod obtained by removing the leg L from X. By the previous example, $F_2(T)$ is a convex triangle with three wings. Note that $F_2(X) = F_2(T) \cup F_2(L) \cup \langle T, L\rangle$, where $\langle T, L\rangle = \{\{p, q\} \in F_2(X) : p \in T \text{ and } q \in L\}$. Observe that $\langle T, L\rangle$ is homeomorphic to $T \times L$, $\langle T, L\rangle \cap F_2(T) = \{\{v, p\} \in F_2(X) : p \in T\}$ is a simple triod located on the convex triangle and $\langle T, L\rangle \cap F_2(L) = \{\{v, q\} \in F_2(X) : q \in L\}$ corresponds to the middle arc in $T \times L$. Thus, to obtain a model of $F_2(X)$ we have to put together three pieces, namely the triangle with wings, $T \times L$ and a convex triangle. The final model is illustrated in Fig. 10.25, where the space $(T \times L) \cup F_2(J)$ is attached to the triangle with wings by the simple triod on the triangle.

10.17 F_2(Noose)

Recall that the noose N is the union of a simple closed curve S and an arc J joined by a point v that is an end point of J. As in the previous example, $F_2(N) = F_2(S) \cup F_2(J) \cup \langle S, J\rangle$, where $\langle S, J\rangle = \{\{p, q\} \in F_2(N) : p \in S \text{ and } q \in J\}$. Note that $\langle S, J\rangle$ is homeomorphic to $S \times J$, $\langle S, J\rangle \cap F_2(S) = \{\{v, q\} \in F_2(N) : q \in S\}$ is a simple closed curve in $F_2(N)$ (like the one we constructed in the example of the Moebius strip (Fig. 10.22)) and $\langle S, J\rangle \cap F_2(J) = \{\{v, p\} \in F_2(N) : p \in J\}$ is an arc in the cylinder $S \times J$. Therefore, a model of $F_2(N)$ can be obtained as illustrated

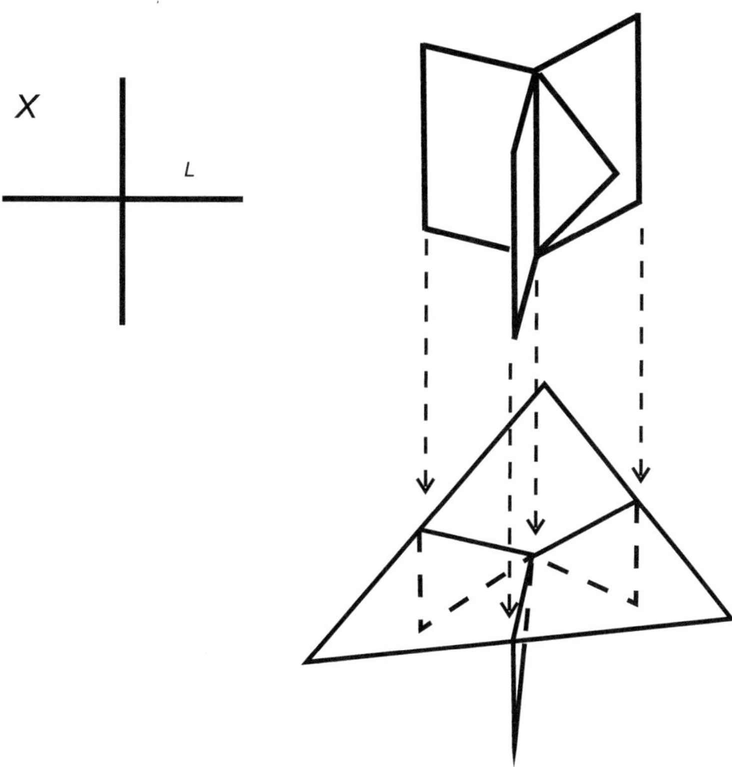

Fig. 10.25 F_2(4-od)

Fig. 10.26 F_2(Noose)

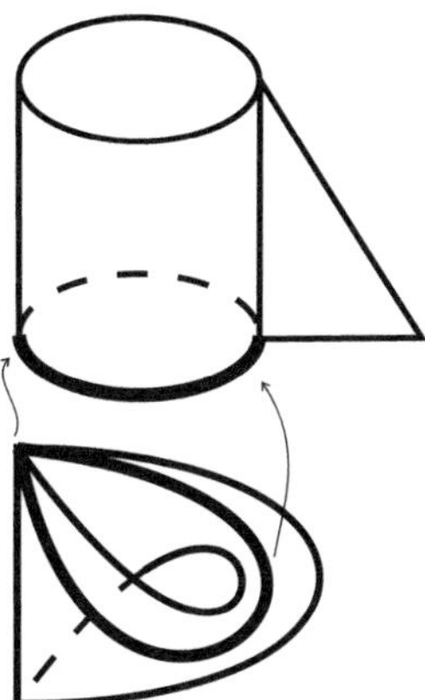

in Fig. 10.26, where the arrows indicate how the simple closed curve in the strip is attached to the base of the cylinder. Notice that, since we can see the curve in the strip, the rest of the strip can be pushed down in such a way that this attachment can be done in the space $\mathbb{R}^3$.

10.18 F_2(Figure Eight Continuum)

Let Y be the "figure eight" continuum. That is, Y is the union of two simple closed curves S_1 and S_2 whose intersection is a one-point set $\{v\}$. Note that $F_2(Y) = F_2(S_1) \cup F_2(S_2) \cup \langle S_1, S_2\rangle$, where $\langle S_1, S_2\rangle = \{\{p, q\} \in F_2(Y) : p \in S_1$ and $q \in S_2\}$. Given a point $p \in S_1$, the set of points $R_p = \{\{p, q\} : q \in S_2\}$ is a simple closed curve. Thus $\bigcup\{R_p : p \in S_1\}$ is a union of pairwise disjoint simple closed curves. Since p runs over the simple closed curve S_1, we can see that $\langle S_1, S_2\rangle$ is homeomorphic to the product $S^1 \times S^1$. Therefore, $\langle S_1, S_2\rangle$ is a torus. Notice that $R_v = \{\{v, q\} : q \in S_2\}$ is an "equator" of the torus $\langle S_1, S_2\rangle$ and the set $R = \{\{p, v\} : p \in S_1\}$ is a "meridian" of the torus. Since $F_2(S_1) \cap \langle S_1, S_2\rangle = R$, $F_2(S_2) \cap \langle S_1, S_2\rangle = R_v$ and $F_2(S_1) \cap F_2(S_2) = \{\{v\}\}$, we have that $F_2(Y)$ is the union of a torus with two Moebius strips attached by R and R_v. A representation of the model of $F_2(Y)$ is illustrated in Fig. 10.27.

Note that, since each of the sets R and R_v in the Moebius strips can be visualized, the final model of $F_2(Y)$ can be constructed in $\mathbb{R}^3$ (in order to see this we can picture the following: in order to attach $F_2(S_1)$ to the torus by the set R we can imagine the curve R in a vertical plane P in front of us in the space $\mathbb{R}^3$, then we can push the rest

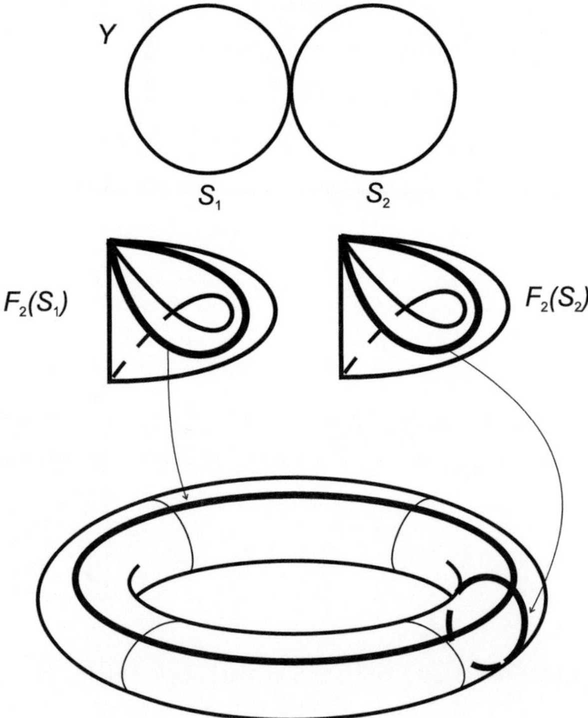

Fig. 10.27 F_2(Figure eight continuum)

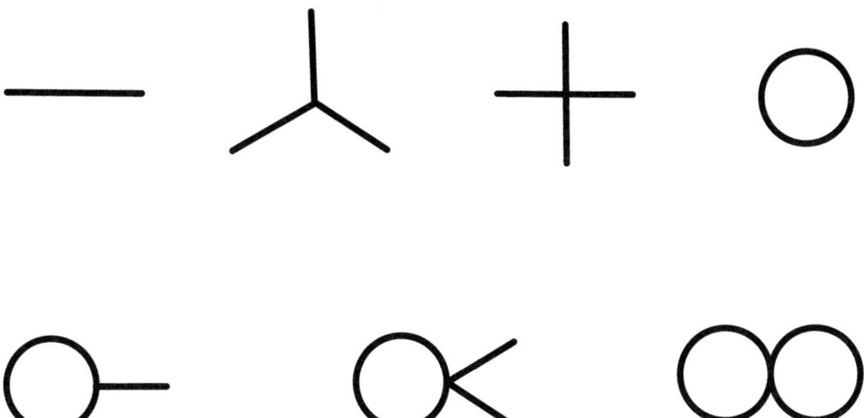

Fig. 10.28 Peano continua X such that $F_2(X)$ is embeddable in $\mathbb{R}^3$

of $F_2(S_1)$ behind P in such a way that $F_2(S_1)$ is in one of the halves in which $\mathbb{R}^3$ is divided by P and that $F_2(S_1) \cap P = R$. Then we can make a rigid transformation to move $F_2(S_1)$ to the upper part of the torus. The Moebius strip $F_2(S_2)$ must be attached to the torus from inside the tube.

We have shown that $F_2(Y)$ can be embedded in $\mathbb{R}^3$. E. Castañeda-Alvarado [21] proved that F_2(simple 5-od) contains a topological copy of the cone over the complete graph K_5 and then, using tools from low-dimensional topology, he showed that this cone is not embeddable in $\mathbb{R}^3$. Thus, F_2(simple 5-od) is not embeddable in $\mathbb{R}^3$. He also found that F_2(figure H-continuum) contains a topological copy of the topological cone of the complete bipartite graph $K_{3,3}$ and then he showed that F_2(figure H-continuum) is not embeddable in $\mathbb{R}^3$. Therefore, if a finite graph G either contains two vertices or it contains a vertex of order at least 5, then $F_2(G)$ is not embeddable in $\mathbb{R}^3$. The two last paragraphs are summarized in the following theorem (Fig. 10.28).

Theorem 10.11 ([21, Theorem 3]) *Given a locally connected continuum X, $F_2(X)$ can be embedded in $\mathbb{R}^3$ if and only if X is embeddable in the figure eight continuum.*

10.19 Hyperspaces $C_n(X)/F_m(X)$, $m \leq n$; and $F_n(X)/F_m(X)$, $m < n$

The theory of hyperspaces has been extended to the study of the so called *hyperspaces suspensions*, which are defined as quotient spaces of the form $C_n(X)/F_m(X)$, $m \leq n$, obtained from $C_n(X)$ by shrinking its subspace $F_m(X)$

to a point; and $F_n(X)/F_m(X)$, $m < n$, defined in a similar way. From the models we have developed in this chapter, it can be observed that

(A) $C([0, 1])/F_1([0, 1])$ is a 2-cell,
(B) $C_2([0, 1])/F_1([0, 1])$ is a 4-cell [99, Theorem 3.3],
(C) $F_2([0, 1])/F_1([0, 1])$ is a 2-cell,
(D) $F_3([0, 1])/F_1([0, 1])$ is homeomorphic to the space obtained from the unit closed ball $B^1 = \{p \in \mathbb{R}^3 : ||p|| \leq 1\}$ by shrinking one diameter of B^1 to a point,
(E) $C(S^1)/F_1(S^1)$ is homeomorphic to the 2-sphere: $\{p \in \mathbb{R}^3 : ||p|| = 1\}$,
(F) $F_2(S^1)/F_1(S^1)$ is homeomorphic to the Klein bottle.

The following results are not observations (the proofs of (H) and (I) are intricate):

(G) $C_2([0, 1])/F_2([0, 1])$ is a 4-cell [100, Theorem 4.6],
(H) $C_2(S^1)/F_2(S^1)$ is homeomorphic to the suspension of the solid torus $B^1 \times S^1$ [68, Theorem 4],
(I) $C_2(S^1)/F_1(S^1)$ is not homeomorphic to the suspension of the solid torus $B^1 \times S^1$ [68, Theorem 14].

10.20 $F_2(\sin(\frac{1}{x})$-Continuum)

In [21, Problem 2] E. Castañeda-Alvarado asked if Theorem 10.11 can be extended for all continua X. In [60] the author constructed a model of $F_2(\sin(\frac{1}{x})$-continuum) and proved that it can be embedded in $\mathbb{R}^3$. In fact, he proved that if X is any compactification of the ray $[0, \infty)$ with an arc as the remainder, then $F_2(X)$ can be embedded in $\mathbb{R}^3$. In [64] it was explained how complex the model of $F_2(\sin(\frac{1}{x})$-continuum) is.

10.21 More Questions

Problem 10.12 ([60]) *Characterize finite graphs G such that $F_2(G)$ is embeddable in $\mathbb{R}^4$.*

Problem 10.13 ([60]) *Is it true that, for a finite graph G, $F_2(G)$ is embeddable in $\mathbb{R}^4$ if and only if G is embeddable in $\mathbb{R}^2$? Is $F_2(K_5)$ embeddable in $\mathbb{R}^4$?*

The sufficiency in the first part of Problem 10.13 is true by the result in [109, Theorem 1] which says that $F_2([0, 1]^2) = [0, 1]^4$. In [109] one can find more results about $F_2([0, 1]^m)$.

10.22 F_n(Hilbert Cube)

Recall that we denote by **Q** the Hilbert cube. V.V. Fedorchuk proved that for each $n \geq 2$, $F_n(\mathbf{Q})$ is homeomorphic to **Q** [38, p. 223]. However, **Q** is not the only continuum X for which $F_n(X)$ is homeomorphic to **Q**. In [70, p. 139] it was shown that if X is the union of two Hilbert cubes joined by a point, then $F_2(X)$ is homeomorphic to **Q**. This is the only case we know in which two different spaces can have the same symmetric product.

Problem 10.14 *Do there exist two non-homeomorphic finite-dimensional continua X and Y such that $F_2(X)$ is homeomorphic to $F_2(Y)$?*

Combining Fedorchuk's Theorem and Theorems 10.5 and 10.9, we can conclude that **Q** has the property that all its hyperspaces $2^{\mathbf{Q}}$, $C_n(\mathbf{Q})$ and for $n > 1$, $F_n(\mathbf{Q})$ are homeomorphic to **Q**. We do not know if **Q** is the only space with this property.

Problem 10.15 *Does there exist a continuum X, non-homeomorphic to the Hilbert cube, such that X is homeomorphic to each of its hyperspaces 2^X, $C_n(X)$ and $F_n(X)$ (for all n)?*

Chapter 11
Irreducible Continua

This chapter is devoted to the classical theorem that says that if X is an irreducible continuum such that each of its indecomposable subcontinua has empty interior, then there exists a monotone mapping $f : X \to [0, 1]$ such that $\text{int}_X(f^{-1}(t)) = \emptyset$ for every $t \in [0, 1]$ (K. Kuratowski [90, Theorem 3, §48, VII, p. 216] and E.S. Thomas, Jr. [120, Theorem 10]).

11.1 Irreducibility

Recall (Definition 4.11) that a connected topological space W is *irreducible between the points a and b* in W if no proper connected closed subset of W contains both points a and b.

Throughout this chapter W will denote a connected space irreducible between its points a and b. Note that we are not asking that W is a continuum.

A point c of W is an *irreducibility point of* W if there exists a point $e \in W$ such that W is irreducible between c and e.

Theorem 11.1 *If A and B are connected closed subsets of W such that $a \in A$ and $b \in B$, then $W \setminus (A \cup B)$ is connected.*

Proof If $A \cap B \neq \emptyset$, then $A \cup B$ is a connected closed subset of W containing $\{a, b\}$, so $A \cup B = W$ and $W \setminus (A \cup B) = \emptyset$ is connected.

Now, suppose that $A \cap B = \emptyset$. By Exercise 11.15, the set $C = W \setminus A$ is connected, and $B \subset C$. Suppose that $W \setminus (A \cup B)$ is not connected. Then there exist disjoint nonempty open subsets U and V of W such that $U \cup V = W \setminus (A \cup B) = C \setminus B$. By Exercise 4.12 applied to the subspace C, we have that $B \cup U$ and $B \cup V$ are connected, so the sets $B \cup \text{cl}_W(U)$ and $B \cup \text{cl}_W(V)$ are also connected. Then $W = A \cup B \cup \text{cl}_W(W \setminus (A \cup B)) = A \cup B \cup \text{cl}_W(U \cup V)$ and $A \cap B = \emptyset$. The connectedness of W implies that $A \cap \text{cl}_W(U \cup V) \neq \emptyset$. If $A \cap \text{cl}_W(U) \neq \emptyset$, then $A \cup B \cup \text{cl}_W(U)$

A. Illanes, *Continuum Theory*, Universitext,
https://doi.org/10.1007/978-3-031-91011-1_11

is a connected closed subset of W containing $\{a, b\}$, so this set coincides with W. This implies that $V \subset W \setminus (A \cup B) \subset \mathrm{cl}_W(U)$ and then $V = \emptyset$. Similarly, if $A \cap \mathrm{cl}_W(V) \neq \emptyset$, we obtain that $U = \emptyset$. In both cases we obtain a contradiction. This ends the proof that $W \setminus (A \cup B)$ is connected. □

Theorem 11.2 *If C is a connected closed subset of W, then* $\mathrm{int}_W(C)$ *is connected.*

Proof If $C = W$, we are done. If $C \neq W$, then $\{a, b\} \nsubseteq C$. We may assume that $a \in W \setminus C$.

If $W \setminus C$ is connected, then $\mathrm{cl}_W(W \setminus C)$ is also connected. By Exercise 11.15, $\mathrm{int}_W(C) = W \setminus (\mathrm{cl}_W(W \setminus C))$ is connected and we are done.

If $W \setminus C$ is not connected, then by Exercise 11.15, there exist disjoint open subsets U and V of W such that $a \in U$, $b \in V$ and $W \setminus C = U \cup V$. By Exercise 4.12, $C \cup V$ is connected, so $C \cup \mathrm{cl}_W(V)$ is closed, connected and contains the point b. By Exercise 11.15, $W \setminus (C \cup \mathrm{cl}_X(V)) = U$ is connected. Similarly, V is connected.

Apply Theorem 11.1 to the sets $A = \mathrm{cl}_W(U)$ and $B = \mathrm{cl}_W(V)$ to obtain that $\mathrm{int}_W(C) = W \setminus (\mathrm{cl}_W(U) \cup \mathrm{cl}_W(V))$ is connected. □

11.2 Closed Domains

Definition 11.3 A subset D of W is a *closed domain* if $D = \mathrm{cl}_W(\mathrm{int}_W(D))$. Define

$$\mathcal{D}(W, a) = \{D \subset W : D \text{ is a connected closed domain of } W \text{ and } a \in D\} \cup \{\emptyset\},$$

$$\mathcal{D}(W, b) = \{D \subset W : D \text{ is a connected closed domain of } W \text{ and } b \in D\} \cup \{\emptyset\}.$$

For each $D \in \mathcal{D}(W, a) \cup \mathcal{D}(W, b)$, define

$$M_D = \mathrm{cl}_W(W \setminus D),$$

and if $D \in \mathcal{D}(W, a)$, define

$$I_D = \{x \in D : D \text{ is irreducible between } a \text{ and } x\}, \text{ and}$$

$$J_D = \{x \in M_D : M_D \text{ is irreducible between } b \text{ and } x\}.$$

Observe that, by Exercise 11.15, each set of the form M_D is connected, and by Exercise 11.18 (a), M_D is a closed domain of W. If $D \in \mathcal{D}(W, a) \setminus \{W, \emptyset\}$, then we have that $b \notin D$, so $M_D \in \mathcal{D}(W, b)$. Similarly, if $D \in \mathcal{D}(W, b) \setminus \{W, \emptyset\}$, then $M_D \in \mathcal{D}(W, a)$.

Theorem 11.4 *If $D \in \mathcal{D}(W, a) \setminus \{W, \emptyset\}$, then D is irreducible between a and each point of* $\mathrm{Fr}_W(D)$.

Proof Since D is a connected closed proper subset of W and contains a, we have that $b \notin D$. By Exercise 11.15, the set $\mathrm{cl}_W(W \setminus D)$ is connected and contains b.

Since $\text{int}_W(D) = W \setminus \text{cl}_W(W \setminus D)$ and D is a closed domain of W, we have that $\text{cl}_W(W \setminus D) \neq W$.

Let $p \in \text{Fr}_W(D)$, and let E be a closed connected subset of D such that $a, p \in E$. Then $E \cup \text{cl}_W(W \setminus D)$ is a closed connected subset of W containing a and b, so $W = E \cup \text{cl}_W(W \setminus D)$. This implies that $\text{int}_W(D) \subset E$. Since D is a closed domain, we conclude that $D \subset E$. Thus $D = E$. This completes the proof that D is irreducible between a and p. □

Theorem 11.5 *If $D, E \in \mathcal{D}(W, a)$ and $E \neq D$, then $E \subset \text{int}_W(D)$ or $D \subset \text{int}_W(E)$.*

Proof We may suppose that $E \neq \emptyset$. Then $a \in E$. Suppose that $E \not\subset \text{int}_W(D)$. Then $D \neq W$ and $\emptyset \neq E \cap (W \setminus \text{int}_W(D)) = E \cap \text{cl}_W(W \setminus D)$. By Exercise 11.15, $W \setminus D$ is a connected subset of W containing b. Then $E \cup \text{cl}_W(W \setminus D)$ is a connected closed subset of W containing $\{a, b\}$, so it is equal to W. Then $W \setminus \text{cl}_W(W \setminus D) \subset E$. Thus $D = \text{cl}_W(\text{int}_W(D)) = \text{cl}_W(W \setminus \text{cl}_W(W \setminus D)) \subset E$. If there exists an element $x \in D \cap \text{Fr}_W(E)$, then D is a closed connected proper subset of E ($E \neq D$, by hypothesis) containing a and x. This contradicts Theorem 11.4. Therefore $D \subset \text{int}_W(E)$. □

As a consequence of Theorem 11.5, $\mathcal{D}(W, a)$ is ordered by inclusion and we have that this order is linear.

Theorem 11.6 *If $\mathcal{D}(W, a) = \mathcal{D}_1 \cup \mathcal{D}_2$ and each element of $\mathcal{D}_1$ is a subset of each element of $\mathcal{D}_2$, then there exists an element D_0 of $\mathcal{D}(W, a)$ such that either D_0 is the minimum element of $\mathcal{D}_2$ or the maximum element of $\mathcal{D}_1$.*

Proof Let $D_0 = \text{cl}_W(\bigcup\{D : D \in \mathcal{D}_1\})$. By Exercise 11.18 (d), D_0 is a closed domain of W. If $\mathcal{D}_1 = \{\emptyset\}$, then $D_0 = \emptyset \in \mathcal{D}(W, a)$. If $\mathcal{D}_1 \neq \emptyset$, then D_0 is the closure of a union of connected sets containing a, so D_0 is connected and $D_0 \in \mathcal{D}(W, a)$. In both cases, $D_0 \in \mathcal{D}(W, a) = \mathcal{D}_1 \cup \mathcal{D}_2$.

In the case that $D_0 \in \mathcal{D}_1$, since it contains all elements of $\mathcal{D}_1$, we have that D_0 is the maximum of $\mathcal{D}_1$. In the case that $D_0 \in \mathcal{D}_2$, given $E \in \mathcal{D}_2$, we have that $D \subset E$ for all $D \in \mathcal{D}_1$. Thus $D_0 \subset E$, so D_0 is the minimum of $\mathcal{D}_2$. □

Theorem 11.7 *Let $A, B, D \in \mathcal{D}(W, a)$. Then the following properties hold*

(a) $D = \text{cl}_W(W \setminus M_D)$, $\text{Fr}_W(D) = \text{Fr}_W(M_D) = D \cap M_D = I_D \cap J_D$,
(b) *if* $A \subsetneq B$, *then* $A \cap I_B = \emptyset = M_B \cap J_A$ *and* $A \cap M_B = \emptyset$,
(c) *if* $A \subsetneq B$, *then* $I_A \cap I_B = \emptyset = J_A \cap J_B$ *and* $I_A \cap J_B = \emptyset$,
(d) *if* $A \subsetneq B \subsetneq D$, *then* $I_D \cap J_A = \emptyset$.

Proof

(a) Since $M_D = \text{cl}_W(W \setminus D)$, $\text{cl}_W(W \setminus M_D) = \text{cl}_W(W \setminus \text{cl}_W(W \setminus D)) = \text{cl}_W(\text{int}_W(D)) = D$, so $D = \text{cl}_W(W \setminus M_D)$. Thus $\text{Fr}_W(D) = D \cap \text{cl}_W(W \setminus D) = D \cap M_D = \text{cl}_W(W \setminus M_D) \cap M_D = \text{Fr}_W(M_D)$. If $\emptyset \neq D \neq W$, by Theorem 11.4, $\text{Fr}_W(D) \subset I_D$ and $\text{Fr}_W(M_D) \subset J_D$. If $D \in \{\emptyset, W\}$, then

$\text{Fr}_W(D) = \emptyset = \text{Fr}_W(M_D)$. This implies that $\text{Fr}_W(D) = \text{Fr}_W(M_D) \subset I_D \cap J_D \subset D \cap M_D = \text{Fr}_W(D)$. This finishes the proof of (a).

(b) If $x \in I_B$, then x cannot be in any proper closed connected subset of B that contains a, so $x \notin A$. Hence $A \cap I_B = \emptyset$. Since $A \subsetneq B$, $M_B = \text{cl}_W(W \setminus B) \subset \text{cl}_W(W \setminus A) = M_A$. If $M_B = M_A$, then $\text{cl}(W \setminus B) = \text{cl}_W(W \setminus A)$. Thus $\text{int}_W(B) = W \setminus \text{cl}_W(W \setminus B) = W \setminus \text{cl}_W(W \setminus A) = \text{int}_W(A)$. Hence $B = A$, a contradiction. Therefore either $M_B = \emptyset$ or M_B is a proper closed connected subset of M_A. By the definition of J_A, $M_B \cap J_A = \emptyset$. The last claim in (b) holds since, by Theorem 11.5, $A \subset \text{int}_W(B) = W \setminus \text{cl}_W(W \setminus B) = W \setminus M_B$.

(c) By (b), $I_A \cap I_B \subset A \cap I_B = \emptyset$, $J_A \cap J_B \subset J_A \cap M_B = \emptyset$ and $I_A \cap J_B \subset A \cap M_B = \emptyset$.

(d) By (b), we have that $M_B \cap J_A = \emptyset$ and $B \cap I_D = \emptyset$. Then $J_A \subset W \setminus M_B \subset B \subset W \setminus I_D$. Therefore $I_D \cap J_A = \emptyset$.

□

Theorem 11.8 *The family $\{I_D \cup J_D : D \in \mathcal{D}(W, a)\}$ covers W.*

Proof Let $p \in W$. Define $\mathcal{D}_1 = \{D \in \mathcal{D}(W, a) : p \in W \setminus D\}$ and $\mathcal{D}_2 = \{D \in \mathcal{D}(W, a) : p \in D\}$. Then the families $\mathcal{D}_1$ and $\mathcal{D}_2$ are disjoint and their union is $\mathcal{D}(W, a)$.

Given $A \in \mathcal{D}_1$ and $B \in \mathcal{D}_2$, let $p \in B \setminus A$. By Theorem 11.5, we have that $A \subset B$. Then Theorem 11.6 implies that there exists a $B_0 \in \mathcal{D}(W, a)$ such that either B_0 is the minimum of $\mathcal{D}_2$ or it is the maximum of $\mathcal{D}_1$. We analyze two cases.

Case 1. $B_0 \in \mathcal{D}_2$.

In this case B_0 is the minimum of $\mathcal{D}_2$, $p \in B_0$ and $B_0 \subset D$, for every $D \in \mathcal{D}_2$.

If B_0 is irreducible between a and p, then $p \in I_{B_0}$ and we are done. Suppose then that B_0 is not irreducible between a and p. Then there exists a connected closed proper subset C of B_0 (and therefore closed in W) such that $a, p \in C$. Then $b \notin C$. By Theorem 11.2, $\text{int}_W(C)$ is connected and so, by Exercise 11.18 (a), the set $D = \text{cl}_W(\text{int}_W(C))$ is a connected closed domain of W.

We check that $D \in \mathcal{D}(W, a)$. If $a \in \text{Fr}_W(C)$, then by Exercise 11.16, $\text{int}_W(C) = \emptyset$. Thus $D \in \mathcal{D}(W, a)$. If $a \in C \setminus \text{Fr}_W(C)$, then $a \in \text{int}_W(C)$ and $D \in \mathcal{D}(W, a)$. Therefore $D \in \mathcal{D}(W, a)$.

Since B_0 is the minimum of $\mathcal{D}_2$ and $D \subset C \subsetneq B_0$, we have that $D \in \mathcal{D}_1$. So, $p \in W \setminus D = W \setminus \text{cl}_W(W \setminus \text{cl}_W(W \setminus C)) = \text{int}_W(\text{cl}_W(W \setminus C)) \subset \text{cl}_W(W \setminus C)$. Thus $p \in C \cap \text{cl}_W(W \setminus C) = \text{Fr}_W(C)$. Let $E = \text{cl}_X(W \setminus C)$. By Exercises 11.15 and 11.18 (a), E is a connected closed domain of W. Then $E \in \mathcal{D}(W, b)$. By Exercise 11.17, E is irreducible between b and p. Set $F = \text{cl}_X(W \setminus E)$. By Exercise 11.18 (e), $F \in \mathcal{D}(W, a)$ and $E = M_F$. Therefore $p \in J_F$.

Case 2. $B_0 \in \mathcal{D}_1$.

In this case, B_0 is the maximum of $\mathcal{D}_1$, for each $D \in \mathcal{D}_1$, $D \subset B_0$, $p \in W \setminus B_0 \subset M_{B_0}$ and, by Exercise 11.18 (e), $M_{B_0} \in \mathcal{D}(W, b)$.

In Case 1, we saw that if there is a minimum element in $\mathcal{D}(W, a)$ containing p, then $p \in I_G \cup J_G$ for some $G \in \mathcal{D}(W, a)$. By the symmetry of the roles of a and b, we can conclude that if there exists a minimum element in $\mathcal{D}(W, b)$ containing p, then there exists a $K \in \mathcal{D}(W, b)$ such that either K is irreducible between b and p or $\mathrm{cl}_W(W \setminus K)$ is irreducible between a and p.

If the first statement holds, since $K = \mathrm{cl}_W(W \setminus \mathrm{cl}_W(W \setminus K))$ and $\mathrm{cl}_W(W \setminus K) \in \mathcal{D}(W, a)$ (Exercise 11.18 (e)), we have that $K = M_{\mathrm{cl}_W(W \setminus K)}$ and $M_{\mathrm{cl}_W(W \setminus K)}$ is irreducible between b and p, so $p \in J_{\mathrm{cl}_W(W \setminus K)}$.

If the second statement holds, then $p \in I_{\mathrm{cl}_W(W \setminus K)}$.

Hence we may assume that $\mathcal{D}(W, b)$ does not have a minimum element containing p. In particular, M_{B_0} is not minimum in this sense, so there exists an $N \in \mathcal{D}(W, b)$ such that $p \in N$ and $N \subsetneq M_{B_0}$.

Set $P = M_N = \mathrm{cl}_W(W \setminus N)$. By Exercise 11.18 (e), $P \in \mathcal{D}(W, a)$ and $B_0 = \mathrm{cl}_W(W \setminus M_{B_0}) \subset \mathrm{cl}_W(W \setminus N) = P$. If $B_0 = P$, then $M_{B_0} = M_P = N$ (Exercise 11.18 (e)), a contradiction. Therefore $B_0 \subsetneq P$. Since, in this case, B_0 is the maximum element of $\mathcal{D}_1$, we have that $P \in \mathcal{D}_2$. Thus $p \in P$. Since $p \in N$, we conclude that $p \in \mathrm{cl}_W(W \setminus N) \cap N = \mathrm{Fr}_W(N)$. Again, by symmetry and Theorem 11.4, we infer that N is irreducible between b and p. Since, $N = M_P$, we conclude that $p \in J_P$. □

Lemma 11.9 *If $D, E \in \mathcal{D}(W, a)$ and $D \subsetneq E$, then $\mathrm{cl}_W(E \setminus D)$ is connected.*

Proof If $D = \emptyset$, then $\mathrm{cl}_W(E \setminus D) = E$ is connected. Suppose then that $D \neq \emptyset$. If $E = W$, the conclusion follows from Exercise 11.15 (b). Suppose then that $E \neq W$. Then $\mathrm{Fr}_W(E) \neq \emptyset$. By Theorem 11.4, E is irreducible between a and each point of $\mathrm{Fr}_W(E)$. Then we can apply Exercise 11.15 (b) to the space E and the connected closed set D (which contains the point a) of E. With this, we obtain that $E \setminus D$ is connected. Therefore $\mathrm{cl}_W(E \setminus D)$ is connected. □

Theorem 11.10 *Let X be a continuum irreducible between a and b such that every indecomposable subcontinuum of X has empty interior. Then for every $D, E \in \mathcal{D}(X, a)$ with $D \subsetneq E$, there exists a $C \in \mathcal{D}(X, a)$ such that $D \subsetneq C \subsetneq E$.*

Proof We analyze two cases.

Case 1. $D \neq \emptyset$.

By Theorem 11.5, $D \subset \mathrm{int}_X(E)$. Since X is connected, D is not open, so the set $U = \mathrm{int}_X(E) \setminus D$ is a nonempty open subset of X contained in $E \setminus D$. By Lemma 11.9, the set $M = \mathrm{cl}_X(E \setminus D)$ is a subcontinuum of X with nonempty interior. By hypothesis, M is decomposable, so there exist proper subcontinua A and B of M such that $M = A \cup B$. Then $E = D \cup M = D \cup A \cup B$. By the connectedness of E, we may suppose that $D \cap A \neq \emptyset$. Observe that $E \setminus D \not\subset A$ and $E \setminus D \not\subset B$. Thus $A \cup D$ and $B \cup D$ are proper subcontinua of E. We consider two subcases.

Subcase 1.1. $E \neq X$.

By Theorem 11.4, E is irreducible between a and each element of $\mathrm{Fr}_X(E)$. Observe that $b \notin E$. Since $A \cup D$ is a proper subcontinuum of E, we have that $A \cap \mathrm{Fr}_X(E) = \emptyset$. Thus $A \cup D \subset \mathrm{int}_X(E)$. Since $\emptyset \neq \mathrm{Fr}_X(E) \subset B$, applying Theorem 11.4 again, we obtain that $B \cap D = \emptyset$. Then $(X \setminus B) \cap \mathrm{int}_X(E)$ is an open subset of X that contains D. Since $D \neq X$, we obtain that

$$D \subsetneq (X \setminus B) \cap \operatorname{int}_X(E) \subset D \cup A \subsetneq E.$$

Define $C = \operatorname{cl}_X((X \setminus B) \cap \operatorname{int}_X(E))$. Then $D \subsetneq C \subsetneq E$.

Since $C = \operatorname{cl}_X((X \setminus B) \cap (X \setminus \operatorname{cl}_X(X \setminus E)) = \operatorname{cl}_X(X \setminus (B \cup \operatorname{cl}_X(X \setminus E)))$, Exercise 1.18 (a) implies that C is a closed domain. Since $a \in E$, Exercise 11.15 (b), implies that $B \cup \operatorname{cl}_X(X \setminus E)$ is a subcontinuum of X containing b and also that C is a subcontinuum of X. Hence $C \in \mathcal{D}(X, a)$. Thus we have finished this subcase.

Subcase 1.2. $E = X$.

In this subcase we have that $X = D \cup A \cup B$. Since $a \in D$ and $A \cup D$ is a proper subcontinuum of X, $b \in B$. Since $\{a, b\} \subset D \cup B$ and $E \setminus D \not\subset B$, we have that $D \cup B \neq X$, so $D \cup B$ is not a subcontinuum of X and $D \cap B = \emptyset$. Since $X \setminus B$ is a proper open subset of X, $a \in D \subset X \setminus B \subsetneq \operatorname{cl}_X(X \setminus B)$. Define $C = \operatorname{cl}_X(X \setminus B)$. Then C is a closed domain containing a. By Exercise 11.5 (b), C is a subcontinuum of X, so $C \in \mathcal{D}(X, a)$. Since $A \cup D$ is a proper subcontinuum of X and $C \subset A \cup D$, we conclude that $C \subsetneq X = E$. This ends this subcase.

Case 2. $D = \emptyset$.

We consider two subcases.

Subcase 2.1. $E = X$.

By hypothesis, X is decomposable, so there exist two proper subcontinua A and B of X such that $X = A \cup B$. Suppose that $a \in A$. By the irreducibility, $b \notin A$. Then $b \in B$. By Exercises 11.15 (b) and 11.18 (a), $\operatorname{cl}_X(X \setminus B) \in \mathcal{D}(X, a)$. Since $a \in \operatorname{cl}_X(X \setminus B) \subset A$, we have that $\emptyset \subsetneq \operatorname{cl}_X(X \setminus B) \subsetneq X$. This subcase is finished.

Subcase 2.2. $E \neq X$.

Since E is a closed domain and $E \neq \emptyset$, $\operatorname{int}_X(E) \neq \emptyset$. By hypothesis, there exist two proper subcontinua A and B of E such that $E = A \cup B$. Since $a \in E$, we may assume that $a \in A$.

Define $C = \operatorname{cl}_X(X \setminus (B \cup \operatorname{cl}_X(X \setminus E)))$. By Exercise 11.18 (a), C is a closed domain. Since $E \neq X$ and $a \in E$, we have that $b \notin E$. By Exercise 11.15 (b), $\operatorname{cl}_X(X \setminus E)$ is a subcontinuum of X. If $B \cup \operatorname{cl}_X(X \setminus E) = X$, then

$$E = \operatorname{cl}_X(\operatorname{int}_X(E)) = \operatorname{cl}_X(X \setminus \operatorname{cl}_X(X \setminus E)) \subset B,$$

a contradiction. Thus $B \cup \operatorname{cl}_X(X \setminus E) \neq X$. Similarly, $A \cup \operatorname{cl}_X(X \setminus E) \neq X$. Since $a \in A$ and $b \in \operatorname{cl}_X(X \setminus E)$, we have that $A \cup \operatorname{cl}_X(X \setminus E)$ cannot be a subcontinuum of X. Then $A \cap \operatorname{cl}_X(X \setminus E) = \emptyset$. Since $A \cup B \cup \operatorname{cl}_X(X \setminus E) = X$, the connectedness of X implies that $B \cap \operatorname{cl}_X(X \setminus E) \neq \emptyset$. Thus $B \cup \operatorname{cl}_X(X \setminus E)$ is a proper subcontinuum of X containing b, so $a \in C$ and $D = \emptyset \subsetneq C$. By Exercise 11.15 (b), C is a subcontinuum of X. Hence $C \in \mathcal{D}(X, a)$. Finally, since $X \setminus A \subset B \cup \operatorname{cl}_X(X \setminus E)$, we obtain that $C \subset A \subsetneq E$. Therefore $D \subsetneq C \subsetneq E$. This ends this subcase and the proof of the theorem. □

Theorem 11.11 *Let X be an irreducible continuum between a and b such that every indecomposable subcontinuum of X has empty interior. Then*

$$\bigcap\{M \subset X : M \in \mathcal{D}(X, b) \setminus \{\emptyset\}\} = \{y \in X : X \text{ is irreducible between } a \text{ and } y\}.$$

Proof First, we prove the inclusion ($\supset$).

Let $y \notin \bigcap\{M \subset X : M \in \mathcal{D}(X, b)\setminus\{\emptyset\}\}$. Then there exists an $N \in \mathcal{D}(X, b)\setminus\{\emptyset\}$ such that $y \notin N$. Thus $N \neq X$. The irreducibility of X between a and b implies that $a \notin N$. Since $N \neq \emptyset$ and $N = \mathrm{cl}_X(\mathrm{int}_X(N))$, we have that $\mathrm{int}_X(N) \neq \emptyset$. Thus $\mathrm{cl}_X(X \setminus N) \neq X$. Hence $\mathrm{cl}_X(X \setminus N)$ is a proper subcontinuum (Exercise 11.15 (b)) of X containing $\{a, y\}$. Therefore X is not irreducible between a and y.

Now we prove the inclusion ($\subset$).

Take $y \in X$ such that X is not irreducible between a and y. Then there exists a proper subcontinuum D of X such that $a, y \in D$. Since $b \notin D$, Exercises 11.15 (b) and 11.18 (a) imply that $\mathrm{cl}_X(X \setminus D) \in \mathcal{D}(X, b) \setminus \{\emptyset\}$. By Theorem 11.10, there exists an $M \in \mathcal{D}(X, b)$ such that $\emptyset \subsetneq M \subsetneq \mathrm{cl}_X(X \setminus D)$. Then $M \in \mathcal{D}(X, b) \setminus \{\emptyset\}$. If $y \in M$, then $M \cup D$ is a subcontinuum of X containing $\{a, b\}$. Thus $X = M \cup D$. Hence $X \setminus D \subset M$ and $\mathrm{cl}_X(X \setminus D) \subset M$, which contradicts the choice of M. We have shown that $y \notin M$. □

Theorem 11.12 *Let X be a continuum irreducible between a and b such that every indecomposable subcontinuum of X has empty interior. Let $D \in \mathcal{D}(X, a)$ be such that $\emptyset \neq D \neq X$ and let $p \in \mathrm{Fr}_X(D)$. Then*

$$I_D = \bigcap\{M \subset D : M \in \mathcal{D}(D, p) \setminus \{\emptyset\}\}.$$

Proof By Exercise 11.22, the indecomposable subcontinua of D have empty interior in D. By Theorem 11.4, D is irreducible between a and p. Then the required equality is consequence of Theorem 11.11. □

11.3 Main Theorem

Theorem 11.13 ([90, Theorem 3, §48, VII, p. 216] and [120, Theorem 10]]) *Let X be a continuum irreducible between a and b such that each of its indecomposable subcontinua has empty interior. Then there exists a monotone mapping $f : X \to [0, 1]$ such that $\mathrm{int}_X(f^{-1}(t)) = \emptyset$ for every $t \in [0, 1]$.*

Proof For each $D \in \mathcal{D}(X, a)$, define $T_D = I_D \cup J_D$. In the set

$$\mathcal{T} = \{T_D : D \in \mathcal{D}(X, a)\},$$

define the relation $\leq$ by: $T_D \leq T_E$ if and only if $D \subset E$.

By Theorem 11.5, $\leq$ is a linear order. Then we consider $\mathcal{T}$ with the order topology (see Exercise 5.6). Notice that $T_\emptyset \leq T_D \leq T_X$ for every $D \in \mathcal{D}(X, a)$. Let $g : X \to \mathcal{T}$ be defined by $g(x) = T_D$ if and only if $x \in T_D$.

We are going to prove that g is well defined and monotone, $\mathcal{T}$ is a continuum and $\text{int}_X(g^{-1}(T)) = \emptyset$ for every $T \in \mathcal{T}$.

Given $x \in X$, Theorem 11.8 guarantees that there exists a $D \in \mathcal{D}(X, a)$ such that $x \in T_D$. We are going to show that D is unique.

Take $E \in \mathcal{D}(X, a)$ such that $E \neq D$. By Theorem 11.5, we have that $D \subsetneq E$ or $E \subsetneq D$. We consider the case when $D \subsetneq E$, the other one being similar. By Theorem 11.10 there exists a $C \in \mathcal{D}(X, a)$ such that $D \subsetneq C \subsetneq E$. By Theorem 11.7 (c) and (d), $I_D \cap I_E = J_D \cap J_E = I_D \cap J_E = \emptyset$ and $I_E \cap J_D = \emptyset$. Thus $T_D \cap T_E = (I_D \cap I_E) \cup (I_D \cap J_E) \cup (J_D \cap I_E) \cup (J_D \cap J_E) = \emptyset$. Hence $x \notin T_E$. We have shown that T_D is the unique element of $\mathcal{T}$ containing x. Therefore g is well defined.

We check that g is onto. It is enough to see that each element T_D of $\mathcal{T}$ is nonempty. Take an element $D \in \mathcal{D}(X, a)$.

If $D = \emptyset$, then $T_D = I_\emptyset \cup J_\emptyset = J_\emptyset = \{x \in X : X$ is irreducible between b and $x\}$. Since $a \in T_\emptyset$, we conclude that $T_\emptyset \neq \emptyset$.

If $D = X$, then $T_D = I_X \cup J_X = I_X = \{x \in X : X$ is irreducible between a and $x\}$. Thus $b \in T_X$ and $T_X \neq \emptyset$.

Finally, if $\emptyset \neq D \neq X$, then there exists an $x \in \text{Fr}_X(D)$ and by Theorem 11.4, D is irreducible between a and x. Then $x \in I_D \subset T_D$ and $T_D \neq \emptyset$. Therefore g is onto.

Now, we see that g is continuous. Let $D \in \mathcal{D}(X, a)$ and $x \in X$ be such that $g(x) < T_D$. Then $g(x) = T_E$ for some $E \in \mathcal{D}(X, a)$. Since $T_E < T_D$, we have that $E \subsetneq D$. By Theorem 11.10, there exists $F \in \mathcal{D}(X, a)$ such that $E \subsetneq F \subsetneq D$. By Theorem 11.5, $E \subset \text{int}_X(F)$.

We claim that $x \in \text{int}_X(F)$ and $g(\text{int}_X(F)) \subset [T_\emptyset, T_D)$.

If $x \notin \text{int}_X(F)$, then $\{b, x\} \subset \text{cl}_X(X \setminus F) \subset \text{cl}_X(X \setminus E)$. If $\text{cl}_X(X \setminus F) = \text{cl}_X(X \setminus E)$, we have that $\text{int}_X(F) = \text{int}_X(E)$. This implies that $F = E$, a contradiction. Therefore, $\text{cl}_X(X \setminus F) \subsetneq \text{cl}_X(X \setminus E)$. Set $A = \text{cl}_X(X \setminus F)$ and $B = \text{cl}_X(X \setminus E)$. Then $M_F = A$ and $M_E = B$. If $x \in J_E$, then B is irreducible between x and b. This contradicts the fact that $\{b, x\} \subset A \subsetneq B$. Thus $x \notin J_E$. Since $x \notin \text{int}_X(F)$, we have that $x \notin E$, so $x \notin I_E$. We have shown that $x \notin I_E \cup J_E = T_E$, a contradiction. This completes the proof that $x \in \text{int}_X(F)$.

Now, take $y \in \text{int}_X(F)$ and let $G \in \mathcal{D}(X, a)$ be such that $y \in T_G = I_G \cup J_G$. In the case when $y \in I_G$, we have that G is irreducible between a and y. Since $a, y \in F$, we have that F is not a proper subcontinuum of G. By Theorem 11.5, we have that $G \subset F$. Thus $G \subsetneq D$. Hence $g(y) = T_G < T_D$. In the case when $y \in J_G$, we have that $y \in \text{cl}_X(X \setminus G)$. If $F \subset G$, then $y \in \text{cl}_X(X \setminus G) \subset \text{cl}_X(X \setminus F)$, and then $y \notin \text{int}_X(F)$, a contradiction. By Theorem 11.5, we conclude that $G \subsetneq F \subsetneq D$. Thus $g(y) = T_G < T_D$. We have shown that $g(\text{int}_X(F)) \subset [T_\emptyset, T_D)$.

We have established that if $g(x) < T_D$, then x has the neighborhood $\text{int}_X(F)$, in X such that $\text{int}_X(F) \subset g^{-1}([T_\emptyset, T_D))$. It follows that $g^{-1}([T_\emptyset, T_D))$ is open in X.

By the symmetry of the roles of a and b, the roles of $\mathcal{D}(X, a)$ and $\mathcal{D}(X, b)$ are also symmetric. Thus, we also can conclude that for each $D \in \mathcal{D}(X, a)$,

$g^{-1}((T_D, T_X])$ is also open. Since the family $\{[T_\emptyset, T_D) : D \in \mathcal{D}(X, a)\} \cup \{(T_D, T_X] : D \in \mathcal{D}(X, a)\}$ is a subbase for the topology of $\mathcal{T}$, we conclude that g is continuous.

Given $T \in \mathcal{T}$, there exists a $D \in \mathcal{D}(X, a)$ such that $T = T_D$. By Exercises 11.23 and 11.24, T is a subcontinuum of X and $\text{int}_X(T) = \emptyset$. Since $T = g^{-1}(T)$, we obtain that g is monotone and its fibers have empty interior.

Since $\mathcal{T}$ has the topology induced by a linear order, it is easy to show that $\mathcal{T}$ is a Hausdorff space. By Exercise 1.23, $\mathcal{T}$ is a continuum. By Exercise 5.9, there exists a homeomorphism $h : \mathcal{T} \to [0, 1]$. Therefore $f = h \circ g : X \to [0, 1]$ has the required properties. □

11.4 Exercises

Exercise 11.14 Show that there exists a continuum X irreducible between two points a and b and a proper connected subset C of X containing a and b.

Exercise 11.15 Let C be a connected closed subset of W. Prove that

(a) if $W \setminus C$ is not connected, then there exist disjoint open subsets U and V of W such that $W \setminus C = U \cup V$, $a \in U$ and $b \in V$, (Hint: use Exercise 4.12.)
(b) if $a \in C$ or $b \in C$, then $W \setminus C$ is connected.

Exercise 11.16 Let C be a connected closed subset of W and $a \in \text{Fr}_W(C)$. Prove that $\text{int}_W(C) = \emptyset$.

Exercise 11.17 Let C be a connected closed proper subset of W with $a \in C$. Show that $\text{cl}_W(W \setminus C)$ is irreducible between b and each point of $\text{Fr}_W(C)$. In particular, if $\text{int}_W(C) = \emptyset$, then W is irreducible between b and each point of C.

Exercise 11.18 Prove the following.

(a) D is a closed domain of W if and only if there exists an open subset G of W such that $D = \text{cl}_W(G)$,
(b) D is a closed domain of W if and only if D is closed in W and $\text{Fr}_W(D) \subset \text{cl}_W(\text{int}_W(D))$,
(c) if D is a closed domain of W, then $\text{Fr}_W(D) = \text{Fr}_W(\text{cl}_W(W \setminus D))$,
(d) if $\{D_\alpha : \alpha \in A\}$ is a family of closed domains of W, then $\text{cl}_W(\bigcup\{D_\alpha : \alpha \in A\})$ is a closed domain of W,
(e) if $D \in \mathcal{D}(W, a)$, then $M_D \in \mathcal{D}(W, b)$; and if $N \in \mathcal{D}(W, b)$, then $\text{cl}_W(W \setminus N) \in \mathcal{D}(W, a)$ and $N = M_{\text{cl}_W(W \setminus N)}$.

Exercise 11.19 Show that $\mathcal{D}(W, b) = \{M_D \subset W : D \in \mathcal{D}(W, a)\}$.

Exercise 11.20 Let c be a irreducibility point of W. Prove that either W is irreducible between a and c or W is irreducible between b and c. Moreover, $\mathcal{D}(W, a) = \mathcal{D}(W, c)$ or $\mathcal{D}(W, b) = \mathcal{D}(W, c)$.

Exercise 11.21 Give an example of a continuum W, irreducible between two points a and b, and two families $\mathcal{D}_1$ and $\mathcal{D}_2$, as in Theorem 11.6, satisfying that the closure of the union of the elements of $\mathcal{D}_1$ is not the intersection of the elements of $\mathcal{D}_2$.

Exercise 11.22 Let E be a closed domain of W and K a subset of E such that $\text{int}_W(K) = \emptyset$. Prove that $\text{int}_E(K) = \emptyset$. Moreover, if L is a closed domain of E, then L is a closed domain of W.

Exercise 11.23 Suppose that W is a continuum such that every indecomposable subcontinuum of W has empty interior. Prove that for each $D \in \mathcal{D}(W, a)$, $I_D \cup J_D$ is a subcontinuum of W. (Hint: in the case when $D \notin \{\emptyset, W\}$, and use Theorem 11.12 to prove that I_D and J_D are subcontinua of W and use also Theorem 11.7 (a).)

Exercise 11.24 Prove that for each $D \in \mathcal{D}(W, a)$, $\text{int}_W(I_D) = \emptyset$ and $\text{int}_W(I_D \cup J_D) = \emptyset$. (Hint: suppose that there exists a nonempty open subset U of W such that $\text{cl}_X(U) \subset I_D$, use Theorem 3.4 to obtain a proper subcontinuum of D containing a and a point in I_D.)

Exercise 11.25 Let W and $f : W \to [0, 1]$ be as in Theorem 11.13. Suppose that A is a subcontinuum of W. Prove that either there exists a $t \in [0, 1]$ such that $A \subset f^{-1}(t)$ or there exist $r, s \in [0, 1]$ such that $0 \leq r < s \leq 1$ and $f^{-1}((r, s)) \subset A \subset f^{-1}([r, s])$.

Exercise 11.26 Let W and $f : W \to [0, 1]$ be as in Theorem 11.13. Prove that for every $x \in f^{-1}(0)$ and $y \in f^{-1}(1)$, W is irreducible between x and y.

Exercise 11.27 Give an example of a continuum X and a monotone mapping $f : X \to [0, 1]$ such that for each $t \in [0, 1]$, $\text{int}_X(f^{-1}(t)) = \emptyset$, but X is not irreducible.

Exercise 11.28 Let X be an irreducible continuum for which there exists a monotone mapping $f : X \to [0, 1]$ such that for each $t \in [0, 1]$, $\text{int}_X(f^{-1}(t)) = \emptyset$. Prove that each indecomposable subcontinuum of X has empty interior.

Exercise 11.29 Give an example of an irreducible continuum W and a mapping $f : X \to [0, 1]$ as in Theorem 11.13, such that for each $t \in [0, 1]$, $f^{-1}(t)$ is an arc.

Chapter 12
Unicoherence

A connected topological space Z is *unicoherent* if $A \cap B$ is connected for every pair of connected closed subsets A and B of Z such that $Z = A \cup B$. This notion was independently introduced by L. Vietoris and K. Kuratowski (who first introduced the word 'unicoherence'). Using unicoherence to study the topological properties of subspaces of the plane, Kuratowski obtained the following remarkable theorem [88, p. 313].
The sphere S^2 is the unique locally connected continuum X satisfying the properties:

(a) *for each $p \in X$, $X \setminus \{p\}$ is connected, and*
(b) *if A is a subcontinuum of X such that $X \setminus A$ is connected, then A is unicoherent.*

In this chapter, we present the basic theory of unicoherence, following S. Eilenberg [35]. As a sample of the many applications that unicoherence has in Continuum Theory, we use the unicoherence of 2-cells to prove the Mountain Climbing Theorem (Theorem 12.17).

The curious reader can find more information about unicoherence in [42], where an extensive bibliography on this topic is included.

12.1 Unicoherence and Property (b)

Definition 12.1 A *region* in a topological space Z is an open connected subset of Z. A connected space Z is *open unicoherent* if $U \cap V$ is connected for every pair of regions of Z such that $Z = U \cup V$.

Definition 12.2 Let $\mathbb{R}$ denote the real line and let S^1 be the unit circle in the Euclidean (or complex) plane.

A. Illanes, *Continuum Theory*, Universitext,
https://doi.org/10.1007/978-3-031-91011-1_12

Let $e : \mathbb{R} \to S^1$ be the exponential mapping. That is, $e(t) = (\cos(t), \sin(t))$; or with complex notation: $e(t) = \mathrm{e}^{it}$, or $e(t) = \cos(t) + i \sin(t)$, where as usual i denotes the square root of the complex number -1.

Given a topological space Z a mapping $f : Z \to S^1$ *can be lifted* if there exists a mapping $g : Z \to \mathbb{R}$ such that $e \circ g = f$. The mapping g is a *lifting* of f.

The space Z has *property (b)* if every mapping $f : Z \to S^1$ can be lifted.

Two mappings $f, g : Z \to Y$ are *homotopic* if there exists a mapping $G : Z \times [0, 1] \to Y$ such that for every $z \in Z$, $G(z, 0) = g(z)$ and $g(z, 1) = f(z)$.

In Exercise 12.21, we ask the reader to prove some basic properties of the exponential mapping. We will use these properties in this chapter.

Lemma 12.3 *Let $f : Z \to S^1$ be a mapping, where Z is a connected space. If the mappings g_1 and g_2 are liftings of f and there exists a point $z_0 \in Z$ such that $g_1(z_0) = g_2(z_0)$, then $g_1 = g_2$.*

Proof By hypothesis, $e \circ g_1 = f = e \circ g_2$. Then for every element $z \in Z$, we have that $\mathrm{e}^{ig_1(z)} = \mathrm{e}^{ig_2(z)}$.

Define $h : Z \to S^1$ by

$$h(z) = \mathrm{e}^{i(g_1(z) - g_2(z))}.$$

Note that h is the constant mapping which takes only the value 1, so h is continuous. Since $\mathrm{e}^{-1}(1) = \{\ldots, -4\pi, -2\pi, 0, 2\pi, 4\pi, \ldots\}$, we have that $(g_1 - g_2)(Z) \subset \{\ldots, -4\pi, -2\pi, 0, 2\pi, 4\pi, \ldots\}$. Thus the mapping $g_1 - g_2$ sends the connected space Z into a discrete space. This implies that $g_1 - g_2$ is a constant mapping.

But $(g_1 - g_2)(z_0) = 0$, so $g_1 - g_2$ is the constant mapping that takes only the value 0. Therefore $g_1 = g_2$. □

Theorem 12.4 *If Z is a non-unicoherent connected normal space, then Z does not have property (b).*

Proof Let A and B be connected closed subsets of Z such that $Z = A \cup B$ and $A \cap B$ is not connected. Then there exist two disjoint nonempty closed subsets K and L of Z such that $A \cap B = K \cup L$. By Urysohn's Lemma, there exist mappings $g_A : A \to [0, \pi]$ and $g_B : B \to [-\pi, 0]$ such that $g_A(K) = \{0\}$, $g_A(L) = \{\pi\}$, $g_B(K) = \{0\}$ and $g_B(L) = \{-\pi\}$.

Let $f : Z \to S^1$ be given by:

$$f(z) = \begin{cases} e(g_A(z)), & \text{if } z \in A, \\ e(g_B(z)), & \text{if } z \in B. \end{cases}$$

If $z \in K$, then $\mathrm{e}^{ig_A(z)} = \mathrm{e}^{i0} = 1$ and $\mathrm{e}^{ig_B(z)} = \mathrm{e}^{i0} = 1$. If $z \in L$, then $\mathrm{e}^{ig_A(z)} = \mathrm{e}^{i\pi} = -1$ and $\mathrm{e}^{ig_B(z)} = \mathrm{e}^{i(-\pi)} = -1$. Thus f is well defined and continuous.

Suppose that f can be lifted. Then there exists a mapping $h : Z \to \mathbb{R}$ such that $e \circ h = f$. Fix $z_0 \in K$ and define $\varphi : Z \to \mathbb{R}$ by

$$\varphi(z) = h(z) - h(z_0).$$

Note that $\varphi(z_0) = 0$.

We claim that $\varphi|_A : A \to \mathbb{R}$ is a lifting of $f|_A$. To prove this, let $a \in A$, then

$$\mathrm{e}^{i\varphi|_A(a)} = \mathrm{e}^{i(h(a)-h(z_0))} = \frac{\mathrm{e}^{ih(a)}}{\mathrm{e}^{ih(z_0)}} = \frac{f(a)}{f(z_0)} = \frac{f(a)}{\mathrm{e}^{ig_A(z_0)}} = \frac{f(a)}{1} = f|_A(a).$$

Therefore $\varphi|_A$ is a lifting of $f|_A$.

Now we prove that $g_A : A \to \mathbb{R}$ also is a lifting of $f|_A$. Given $a \in A$, we have that

$$\mathrm{e}^{ig_A(a)} = f(a) = f|_A(a).$$

Therefore g_A is a lifting of $f|_A$.

Observe that $\varphi|_A(z_0) = \varphi(z_0) = 0$ and $g_A(z_0) = 0$, so $\varphi|_A(z_0) = g_A(z_0)$.

We have shown that $\varphi|_A$ and g_A are liftings of $f|_A$ and they take the same value at the point z_0. By Lemma 12.3, $\varphi|_A = g_A$. In a similar way it can be proved that $\varphi|_B = g_B$.

Given an element $y \in L$, we have that

$$g_A(y) = \pi = \varphi|_A(y) \text{ and}$$
$$g_B(y) = -\pi = \varphi|_B(y).$$

Since $L \subset A \cap B$, we have that $y \in A \cap B$. This implies that $\varphi|_A(y) = \varphi|_B(y)$, a contradiction. Therefore f cannot be lifted. We have shown that Z does not have property (b). □

Lemma 12.5 *Let A and B be closed (or open) connected subsets of a space Z such that $A \cap B$ is connected and nonempty. Let $f : A \cup B \to S^1$ be a mapping such that $f|_A$ and $f|_B$ can be lifted. Then f can be lifted.*

Proof By hypothesis there are two mappings $h_A : A \to \mathbb{R}$ and $h_B : B \to \mathbb{R}$ such that $\mathrm{e}^{ih_A} = f|_A$ and $\mathrm{e}^{ih_B} = f|_B$. Choose a point $z_0 \in A \cap B$ and define $g_B : B \to \mathbb{R}$ as:

$$g_B(z) = h_B(z) - h_B(z_0) + h_A(z_0).$$

Clearly, g_B is continuous and we have that

$$\mathrm{e}^{ig_B} = \frac{\mathrm{e}^{ih_B} \cdot \mathrm{e}^{ih_A(z_0)}}{\mathrm{e}^{ih_B(z_0)}} = \frac{f|_B \cdot f(z_0)}{f(z_0)} = f|_B.$$

Hence g_B is a lifting of $f|_B$. Note that $g_B(z_0) = h_A(z_0)$. Then $g_B|_{A\cap B}$ and $h_A|_{A\cap B}$ are liftings of $f|_{A\cap B}$ and they coincide at the point z_0. By Lemma 12.3, $g_B|_{A\cap B} = h_A|_{A\cap B}$.

Define $k : A \cup B \to \mathbb{R}$ by:

$$f(z) = \begin{cases} h_A(z), & \text{if } z \in A, \\ g_B(z), & \text{if } z \in B. \end{cases}$$

Clearly, k is continuous and it is a lifting of f. □

Theorem 12.6 *Let Z be a connected metric space, A a subset of Z and $f : Z \to S^1$ a mapping such that $f|_A$ can be lifted. Then there exists an open subset U of Z such that $A \subset U$ and $f|_U$ can be lifted.*

Proof Let d be a metric for Z. Let $g : A \to \mathbb{R}$ be a mapping such that $f|_A = e \circ g$.

Given $a \in A$, by the continuity of g and f, we can choose a number $\varepsilon_a > 0$ such that diameter$(g(B(a, 2\varepsilon_a) \cap A)) < \frac{1}{2}$ and diameter$(f(B(a, 2\varepsilon_a))) < \frac{1}{4}$. Set $U_a = B(a, \varepsilon_a)$.

Fix a subarc J_a of S^1 such that $f(U_a) \subset J_a$ and diameter$(J_a) < \frac{1}{4}$.

By Exercise 12.21 (b), there exists a countable family of arcs $\{L_z^{(a)} \subset \mathbb{R} : z \in \mathbb{Z}\}$ such that $\bigcup\{L_z^{(a)} : z \in \mathbb{Z}\} = \mathrm{e}^{-1}(J_a)$, for every $z \in \mathbb{Z}$, $L_z^{(a)} = z + L_0^{(a)}$ and $e|_{L_z^{(a)}} : L_z^{(a)} \to J_a$ is a homeomorphism.

Since $e(g(a)) = f(a) \in J_a$, we have that there exists a $z_a \in \mathbb{Z}$ such that $g(a) \in L_{z_a}^{(a)}$.

Define $h_a : U_a \to \mathbb{R}$ in the following way:

$$h_a = (e|_{L_{z_a}^{(a)}})^{-1} \circ f|_{U_a}.$$

Since $f(U_a) \subset J_a$, we have that h_a is well defined and continuous.

Note that $e \circ h_a = f|_{U_a}$ and $h_a(a) = (e|_{L_{z_a}^{(a)}})^{-1}(f(a)) = (e|_{L_{z_a}^{(a)}})^{-1}(e(g(a)))$. Since $g(a) \in L_{z_a}^{(a)}$, we have that $e(g(a)) = e|_{L_{z_a}^{(a)}}(g(a))$. Thus $h_a(a) = g(a)$. Hence h_a is a lifting of $f|_{U_a}$ such that $h_a(a) = g(a)$.

We want to show that there exists a continuous common extension of all mappings h_a. In order to do this, it is enough to show that if $a, b \in A$ satisfy $U_a \cap U_b \neq \emptyset$, then $h_a|_{U_a} = h_b|_{U_b}$.

Suppose then that $a, b \in A$ and that we can take a point $x \in U_a \cap U_b$. Then $d(a, x) < \varepsilon_a$ and $d(b, x) < \varepsilon_b$. Without loss of generality, we may suppose that $\varepsilon_b \leq \varepsilon_a$. Then $d(a, b) < 2\varepsilon_a$. Thus $|g(a) - g(b)| < \frac{1}{2}$.

Since diameter$(J_a) < \frac{1}{4}$, diameter$(J_b) < \frac{1}{4}$ and $f(x) \in J_a \cap J_b$, we have that diameter$(J_a \cup J_b) < \frac{1}{2}$. This implies that $J_a \cup J_b$ is a subarc of S^1.

By Exercise 12.21 (b), there exists a countable family $\{L_z \subset \mathbb{R} : z \in \mathbb{Z}\}$ such that $\bigcup\{L_z : z \in \mathbb{Z}\} = \mathrm{e}^{-1}(J_a \cup J_b)$, for every $z \in \mathbb{Z}$, $L_z = z + L_0$ and $e|_{L_z} : L_z \to J_a \cup J_b$ is a homeomorphism.

Since $e(g(a)) = f(a) \in J_a$, we have that there exists a $z \in \mathbb{Z}$ such that $g(a) \in L_z$.

Note that the following two functions:

$$(e|_{L_{za}^{(a)}})^{-1}, (e|_{L_z})^{-1}|_{J_a} : J_a \to L_z$$

are continuous and both are liftings of the identity defined on J_a. Moreover, since $g(a)$ belongs to each of the intervals $L_{za}^{(a)}$ and L_z, and $e(g(a)) = f(a)$, we have that both functions $(e|_{L_{za}^{(a)}})^{-1}$ and $(e|_{L_z})^{-1}|_{J_a}$ take the value $g(a)$ at the point $f(a)$. Then, Lemma 12.3, implies that $(e|_{L_{za}^{(a)}})^{-1} = (e|_{L_z})^{-1}|_{J_a}$. Therefore $h_a = (e|_{L_z})^{-1}|_{J_a} \circ f|_{U_a}$.

Since $e|_{L_z} : L_z \to J_a \cup J_b$, is a homeomorphism, L_z is a closed subinterval of $\mathbb{R}$. Since the arc $J_a \cup J_b$ has diameter less than $\frac{1}{2}$, this arc covers less than one fourth of the circle S^1. Then each of the intervals of the form L_w has length less than $\frac{\pi}{2}$, and two consecutive such intervals have distance at least $\frac{3\pi}{2}$.

Since $f(b) \in J_b$ and $e(g(b)) = f(b)$, we have that $g(b) \in \mathrm{e}^{-1}(J_a \cup J_b)$. Then $g(b)$ belongs to some interval of the form L_w. If $z \neq w$, then the distance from L_w to L_z is at least $\frac{3\pi}{2}$. But $|g(a) - g(b)| < \frac{1}{2}$. This implies that $w = z$. Therefore $g(b) \in L_z$.

Proceeding as we did with a, we conclude that $(e|_{L_{zb}^{(b)}})^{-1} = (e|_{L_z})^{-1}|_{J_b}$ and that $h_b = (e|_{L_z})^{-1}|_{J_b} \circ f|_{U_b}$. Since $f(U_a \cap U_b) \subset J_a \cap J_b$, we have that $h_a|_{U_a \cap U_b} = (e|_{L_z})^{-1}|_{J_a} \circ f|_{U_a \cap U_b} = (e|_{L_z})^{-1}|_{J_b} \circ f|_{U_a \cap U_b} = h_b|_{U_a \cap U_b}$.

We have shown that there exists a common extension of all functions of the form h_a. We call this function h. Define $U = \bigcup\{U_a : a \in A\}$. So we have defined $h : U \to \mathbb{R}$ by

$$h(x) = h_a(x), \text{ if } x \in U_a \text{ and } a \in A.$$

Then h is a continuous function. If $x \in U_a$ and $a \in A$, then $e(h(x)) = e((e|_{L_{za}^{(a)}})^{-1} \circ f(x)) = f(x)$. Thus h is a lifting of $f|_U$. Since $A \subset U$, we have finished the proof of the theorem. □

The following theorem is valid for all topological spaces (see [103]).

Theorem 12.7 *Let Z be a metric space and let $f : Z \to S^1$ be a mapping. Then f can be lifted if and only if f is homotopic to a constant mapping.*

Proof (Necessity.) Suppose that f can be lifted. Then there exists a mapping $h : Z \to \mathbb{R}$ such that $f = e \circ h$. Define $G : Z \times [0, 1] \to S^1$ by

$$G(z, t) = e((1 - t)h(z)).$$

Clearly, G is a continuous function such that for each $z \in Z$, $G(z, 0) = e(h(z)) = f(z)$ and $f(z, 1) = e(0) = (1, 0)$. Thus f is homotopic to a constant mapping.

(Sufficiency) Now, suppose that f is homotopic to a constant mapping. Then there exist a mapping $G : Z \times [0, 1] \to S^1$ and $w \in S^1$ such that for each $z \in Z$, $G(z, 0) = f(z)$ and $G(z, 1) = w$. Choose $s \in \mathbb{R}$ such that $e(s) = w$.

Given $z \in Z$, by Exercise 12.29, there exists a lifting $k_z : \{z\} \times [0, 1] \to \mathbb{R}$ of $G|_{\{z\}\times[0,1]}$. Define $h_z : \{z\} \times [0, 1] \to \mathbb{R}$ by $h_z(z, t) = k_z(z, t) - k_z(z, 1) + s$. Observe that h_z is a continuous function such that for each $t \in [0, 1]$, $e(h_z(z, t)) = e(s) \cdot e(k_z(z, t))/e(k_z(z, 1)) = w \cdot G(z, t)/w = G(z, t)$. Hence h_z is also a lifting of $G|_{\{z\}\times[0,1]}$. Note that $h_z(z, 1) = s$. The mappings h_z have been defined in such a way that all of them take the same value at all points of the form $(z, 1)$.

Define $F : Z \times [0, 1] \to \mathbb{R}$ by

$$F(z, t) = h_z(z, t).$$

Note that for every $(z, t) \in X \times [0, 1]$, $e(F(z, t)) = e(h_z(z, t)) = G(z, t)$.

We check that F is continuous. Take $(z, t) \in Z \times [0, 1]$.

Set $A = (\{z\} \times [0, 1]) \cup (Z \times \{1\})$ and define $\lambda : A \to \mathbb{R}$ by

$$\lambda(u, r) = \begin{cases} h_z(z, r), & \text{if } u = z, \\ s, & \text{if } r = 1. \end{cases}$$

Since $h_z(z, 1) = s$, we have that λ is well defined and continuous.

Given $(u, 1) \in Z \times \{1\}$, we have that $e(\lambda(u, 1)) = e(s) = w = G(u, 1)$. Moreover, given $r \in [0, 1]$, $e(\lambda(z, r)) = e(h_z(z, r)) = G(z, r)$. Thus λ is a lifting of $G|_A$.

Since $Z \times [0, 1]$ is a metric space, by Theorem 12.6, there exists an open subset U of $Z \times [0, 1]$ such that $A \subset U$ and there exists a lifting $\Lambda : U \to \mathbb{R}$ of the mapping $G|_U$ such that $\Lambda|_A = \lambda$.

Since $[0, 1]$ is compact, there exists and open subset V of Z such that $V \times [0, 1] \subset U$.

We claim that $F|_{V\times[0,1]} = \Lambda|_{V\times[0,1]}$.

In order to check this, take $(v, r) \in V \times [0, 1]$. Then $F|_{\{v\}\times[0,1]} = h_v$ and $\Lambda|_{\{v\}\times[0,1]}$ are liftings of $G|_{\{v\}\times[0,1]}$ such that $h_v(v, 1) = s$ and $\Lambda(v, 1) = s$. According to Lemma 12.3, $F|_{\{v\}\times[0,1]} = \Lambda|_{\{v\}\times[0,1]}$. Thus $F(v, r) = \Lambda(v, r)$. We have shown that $F|_{V\times[0,1]} = \Lambda|_{V\times[0,1]}$.

Since Λ is continuous, we have that $F|_{V\times[0,1]}$ is continuous, and since $V \times [0, 1]$ contains (z, t) in its interior, we conclude that F is continuous in (z, t).

We have shown that F is a continuous function.

Define $\varphi : Z \to \mathbb{R}$ by $\varphi(z) = F(z, 0)$. Then φ is continuous. If $z \in Z$, then $e(\varphi(z)) = e(h_z(z, 0)) = G(z, 0) = f(z)$. Therefore φ is a lifting of f. □

12.2 Open Unicoherence

In the following theorem we prove that in connected and locally connected spaces, unicoherence and open unicoherence are equivalent. The proof of this result is much easier if we impose the additional hypothesis of normality (Exercise 12.28). Answering a question by A.H. Stone, in [42, Example 1.7], it was shown that there exists a non-open unicoherent, unicoherent, connected plane space.

Theorem 12.8 *Let Z be a connected and locally connected space. Then the following are equivalent:*

(a) Z is unicoherent,
(b) if U and V are regions of Z with disjoint boundaries, then $U \cap V$ is connected, and
(c) Z is open unicoherent.

Proof (a) $\Rightarrow$ (c). Suppose that Z is unicoherent and that there exist two regions U and V of Z such that $Z = U \cup V$ and $U \cap V = W \cup Y$, where W and Y are disjoint nonempty open subsets of Z. Choose points $w \in W$ and $y \in Y$. Let R be the component of W containing w and let S be the component of $Z \setminus R$ containing y. By Exercise 2.13, R is open, then S is closed.

By Exercise 12.24, $Z \setminus S$ is connected. Since we are assuming that Z is unicoherent, we have that $\mathrm{Fr}_Z(S) = \mathrm{cl}_Z(Z \setminus S) \cap S$ is connected. By Exercise 2.37, $\mathrm{Fr}_Z(S) \subset \mathrm{Fr}_Z(Z \setminus R) = \mathrm{Fr}_Z(R) \subset \mathrm{Fr}_Z(W)$.

Since $y \in U \cap V \cap S$ and $w \in U \cap V \setminus S$, the connectedness of U implies that the set $K = \mathrm{Fr}_Z(S) \cap U$ is nonempty. Since W and Y are nonempty and disjoint, $\mathrm{Fr}_Z(S)$ does not intersect $W \cup Y = U \cap V$, and since $Z = U \cup V$, we obtain that $\mathrm{Fr}_Z(S) \cap U = \mathrm{Fr}_Z(S) \cap (Z \setminus V)$. Thus K is closed in Z. Similarly, the set $L = \mathrm{Fr}_Z(S) \cap V = \mathrm{Fr}_Z(S) \cap (Z \setminus U)$ is closed and nonempty. Therefore K and L are disjoint, closed and nonempty.

Note that $\mathrm{Fr}_Z(S) \subset K \cup L$. Therefore $\mathrm{Fr}_Z(S) = K \cup L$. This contradicts the connectedness of $\mathrm{Fr}_Z(S)$ and finishes the proof that (a) $\Rightarrow$ (c).

(c) $\Rightarrow$ (b). Let U and V be regions of Z such that $\mathrm{Fr}_Z(U) \cap \mathrm{Fr}_Z(V) = \emptyset$. We suppose that $U \cap V \neq \emptyset$. Let

$$U_0 = U \cup (\bigcup\{D \subset Z : D \text{ is a component of } Z \setminus \mathrm{cl}_Z(V) \text{ and } D \cap U \neq \emptyset\}), \text{ and}$$

$$V_0 = V \cup (\bigcup\{D \subset Z : D \text{ is a component of } Z \setminus \mathrm{cl}_Z(U) \text{ and } D \cap V \neq \emptyset\}).$$

By Exercise 2.13, U_0 and V_0 are regions of Z.

We are going to show that $Z = U_0 \cup V_0$. Take $z \in Z \setminus (U \cup V)$. When $z \in \mathrm{Fr}_Z(U)$, we have that $z \notin \mathrm{Fr}_Z(V)$. Thus $z \notin \mathrm{cl}_Z(V)$. Let D be the component of $Z \setminus \mathrm{cl}_Z(V)$ such that $z \in D$. Since D is open, $D \cap U \neq \emptyset$. Thus $z \in U_0$. In a similar way it can be proved that if $z \in \mathrm{Fr}_Z(V)$, then $z \in V_0$. Finally we suppose that $z \notin \mathrm{cl}_Z(U \cup V)$. Let E be the component of $Z \setminus \mathrm{cl}_Z(U \cup V)$ such that $z \in E$. By the connectedness of Z and Exercise 2.37, we can take a point $q \in \mathrm{Fr}_Z(E) \subset \mathrm{Fr}_Z(\mathrm{cl}_Z(U \cup V)) \subset$

$\mathrm{Fr}_Z(U) \cup \mathrm{Fr}_Z(V)$. We may assume that $q \notin \mathrm{Fr}_Z(V)$. Then $q \in \mathrm{Fr}_Z(U)$. Let D be component of $Z \setminus \mathrm{cl}_Z(V)$ such that $q \in D$. Since $E \cup \{q\}$ is a connected subset of $Z \setminus \mathrm{cl}_Z(V)$, we obtain that $E \cup \{q\} \subset D$. Since D is open in Z and $q \in D$, we conclude that $D \cap U \neq \emptyset$. Therefore $z \in D \subset U_0$. This completes the proof that $Z = U_0 \cup V_0$.

Since Z is open unicoherent, we have that $U_0 \cap V_0$ is connected. Finally, we prove that $U_0 \cap V_0 = U \cap V$.

Given a point $p \in (U_0 \cap V_0) \setminus (U \cap V)$, if $p \notin V$, then there exists a component D of $Z \setminus \mathrm{cl}_Z(U)$ such that $p \in D$. Thus $p \notin U$, so there exists a component E of $Z \setminus \mathrm{cl}_Z(V)$ such that $p \in E$. Hence $p \notin \mathrm{cl}_Z(U \cup V)$. If $p \notin U$, then by a symmetric argument, we conclude that $p \notin \mathrm{cl}_Z(U \cup V)$. We have shown that $(U_0 \cap V_0) \setminus (U \cap V) \subset Z \setminus \mathrm{cl}_Z(U \cup V)$. This proves that $U_0 \cap V_0 \subset (U \cap V) \cup (Z \setminus \mathrm{cl}_Z(U \cup V))$. Since $U_0 \cap V_0$ is connected and $\emptyset \neq U \cap V \subset U_0 \cap V_0$, we conclude that $U_0 \cap V_0 = U \cap V$. This ends the proof of (c) $\Rightarrow$ (b).

(b) $\Rightarrow$ (a). Suppose (b) and suppose that Z is not unicoherent. Then there exist closed connected subsets A and B of Z such that $Z = A \cup B$ and $A \cap B$ is disconnected. Let K and L be nonempty closed disjoint subsets of Z such that $A \cap B = K \cup L$.

We claim that there exists a component R of $X \setminus A$ such that $\mathrm{Fr}_Z(R)$ intersects both sets K and L. Suppose to the contrary that no such component R exists. Let

$$P = K \cup \left(\bigcup\{S : S \text{ is a component of } Z \setminus A \text{ and } \mathrm{Fr}_Z(S) \subset K\}\right), \text{ and}$$

$$Q = L \cup \left(\bigcup\{S : S \text{ is a component of } Z \setminus A \text{ and } \mathrm{Fr}_Z(S) \subset L\}\right).$$

Given a component S of $Z \setminus A$, the connectedness of Z implies that $\mathrm{Fr}_Z(S) \neq \emptyset$. Since $S \subset B$, by Exercise 2.37, $\mathrm{Fr}_Z(S) \subset A \cap B = K \cup L$. By our assumption either, $\mathrm{Fr}_Z(S) \subset K$ or $\mathrm{Fr}_Z(S) \subset L$. Then $S \subset P$ or $S \subset Q$. This implies that $B \subset P \cup Q$. Therefore $B = P \cup Q$. By Exercise 12.25, $\mathrm{Fr}_Z(\bigcup\{S : S \text{ is a component of } Z \setminus A \text{ and } \mathrm{Fr}_Z(S) \subset K\}) \subset K$. Then P is closed in Z. Similarly, Q is closed in Z. Clearly, $P \cap Q = \emptyset$. Since P and Q are nonempty, we conclude that B is disconnected. This contradiction proves the existence of the component R.

By Exercise 12.24, $Z \setminus R$ is connected. Then $Z \setminus R$ and $\mathrm{cl}_Z(R)$ are closed connected subsets of Z such that $Z = (Z \setminus R) \cup \mathrm{cl}_Z(R)$ and $(Z \setminus R) \cap \mathrm{cl}_Z(R) = \mathrm{Fr}_Z(R) \subset A \cap B = K \cup L$ (see Exercise 2.37), and the sets $K_0 = \mathrm{Fr}_Z(R) \cap K$ and $L_0 = \mathrm{Fr}_Z(R) \cap L$ form a separation of $\mathrm{Fr}_Z(R)$. Thus $Z \setminus R$ and $\mathrm{cl}_Z(R)$ play similar roles to those of the sets A and B. Therefore with an analogous argument we can obtain a component T of $Z \setminus \mathrm{cl}_Z(R)$ such that $\mathrm{cl}_Z(T) \cap K_0 \neq \emptyset$ and $\mathrm{cl}_Z(T) \cap L_0 \neq \emptyset$. By Exercise 2.13, R is open in Z, so $(R \cup T) \cap (L_0 \cup K_0) = \emptyset$.

Let U (respectively, V) be the component of $Z \setminus K_0$ (respectively, $Z \setminus L_0$) containing T. By Exercise 2.13, U and V are open. By Exercise 2.37, $\mathrm{Fr}_Z(U) \subset K_0$ and $\mathrm{Fr}_Z(V) \subset L_0$. Thus U and V are regions in Z with disjoint boundaries. By hypothesis, $U \cap V$ is connected.

Fix a point $p_0 \in \mathrm{cl}_Z(T) \cap K_0$. Then $p_0 \in \mathrm{cl}_Z(R) \cap \mathrm{cl}_Z(T) \setminus L_0$. Since $\{p_0\} \cup R \cup T$ is a connected subset of $Z \setminus L_0$, we conclude that $R \subset V$. Similarly, $R \subset U$. Therefore $R \cup T \subset U \cap V$. Since $U \cap V$ is connected and intersects R and $T \subset Z \setminus R$, we infer that $\emptyset \neq \mathrm{Fr}_Z(R) \cap U \cap V = (K_0 \cup L_0) \cap U \cap V$. This contradicts the choice of U and V and proves that Z is unicoherent. Therefore (b) implies (a), and the proof of the theorem is complete. □

Theorem 12.9 *Let Z be a unicoherent, connected and locally connected space. Then Z has property (b).*

Proof Let $f : Z \to S^1$ be a mapping. If f is not onto, by Exercise 12.21 (c), it can be lifted and we are done. Suppose then that f is onto. Let $V_1 = S^1 \cap ((-2, 2) \times (-\frac{1}{2}, 2))$ and $V_2 = S^1 \cap ((-2, 2) \times (-2, \frac{1}{2}))$. Then $\{V_1, V_2\}$ is an open covering of S^1. For each $j \in \{1, 2\}$, set $U_j = f^{-1}(V_j)$. Then $\{U_1, U_2\}$ is an open covering of Z. Set

$$C = \{W \subset Z : W \text{ is component of } U_j \text{ for some } j \in \{1, 2\}\}.$$

Claim 1. Let $W, U \in C$. Then $W \cap U$ is connected.

We prove Claim 1. Suppose that $W \neq U$ and $W \cap U \neq \emptyset$. By Theorem 12.8, it is enough to show that $\mathrm{Fr}_Z(W) \cap \mathrm{Fr}_Z(U) = \emptyset$. Suppose to the contrary that there exists a point $p \in \mathrm{Fr}_Z(W) \cap \mathrm{Fr}_Z(U)$. Since C covers Z, there exists an $R \in C$ such that $p \in R$. Then $W \neq R \neq U$ and $W \cap R \neq \emptyset \neq R \cap U$. We may assume that R is a component of U_1. Then W and U are components of U_2. This is impossible since $W \neq U$ and $W \cap U \neq \emptyset$. This ends the proof of Claim 1.

Claim 2. Let $W \in C$ and $\mathcal{D}$ be a nonempty subset of $C \setminus \{W\}$ such that the set $D = \bigcup\{R : R \in \mathcal{D}\}$ is connected. Let U be the component of the set $V = \bigcup\{R : R \in C \setminus \{W\}\}$ containing D. Then $W \cap U$ is connected.

We prove Claim 2. We suppose that $W \cap U \neq \emptyset$. By Exercise 2.13, W and U are regions of Z. By Theorem 12.8, in order to prove Claim 2, we only need to show that $\mathrm{Fr}_Z(W) \cap \mathrm{Fr}_Z(U) = \emptyset$. Suppose to the contrary that there exists a point $p \in \mathrm{Fr}_Z(W) \cap \mathrm{Fr}_Z(U)$. Since C is a cover of Z, there exists an $R \in C$ such that $p \in R$. Then R is open, $R \neq W$ and $R \cap U \neq \emptyset$. Then $R \subset V$. Since R is connected, $R \subset U$. This contradicts the fact that $p \in \mathrm{Fr}_Z(U)$. We have shown that $\mathrm{Fr}_Z(W) \cap \mathrm{Fr}_Z(U) = \emptyset$. Therefore $W \cap U$ is connected.

Claim 3. Let $W \in C$ and $\mathcal{D}$ be a nonempty subset of $C \setminus \{W\}$ such that the set $D = \bigcup\{R : R \in \mathcal{D}\}$ is connected. Then $W \cap D$ is connected.

We prove Claim 3. Suppose to the contrary that there exist two nonempty disjoint open subsets V_1 and V_2 of Z such that $W \cap D = V_1 \cup V_2$. Let U be the component of $\bigcup\{R : R \in C \setminus \{W\}\}$ containing D. By Claim 2, $W \cap U$ is connected.

Let

$$V_3 = \bigcup\{R \in C \setminus \{W\} : R \subset U \text{ and } R \cap V_1 \neq \emptyset\}, \text{ and}$$

$$V_4 = \bigcup\{R \in C \setminus \{W\} : R \subset U \text{ and } R \cap V_1 = \emptyset\}.$$

Note that V_3 and V_4 are open in Z. Given a point $p \in U$, there exists an $R \in C\setminus\{W\}$ such that $p \in R$. Since R is connected and intersects U, we have that $R \subset U$. Then $R \subset V_3$ or $R \subset V_4$. We have shown that $U = V_3 \cup V_4$. In particular,

$$W \cap U = (W \cap V_3) \cup (W \cap V_4).$$

Choose a point $p \in V_1$. Then there exists an $R \in \mathcal{D}$ such that $p \in R \subset D \subset U$. Then $R \subset V_3$. Thus $W \cap V_3 \neq \emptyset$.

Choose a point $q \in V_2$. Let $R \in \mathcal{D}$ be such that $q \in R \subset D \subset U$. Note that $W \cap R \subset W \cap D = V_1 \cup V_2$. By Claim 1, $W \cap R$ is a connected subset of $V_1 \cup V_2$. Since $q \in W \cap R \cap V_2$, we conclude that $W \cap R \subset V_2$, so $R \cap V_1 = W \cap R \cap V_1 = \emptyset$. Thus $R \subset V_4$ and $q \in V_4$. Therefore $W \cap V_4 \neq \emptyset$.

Finally, we check that $W \cap V_3 \cap V_4 = \emptyset$. Suppose the contrary and take a point p in this intersection. Then there exist $R_1, R_2 \in C \setminus \{W\}$ such that $p \in R_1 \cap R_2$, $R_1 \cup R_2 \subset U$, $R_1 \cap V_1 \neq \emptyset$ and $R_2 \cap V_1 = \emptyset$. Without loss of generality, we may suppose that W is component of U_1. Since R_1 and R_2 intersect W and $R_1, R_2 \in C \setminus \{W\}$, R_1 and R_2 are components of U_2. This is impossible since $R_1 \neq R_2$ and $R_1 \cap R_2 \neq \emptyset$. We conclude then that $W \cap V_3 \cap V_4 = \emptyset$.

We have obtained then a separation of the connected set $W \cap U$. This contradiction ends the proof of Claim 3.

Fix a point $p_0 \in Z$ such that $f(p_0) = (0, 1)$. Then $p_0 \in U_1 \setminus U_2$. Thus there exists a unique element $R_0 \in C$ such that $p_0 \in R_0$.

Given a point $p \in Z$, by Exercise 2.20, there exists a chain

$$\mathcal{W}^{(p)} = \{W_1^{(p)}, \ldots, W_{m_p}^{(p)}\}$$

of elements of C such that $W_1^{(p)} = R_0$ and $p \in W_{m_p}^{(p)}$. We ask that m_p be minimum. Then $p \notin W_1^{(p)} \cup \cdots \cup W_{m_p-1}^{(p)}$.

Claim 4. For each $p \in Z$, the chain $\mathcal{W}^{(p)}$ is unique.

Suppose, contrary to Claim 4, that there exist two distinct chains $\mathcal{W}^{(p)} = \{W_1^{(p)}, \ldots, W_{m_p}^{(p)}\}$ and $\mathcal{T}^{(p)} = \{T_1, \ldots, T_{m_p}\}$, with $W_1^{(p)} = R_0 = T_1$ and $p \in T_{m_p}$. Let $k = \min\{j \in \{1, \ldots, m_p\} : W_j^{(p)} \neq T_j\}$. Then $k > 1$. Without loss of generality, we may suppose that $W_{k-1}^{(p)} = T_{k-1}$ is a component of U_1. Since $W_k^{(p)}$ and T_k are distinct from $W_{k-1}^{(p)}$, we obtain that $W_k^{(p)}$ and T_k are components of U_2. Then $W_k^{(p)} \cap T_k = \emptyset$. Let $D = W_k^{(p)} \cup \cdots \cup W_{m_p}^{(p)} \cup V_k \cup \cdots \cup V_{m_p}$. Then D is a region. By the definition of chain, the disjoint sets $W_{k-1}^{(p)} \cap W_k^{(p)}$ and $T_{k-1} \cap T_k$ are open, nonempty and their union is equal to $W_{k-1}^{(p)} \cap D$. This contradicts Claim 3 and finishes the proof of Claim 4.

Given $R \in C$, we have that $f(R) \subset V_1$ or $f(R) \subset V_2$, so $f(R)$ is a proper subset of S^1. By Exercise 12.21 (c), $f|R$ can be lifted.

Given a point $p \in Z$, take the chain $\mathcal{W}^{(p)} = \{W_1^{(p)}, \ldots, W_{m_p}^{(p)}\}$, of elements of C, associated to p. By the previous paragraph, $f|W_1^{(p)}$ can be lifted. By Claim 1, $W_1^{(p)} \cap W_2^{(p)}$ is connected, so Lemma 12.5 implies that $f|W_1^{(p)} \cup W_2^{(p)}$ can be lifted. By Claim 3, $(W_1^{(p)} \cup W_2^{(p)}) \cap W_3^{(p)}$ is connected. By Lemma 12.5, $f|W_1^{(p)} \cup W_2^{(p)} \cup W_3^{(p)}$ can be lifted. Proceeding in this way, it is possible to conclude that $f|W_1^{(p)} \cup \cdots \cup W_{m_p}^{(p)}$ can be lifted. Let $h_1 : W_1^{(p)} \cup \cdots \cup W_{m_p}^{(p)} \to \mathbb{R}$ be a lifting of this mapping. Fix a number $t_0 \in \mathbb{R}$ such that $e(t_0) = (0, 1)$. Let $h_p = h_1 - h_1(p_0) + t_0$. Then

$$\mathrm{e}^{h_p} = \frac{\mathrm{e}^{h_1} \cdot \mathrm{e}^{h_1(p_0)}}{i} = \frac{\mathrm{e}^{h_1} \cdot f(p_0)}{i} = \frac{\mathrm{e}^{h_1} \cdot i}{i} = f|W_1^{(p)} \cup \cdots \cup W_{m_p}^{(p)}.$$

Hence h_p is the unique (see Lemma 12.3) lifting of $f|W_1^{(p)} \cup \cdots \cup W_{m_p}^{(p)}$ satisfying $h_p(p_0) = t_0$.

Define $h : Z \to S^1$ by

$$h(p) = h_p(p).$$

Since there exists a unique way to choose the chain $\mathcal{W}^{(p)}$ and there exists a unique lifting h_p, we infer that h is well defined and $\mathrm{e}^{h(p)} = f(p)$.

In order to prove that h is continuous, take a point $q \in W_{m_p}^{(p)}$. We claim that $h(q) = h_p(q)$.

Since the sequences $\mathcal{W}^{(p)}$ and $\mathcal{W}^{(q)} \cup \{W_{m_p}^{(p)}\}$ are weak chains containing, respectively, the pairs of points $\{p_0, q\}$ and $\{p_0, p\}$, it follows that $m_q \leq m_p$ and $m_p \leq m_q + 1$. Then $m_p - 1 \leq m_q \leq m_p$.

If $m_q = m_p - 1$, the minimality of m_p implies that $\mathcal{W}^{(p)} = \mathcal{W}^{(q)} \cup \{W_{m_p}^{(p)}\}$ and $\mathcal{W}^{(q)} = W_1^{(p)} \cup \cdots \cup W_{m_p-1}^{(p)}$. Thus $h_q = h_p|(W_1^{(p)} \cup \cdots \cup W_{m_p-1}^{(p)}$ and $h(q) = h_q(q) = h_p(q)$.

If $m_q = m_p$, then we have that $q \notin W_1^{(p)} \cup \cdots \cup W_{m_p-1}^{(p)}$, and by the uniqueness of the chain $\mathcal{W}^{(q)}$, we obtain that $\mathcal{W}^{(q)} = \mathcal{W}^{(p)}$. Hence $h(q) = h_q(q) = h_p(q)$.

We have shown that h coincides with h_p in the open set $W_{m_p}^{(p)}$. Since h_p is continuous at p, we conclude that h is continuous at p. This ends the proof of the theorem. □

Corollary 12.10 *Let Z be a connected, locally connected and normal space. Then the following are equivalent.*

(a) Z is unicoherent,
(b) Z is open unicoherent,
(c) Z has property (b).

12.3 The Disk

Theorem 12.11 *Let Z be a contractible metric space. Then Z is unicoherent.*

Proof Let $G : Z \times [0, 1] \to Z$ be a mapping and $z_0 \in Z$ such that for every $z \in Z$, $G(z, 0) = z$ and $G(z, 1) = z_0$. Given a mapping $f : Z \to S^1$, the mapping $K : Z \times [0, 1] \to S^1$ given by $K = f \circ G$ satisfies that for each $z \in Z$, $K(z, 0) = f(z)$ and $K(z, 1) = f(z_0)$. Then f is homotopic to a constant mapping, and by Theorem 12.7, f can be lifted. We have shown that Z has property (b). By Theorem 12.4, Z is unicoherent. □

Corollary 12.12 *The square $[0, 1]^2$ is unicoherent and it has property (b).*

Theorem 12.13 *Let Z be a connected locally connected unicoherent metric space and $Y \subset Z$ a retract of Z. Then Y is unicoherent.*

Proof By Exercise 12.26, Y is locally connected. By Theorem 12.4, it is enough to prove that Y has property (b). Let $f : Y \to S^1$ be a mapping and $r : Z \to Y$ a retraction. Since Z has property (b) (Corollary 12.10), $f \circ r : Z \to S^1$ can be lifted. Let $g : Z \to \mathbb{R}$ be a mapping such that $f \circ r = e \circ g$. Then for each $y \in Y$, $e(g(y)) = f(r(y)) = f(y)$. Therefore, $g|_Y$ is a lifting of f, and hence Y is unicoherent. □

Corollary 12.14 *S^1 is not a retract of the unit disk.*

12.4 The Mountain Climbing Theorem

The Mountain Climbing Theorem informally says that two alpinists can escalate a mountain, starting at opposite positions from the bottom and finishing at the top, coordinating their movements in order to always stay at the same height. In mathematics, this theorem helps us to synchronize two functions in such a way that they take the same value at each moment (see Theorem 12.17). This theorem was proved by T. Homma in 1952 [49] and it has been rediscovered several times. It has been useful to solve problems in several branches of mathematics. The author of this book has applied it in various results or theorems, particularly in problems related to chainable continua and inverse limits with set-valued functions. For example, this result was used to prove that, in the product of two pseudo-arcs, there exists a non-degenerate subcontinuum containing no pseudo-arcs [65]; and that a circle is not the generalized inverse limit of a subset of $[0, 1]^2$ [63]. The proof of Theorem 12.17 is an application of the unicoherence of the square. A. Tucker has given an alternative and very nice proof of this theorem by using Graph Theory [121].

Lemma 12.15 *Let A be a closed subset of $[0, 1]^2$ containing $(0, 0)$ and $(1, 1)$. Let E be the component of A containing $(0, 0)$. Suppose that $(1, 1) \notin E$. Then there exists a piecewise linear arc $\alpha \subset [0, 1]^2 \setminus A$ that separates $(0, 0)$ and $(1, 1)$ in*

$[0,1]^2$, *and* α *joins one point* $p \in ([0,1] \times \{0\}) \cup (\{1\} \times [0,1])$ *to one point* $q \in (\{0\} \times [0,1]) \cup ([0,1] \times \{1\})$.

Proof We apply Theorem 3.3 to the space A, and the closed subsets E and $\{(1,1)\}$ of A. Then there exist disjoint closed subsets K and L of A such that $A = K \cup L$, $E \subset K$ and $(1,1) \in L$. Let U and V be disjoint open subsets of $[0,1]^2$ such that $K \subset U$ and $L \subset V$. Let $M = \mathrm{Fr}_{[0,1]^2}(U)$. Then M is a closed subset of $[0,1]^2$ that separates E and $(1,1)$ in $[0,1]^2$. By Exercise 12.25 (a), there exists a closed connected subset B of M that separates $(0,0)$ and $(1,1)$. Since $A \subset U \cup V$, we have that $A \cap B = \emptyset$. Since the set $J = ([0,1] \times \{0\}) \cup (\{1\} \times [0,1])$ is an arc in $[0,1]^2$ joining $(0,0)$ and $(1,1)$, we have that there exists a point $p \in B \cap J$. Similarly, there exists a point $q \in B \cap L$, where $L = (\{0\} \times [0,1]) \cup ([0,1] \times \{1\})$. Since B is connected, there exists an open connected subset W of $[0,1]^2$ such that $B \subset W$ and $W \cap A = \emptyset$.

Since $[0,1]^2$ has a basis of neighborhoods connected by piecewise linear arcs, each open connected subset of A is connected by piecewise linear arcs. Thus there exists a piecewise linear arc α in W connecting p and q and we can ask that $\alpha \cap \mathrm{Fr}_{\mathbb{R}^2}([0,1]^2) = \{p,q\}$. Then, by the Theta Curve Theorem, see [90, Theorem 2, §61 II, ch. X], α separates $(0,0)$ and $(1,1)$ in $[0,1]^2$. □

Definition 12.16 A mapping $f : [0,1] \to [0,1]$ is a *PL mapping (piecewise linear mapping)*, if there exists a partition $P : 0 = t_0 < t_1 < \cdots < t_n = 1$ of $[0,1]$ such that, for each $i \in \{1,\ldots,n\}$ and each $t \in [t_{i-1}, t_i]$, $f(t) = \frac{t - t_{i-1}}{t_i - t_{i-1}} f(t_i) + \frac{t_i - t}{t_i - t_{i-1}} f(t_{i-1})$. In this case we say that f is *supported* by P. A PL mapping f is a *jumping mapping* if $f(0) = 0$ and $f(1) = 1$.

Theorem 12.17 (Mountain Climbing Theorem) *If* f *and* g *are jumping mappings, then there exist jumping mappings* α *and* β *such that* $f \circ \alpha = g \circ \beta$.

Proof Given points $p, q \in \mathbb{R}^2$, let pq denote the convex segment in the plane joining p and q.

Let $A = \{(x,y) \in [0,1]^2 : f(x) = g(y)\}$. Then A is a compact subset of $[0,1]^2$ such that $(0,0), (1,1) \in A$. Let E be the component of A containing $(0,0)$.

We check that $(1,1) \in E$. Suppose to the contrary that $(1,1) \notin E$. By Theorem 12.15, there exists a piecewise linear arc $\alpha \subset [0,1]^2 \setminus A$ that separates $(0,0)$ and $(1,1)$ in $[0,1]^2$, and α joins one point $p = (r,s) \in ([0,1] \times \{0\}) \cup (\{1\} \times [0,1])$ and one point $q = (t,u) \in (\{0\} \times [0,1]) \cup ([0,1] \times \{1\})$.

If $s = 0$, then $f(r) \geq 0 = g(s)$. If $r = 1$, then $f(r) = 1 \geq g(s)$. In both cases, $f(r) \geq g(s)$. Similarly, $f(t) \leq g(u)$. By the connectedness of α and the Intermediate Value Theorem, we conclude that there exists a $z = (x,y) \in \alpha$ such that $f(x) = g(y)$. Therefore $z \in \alpha \cap A$. This contradicts the properties of α and proves that $(1,1) \in E$.

Since f and g are PL mappings, by Exercise 12.31, we have that there exists a partition $P : 0 = t_0 < t_1 < \cdots < t_n = 1$ of $[0,1]$ such that f and g are supported by P.

Given $r \in [0, 1]$, there exists an $\varepsilon_r > 0$ such that if $s \in [0, 1]$ and $|r - s| < \varepsilon_r$ then there exists a $j \in \{1, \ldots, n\}$ such that $\{r, s\} \subset [t_{j-1}, t_j]$. Given $p = (x, y) \in E$, let $\delta_p = \min\{\varepsilon_x, \varepsilon_y\}$. Set $D_p = \{(u, v) \in E : |x - u| < \delta_p$ and $|y - v| < \delta_p\}$. Then D_p is a neighborhood of p in E. By Exercise 2.20, there exist $m \in \mathbb{N}$ and points $p_1 = (x_1, y_1), \ldots, p_m = (x_m, y_m)$ in E such that $p_1 = (0, 0)$, $p_m = (1, 1)$ and for each $j \in \{2, \ldots, m\}$, $D_{p_{j-1}} \cap D_{p_j} \neq \emptyset$. Given $j \in \{2, \ldots, m\}$, take a point $q_j = (u_j, v_j) \in D_{p_{j-1}} \cap D_{p_j}$.

We see that the convex segments $p_1q_2, q_2p_2, p_2q_3, q_3p_3, \ldots, p_{m-1}q_m, q_mp_m$ are contained in E. We prove that $p_1q_2 \subset E$ (the rest of the cases can be treated in a similar way). Since $q_2 \in D_{p_1}$, we have that $|x_1 - u_2| < \varepsilon_{x_1}$ and $|y_1 - v_2| < \varepsilon_{y_1}$. Then there exist $j, k \in \{1, \ldots, m\}$ such that $\{x_1, u_2\} \subset [t_{j-1}, t_j]$ and $\{y_1, v_2\} \subset [t_{k-1}, t_k]$. Since $\{p_1, q_2\} \subset E \subset A$, we have that $f(x_1) = g(y_1)$ and $f(u_2) = g(v_2)$. The we can apply Exercise 12.32 and obtain that for each $t \in [0, 1]$, $f(tx_1 + (1 - t)u_2) = g(ty_1 + (1 - t)v_2)$. Thus for each $t \in [0, 1]$, $(tx_1 + (1 - t)u_2, ty_1 + (1 - t)v_2) \in E$. Therefore, $p_1q_2 \subset E$.

We put together the points $p_1, q_2, p_2, q_3, \ldots, p_{m-1}, q_m, p_m$ in a sequence by defining: $k = 2(m - 1)$ and $e_0 = p_1, e_1 = q_2, e_2 = p_2, e_3 = q_3, \ldots, e_{k-2} = p_{m-1}, e_{k-1} = q_m$ and $e_k = p_m$. Then, for each $j \in \{1, \ldots, l\}$, $e_{j-1}e_j \subset E$.

For each $j \in \{0, \ldots, k\}$, let $e_j = (w_j, z_j)$. Define the partition Q of $[0, 1]$ given by $Q : 0 = \frac{0}{k} < \frac{1}{k} < \cdots < \frac{k}{k} = 1$.

Define $\alpha, \beta : [0, 1] \to [0, 1]$ in the following way. Given $j \in \{1, \ldots, k\}$ and $t \in [\frac{j-1}{k}, \frac{j}{k}]$, let

$$\alpha(t) = (kt - j + 1)w_j + (j - kt)w_{j-1}, \text{ and}$$

$$\beta(t) = (kt - j + 1)z_j + (j - kt)z_{j-1}.$$

Given $j \in \{1, \ldots, k - 1\}$, $\alpha(\frac{j}{k}) = w_j$ in both cases, when we consider $\frac{j}{k}$ either in the interval $[\frac{j-1}{k}, \frac{j}{k}]$ or in the interval $[\frac{j}{k}, \frac{j+1}{k}]$. This shows that α is well defined and continuous. Similarly, β is well defined and continuous. Clearly, α and β are PL mappings.

Given $j \in \{1, \ldots, k\}$ and $t \in [\frac{j-1}{k}, \frac{j}{k}]$, the point $(\alpha(t), \beta(t))$ belongs to $e_{j-1}e_j \subset E$. Thus $f(\alpha(t)) = g(\beta(t))$. Therefore $f \circ \alpha = g \circ \beta$.

Since $\alpha(0) = w_0 = 0$, $\beta(0) = z_0 = 0$, $\alpha(1) = w_k = 1$ and $\beta(1) = z_k = 1$, we conclude that α and β are jumping mappings. □

12.5 The Fundamental Theorem of Algebra

The Fundamental Theorem of Algebra (TFA) says that every polynomial with complex coefficients of degree at least 1 has at least one complex root. In 1746 J. le R. D'Alembert made a first attempt to prove it and produced an incomplete proof. L. Euler (1749) and J.-L. Lagrange (1772) also tried to prove it without success. In 1799, using geometric arguments, C.F. Gauss gave a proof in his doctoral dissertation (he was only 22 years old), but motivated by the fact that his contemporaries wanted an entirely algebraic proof, he gave three more demonstrations of this theorem, the last one when he was 70 years old.

Now we know of many ways to prove the TFA, the simplest perhaps being the one that only uses the fact that a disk (a 2-cell) is compact. In this section we give a beautiful proof as an application of Theorem 12.7.

Lemma 12.18 *For each $n \in \mathbb{N}$, the mapping $\alpha : S^1 \to S^1$ defined by $\alpha(z) = z^n$ cannot be lifted. Thus α is not homotopic to a constant mapping.*

Proof Suppose, to the contrary, that there exists a mapping $g : S^1 \to \mathbb{R}$ such that $\alpha = e \circ g$. Since $\alpha(1) = 1$, adding an integer multiple of 2π to g, we may assume that $g(1) = 0$.

Let $f : [0, 2\pi] \to \mathbb{R}$ be given by $f(t) = nt$. Then for each $t \in [0, 2\pi]$, $(\alpha \circ e)(t) = \alpha(\mathrm{e}^{it}) = \mathrm{e}^{int} = (e \circ f)(t)$ and $(\alpha \circ e)(t) = (e \circ (g \circ e))(t)$. Thus f and $g \circ e$ are liftings of the mapping $\alpha \circ e$. Since $(g \circ e)(0) = g(\mathrm{e}^{i0}) = g(1) = 0$ and $f(0) = 0$ and $[0, 2\pi]$ is connected, by Lemma 12.3, $f = g \circ e$. Hence $0 = f(0) = g(\mathrm{e}^{i0}) = g(1) = g(\mathrm{e}^{i2\pi}) = f(2\pi) = n2\pi$, a contradiction. We have shown that α cannot be lifted. Therefore, by Theorem 12.7, α is not homotopic to a constant mapping. □

Theorem 12.19 *Every polynomial with complex coefficients of degree at least* 1 *has a root in the set $\mathbb{C}$ of complex numbers.*

Proof Suppose, to the contrary, that

$$f(z) = a_0 + a_1 z + \cdots + a_n z^n,$$

where $a_0, \ldots, a_n \in \mathbb{C}$ and $a_n \neq 0$, has no roots in $\mathbb{C}$.

Define $F : \mathbb{C} \to \mathbb{C}$ by

$$F(z) = a_0 z^n + a_1 z^{n-1} + \cdots + a_{n-1} z + a_n.$$

Then $F(0) = a_n \neq 0$ and for each $z \neq 0$, $F(z) = z^n f(\frac{1}{z}) \neq 0$, so F also has no roots. Let $r : \mathbb{C} \setminus \{0\} \to S^1$ be the retraction $r(z) = \frac{z}{|z|}$.

Define $h : S^1 \times [0, 1] \to S^1$ by

$$h(z, t) = r(f(tz)/F(t/z)).$$

Then h is a well-defined homotopy such that for each $z \in \mathbb{C}$, $h(z, 0) = r(a_0/a_n)$ and $h(z, 1) = r(f(z)/F(1/z)) = r(z^n) = z^n$. Therefore the mapping $z \rightarrowtail z^n$ is homotopic to a constant mapping, a contradiction with Lemma 12.18. This finishes the proof of the theorem. □

12.6 Exercises

Exercise 12.20 Determine which of the following continua are unicoherent.

(a) $[0, 1]$,
(b) S^1,
(c) a finite graph containing a simple closed curve,
(d) the Warsaw circle,
(e) the Euclidean plane minus the one-point set $\{(0, 0)\}$,
(f) a compactification of the ray $[0, \infty)$.

Exercise 12.21 Show that the exponential mapping has the following properties.

(a) e has period 2π,
(b) if $J \subsetneq S^1$, then there exists a countable family $\{L_z \subset \mathbb{R} : z \in \mathbb{Z}\}$ such that $\bigcup\{L_z : z \in \mathbb{Z}\} = \mathrm{e}^{-1}(J)$, for each $z \in \mathbb{Z}$, $L_z = z + L_0$ and $e|_{L_z} : L_z \to J$ is a homeomorphism,
(c) if Z is a topological space and $f : Z \to S^1$ is a non-onto mapping, then f can be lifted.

Exercise 12.22 Find a unicoherent continuum without property (b).

Exercise 12.23 Prove that the following spaces are unicoherent.

(a) an n-cell ($n \in \mathbb{N}$),
(b) a sphere S^n ($n \geq 2$),
(c) the Hilbert cube.

Exercise 12.24 Let Z be a connected space, A a connected subset of Z and C a component of $Z \setminus A$. Prove that $Z \setminus C$ is connected. (Hint: use Exercise 4.12.)

Exercise 12.25 Let Z be a connected and locally connected space.

(a) Suppose that Z is unicoherent, A is a closed subset of Z and A separates the points p and q in Z. Prove that there exists a closed connected subset B of A such that B also separates p and q in Z,
(b) If $\{A_j : j \in J\}$ is a family of subsets of Z, show that

$$\mathrm{Fr}_Z(\bigcup\{A_j : j \in J\}) \subset \mathrm{cl}_Z(\bigcup\{\mathrm{Fr}_Z(A_j) : j \in J\}).$$

(Hint: for (a) use Exercises 2.37 and 12.24.)

Exercise 12.26 Prove that if Z is locally connected and $Y \subset Z$ is a retract of Z, then Y is locally connected.

Exercise 12.27 Find an open unicoherent continuum which is not unicoherent.

Exercise 12.28 Suppose that Z is a connected and locally connected space. Prove that, if we also assume that Z is normal, then it is possible to give a much easier

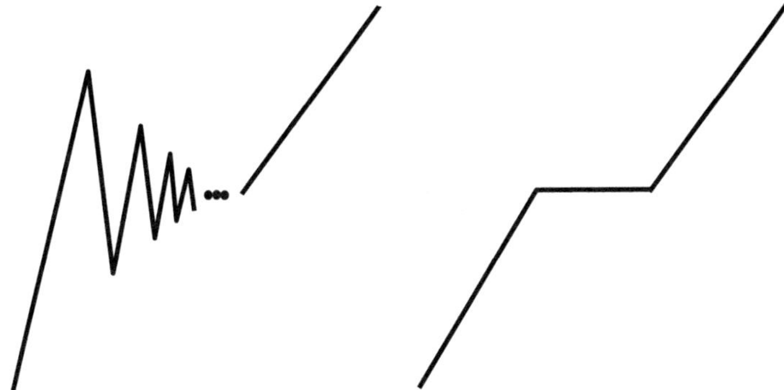

Fig. 12.1 Example related to the mountain climbing theorem

proof, than the one given in Theorem 12.8, of the fact that open unicoherence implies unicoherence for Z.

Exercise 12.29 Prove that every chainable continuum has property (b).

Exercise 12.30 Prove that the composition of PL-mappings is also a PL-mapping.

Exercise 12.31 Suppose that $f, g : [0, 1] \to [0, 1]$ are PL mappings. Then there exists a partition P of $[0, 1]$ such that both mappings f and g are supported by P.

Exercise 12.32 Suppose that $f, g : [0, 1] \to [0, 1]$ are PL mappings such that both are supported by a partition $P : 0 = t_0 < t_1 < \cdots < t_n = 1$ of $[0, 1]$. Let $i, j \in \{1, \ldots, n\}$, and numbers $x, u \in [t_{i-1}, t_i]$ and $y, v \in [t_{j-1}, t_j]$ be such that $f(x) = g(y)$ and $f(u) = g(v)$. Then for each $t \in [0, 1]$, $f(tx + (1 - t)u) = g(ty + (1 - t)v)$.

Exercise 12.33 Let $f, g : [0, 1] \to [0, 1]$ be mappings represented in Fig. 12.1, where the graph of f, represented by the picture on the left, crosses the line $y = \frac{1}{2}$ infinitely many times. Observe that f and g are continuous, $f(0) = 0 = g(0)$, $f(1) = 1 = g(1)$, and the graph of g contains a convex segment in the line $y = \frac{1}{2}$. Show that it is impossible to find mappings $\alpha, \beta : [0, 1] \to [0, 1]$ such that $f \circ \alpha = g \circ \beta$.

Exercise 12.34 Let $f : [0, 1] \to [0, 1]$ be a PL mapping such that $f(1) = 1$ and $g : [0, 1] \to [0, 1]$ is a jumping mapping. Then there exist a jumping mapping α and a PL mapping β such that $\beta(1) = 1$ and $f \circ \alpha = g \circ \beta$. (Hint: define the jumping mappings $h, k : [0, 1] \to [0, 1]$ by $h(t) = 2tf(0))$, if $t \in [0, \frac{1}{2}]$, and $h(t) = f(2t - 1)$, if $t \in [\frac{1}{2}, 1]$; and $k(t) = 0$, if $t \in [0, \frac{1}{2}]$, and $k(t) = g(2t - 1)$, if $t \in [\frac{1}{2}, 1]$. Obtain jumping mappings γ and δ such that $h \circ \gamma = k \circ \lambda$. Define $s_0 = \max \gamma^{-1}(\frac{1}{2})$, $\alpha(t) = 2\gamma(t + (1 - t)s_0) - 1$ and $\beta(t) = \max\{2\lambda(t + (1 - t)s_0) - 1, 0\}$.)

Chapter 13
The Fixed Point Property

13.1 Introduction

Definition 13.1 Let X be a topological space and $f : X \to X$ a mapping. A point $p \in X$ is a *fixed point* for f if $f(p) = p$. The space X has the *fixed point property* (*fpp*) if every mapping $f : X \to X$ has a fixed point.

In the literature, one can find many problems that ask us to characterize the continua that have some topological property. Among these kinds of problems, the ones that seems particularly difficult (almost impossible) to solve are those that refer to the fixed point property. Counterexamples have played a very important role in this area, since many reasonable conjectures have been solved in the negative with examples. This can be corroborated by reading the classical paper: *The elusive fixed point property* written in 1969 by RH Bing [12]. In 2007, C.L. Hagopian [44] and R. Mańka [102] wrote updates of Bing's paper.

Perhaps the most interesting open problem on the topology of the plane is the following.

Problem 13.2 *Suppose that X is a continuum in the plane $\mathbb{R}^2$ such that $\mathbb{R}^2 \setminus X$ is connected. Does X have the fpp?*

Problem 13.2 first appeared in 1930 in a paper by W.L. Ayres, who referred to it as "a well known problem" [6].

This chapter provides a brief introduction to the fpp on continua. We include many basic facts as exercises and we discuss four appealing topics: the technique of "dog chasing rabbit"; the square $[0, 1]^2$ has the fpp; dendroids have the fpp; and the cone over a spiral formed by a ray surrounding a circle does not have the fpp.

A. Illanes, *Continuum Theory*, Universitext,
https://doi.org/10.1007/978-3-031-91011-1_13

13.2 Dog Chasing Rabbit

In the following example we use the so-called "Dog chasing rabbit" technique. At first sight it can be appear to be an informal technique but the steps can be justified by using Exercise 13.9.

Example 13.3 The Warsaw circle has the fpp.

Denote by X the Warsaw circle. Then $X = R \cup J \cup L$, where R is the topologist's curve (the closure in the plane of the graph of the mapping $\sin(\frac{1}{x})$ defined in the interval $(0, 1]$), $J = \{0\} \times [-1, 1]$ and L is an arc joining the points $(0, -1)$ and $(1, \sin(1))$ satisfying $(R \cup J) \cap L = \{(0, -1),(1, \sin(1))\}$ (Fig. 13.1).

Suppose that there exists a fixed-point-free mapping $f : X \to X$. We imagine that the position of the dog is marked by p and the position of the rabbit is marked by $f(p)$. The dog will try to catch the rabbit, so the dog starts running toward the rabbit. We have assumed that f does not have fixed points, so the dog does not catch the rabbit.

Note that the Warsaw circle is uniquely arcwise connected and the only end-point of X is the point $(0, 1)$. Note also that we can give a natural order to X where $(0, 1)$ is the minimum and for two points $w, z \in X$, $w \leq z$ if and only if w belongs to the unique arc in X connecting $(0, 1)$ to z.

The dog starts at the point $e = (0, 1)$. When the dog runs in the only possible direction, the rabbit will always remain ahead of the dog (they cannot cross paths). This means that for each $p \in X$, $p < f(p)$, in the order defined above. In particular, when the dog is at the position $q_n = (\frac{1}{\pi(2n+\frac{3}{2})}, -1)$, the rabbit must be in the rectangle $([0, \frac{1}{\pi(2n+\frac{3}{2})}] \times [-1, 1]) \cap X$. Since $\lim_{n\to\infty} q_n = q = (0, -1)$, by the continuity of f, $f(q)$ must be in the degenerate rectangle $\{0\} \times [-1, 1]$. Thus $f(q)$ is in the arc J. Hence $f(q) \leq q$, which is a contradiction. This completes the proof that f has the fpp.

Fig. 13.1 Warsaw circle

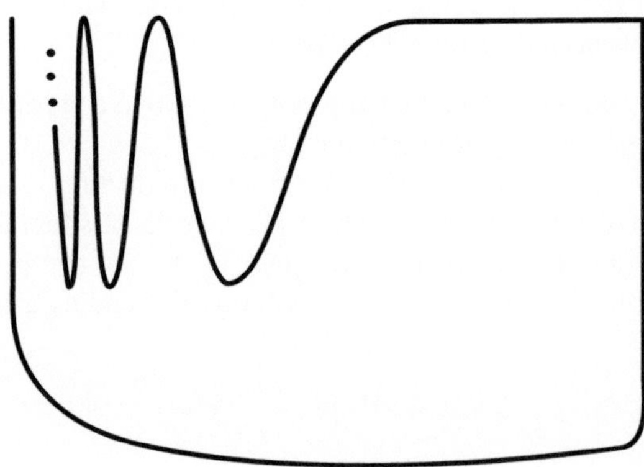

13.3 Cells

Theorem 13.4 $[0, 1]^2$ *has the fpp.*

Proof Let D be the unit disk in the plane $\mathbb{R}^2$ and S^1 its boundary. It is enough to show that D has the fpp. Suppose to the contrary that there exists a mapping $f : D \to D$ without fixed points.

Define $g : D \to D$ in the following way. Given $p \in D$, since $p \neq f(p)$, we may consider the line $L(p)$ in $\mathbb{R}$ determined by the points p and $f(p)$. Inside $L(p)$ we can consider the infinite ray $J_p \subset L(p)$ that starts at $f(p)$ and contains p. Let $g(p)$ be the unique point in $(J_p \setminus \{f(p)\}) \cap S^1$. It is easy to see that g is continuous. If $p \in S^1$, then $p \in J(p) \cap S^1$, so $g(p) = p$. We have shown that S^1 is a retract of D. This contradicts Corollary 12.14 and finishes the proof that $[0, 1]^2$ has the fpp. □

We have proved Theorem 13.4 as a consequence of the unicoherence of the square $[0, 1]^2$. The proof that each n-cell has the fpp needs other techniques. A short and elegant proof of this result, using Sperner's lemma, was given by B. Knaster, K. Kuratowski and S. Mazurkiewicz [86].

Theorem 13.5 ([86]) *For every* $n \in \mathbb{N}$, $[0, 1]^n$ *has the fpp.*

13.4 Dendroids

Theorem 13.6 ([18]) *Dendroids have the fpp.*

Proof Suppose to the contrary that there exists a dendroid X and a mapping $f : X \to X$ without fixed points. Let d be a metric for X. By Exercise 13.16, there exists an $\varepsilon > 0$ such that $d(x, f(x)) > 2\varepsilon$ for every $x \in X$.

Fix a point $a \in X$. Then $d(a, f(a)) > 2\varepsilon$. Let b_1 be the first point in the arc $af(a)$, going from a to $f(a)$, such that $d(a, b_1) = \varepsilon$. Note that $ab_1 \subset \{p \in X : d(a, p) \leq \varepsilon\}$.

We need to prove the following claim.

Claim $b_1 \in af(b_1)$. We prove this claim. Let $z \in af(a)$ be such that $\{z\} = af(a) \cap zf(b_1)$. Since $b_1 \in af(a)$, we have that $af(a) = ab_1 \cup b_1 f(a)$.

We claim that $z \in b_1 f(a)$. Suppose the contrary, then $z \in ab_1$. Then $d(a, z) \leq \varepsilon$. Since $f(ab_1)$ is a subcontinuum (in fact, a dendroid) of X and $f(a), f(b_1) \in f(ab_1)$, we have that $f(a)f(b_1) \subset f(ab_1)$.

Since $\{z\} = af(a) \cap zf(b_1)$, z is in the unique arc joining $f(a)$ and $f(b_1)$. That is, $z \in f(a)f(b_1) \subset f(ab_1)$. Thus there exists an $x \in ab_1$ such that $z = f(x)$. Note that $d(a, x) \leq \varepsilon$. By the triangle inequality, we have that $d(x, f(x)) \leq d(x, a) + d(a, z) \leq 2\varepsilon$, contradicting the choice of ε.

We have shown that $z \in b_1 f(a)$. Thus b_1 is in the arc az. By the definition of z, $z \in af(b_1)$. Thus, when we go from a to $f(b_1)$, we pass through b_1. This proves that $b_1 \in af(b_1)$, and finishes the proof of the claim.

Since $d(b_1, f(b_1)) > 2\varepsilon$, we can take the first point b_2 in the arc $b_1 f(b_1)$, going from b_1 to $f(b_1)$, such that $d(b_1, b_2) = \varepsilon$. Since $b_1 \in af(b_1)$, we have that $af(b_1) = ab_1 \cup b_1 f(b_1)$. Since $b_2 \in b_1 f(b_1)$, we conclude that $ab_1 \subset ab_2$.

Proceeding as in the claim, it is possible to show that $b_2 \in b_1 f(b_2)$.

Repeating this argument, it is possible to find a point $b_3 \in b_2 f(b_2)$ such that $d(b_2, b_3) = \varepsilon$, $ab_2 \subset ab_3$ and $b_3 \in b_2 f(b_3)$.

Proceeding in a similar way, it is possible to construct a sequence $\{b_n\}_{n=1}^{\infty}$ in X such that $d(b_n, b_{n+1}) = \varepsilon$ and $ab_n \subset ab_{n+1}$ for every $n \in \mathbb{N}$.

By Theorem 7.2, there exists an $x \in X$ such that $ab_n \subset ax$ for every $n \in \mathbb{N}$.

Thus the sequence $\{b_n\}_{n=1}^{\infty}$ is a monotone sequence in the arc ax, so this sequence is convergent. This is absurd since for each $n \in \mathbb{N}$, $d(b_n, b_{n+1}) = \varepsilon$.

Since this contradiction comes from the assumption that X does not have the fpp, the proof of the theorem is complete. □

13.5 The Cone of a Spiral

In 1948 K. Borsuk [16, p. 332] asked if the cone over a continuum always has the fixed point property. In 1953 this question was answered in the negative by S. Kinoshita [81] with the example called "the can with a roll of toilet paper". In 1967, R.J. Knill [87] showed that the cone over a circle with a spiral has the fpp. The proof presented in this section is a little simpler than Knill's proof. Another example in this direction appeared in 2007 [59]—a tree-like continuum (the one in Fig. 10.13) whose cone does not have the fpp.

Example 13.7 ([87]) There exists a continuum X whose cone does not have the fpp.

The continuum X in this example is the simplest compactification of the ray $[0, \infty)$ having a simple closed curve as its remainder (Fig. 13.2).

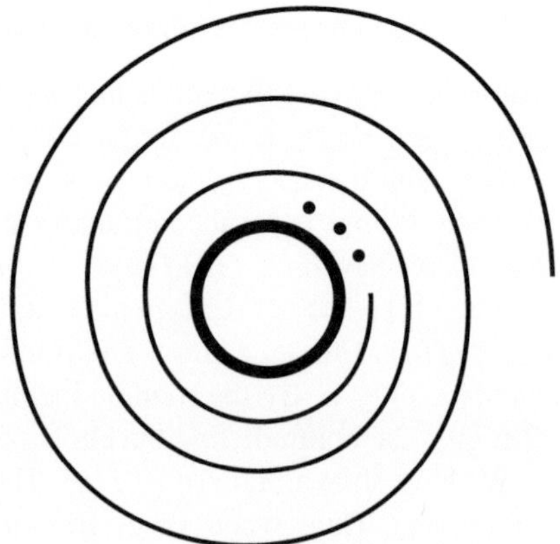

Fig. 13.2 A spiral

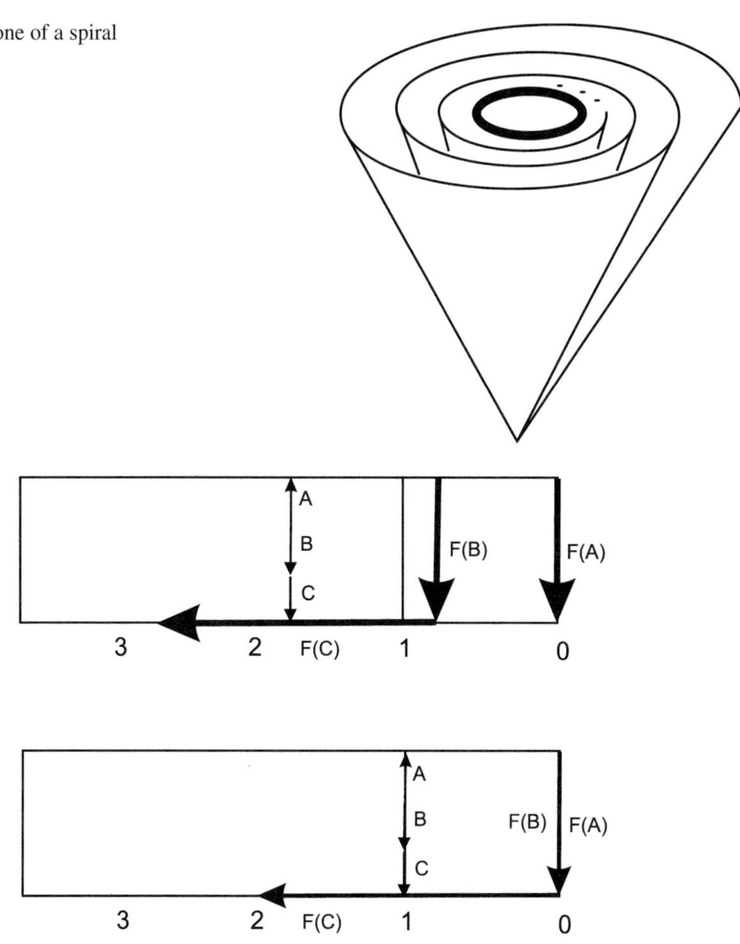

Fig. 13.3 Cone of a spiral

Fig. 13.4 The definition of F

Let Z be the cone of X and v the vertex of Z (Fig. 13.3). We show that Z does not have the fpp.

We show how to define a fixed-point-free mapping $F : Z \to Z$. We describe the ray as the image of the mapping $\varphi : [0, \infty) \to \mathbb{R}^2$ given by

$$\varphi(t) = (1 + \frac{1}{1+t})(\sin(2\pi t), \cos(2\pi t)).$$

As usual, we consider Z as the image of the mapping $\psi : X \times [0, 1] \to Z$, where $\psi(X \times \{1\}) = v$.

The geometric ideas of the definition F can be followed in Figs. 13.4 and 13.5.

We will describe F on the infinite rectangle $R = [0, \infty) \times [0, 1]$. We ask the reader to check that this definition can be extended to $S^1 \times [0, 1]$ and that it is well

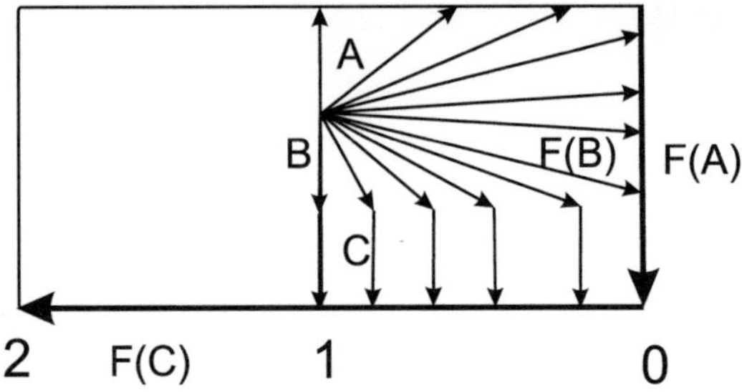

Fig. 13.5 End of the definition of F

defined on Z. Note that the image of the infinite ray $[0, \infty) \times \{1\}$ under ψ is the vertex v of Z. Then the vertical segments of the form $\{t\} \times [0, 1]$ (in R) represent segments in the cone that join a point in the base to v.

We first describe how F acts on segments of the form $R_t = \{t\} \times [0, 1]$, for $t \geq 1$. Given $t \geq 1$, we take a division of R_t into the three segments

$$A_t = \{t\} \times [\frac{2}{3}, 1], \; B_t = \{t\} \times [\frac{1}{3}, \frac{2}{3}] \text{ and } C_t = \{t\} \times [0, \frac{1}{3}],$$

and we consider the segments A_t, B_t and C_t as arrows as in Fig. 13.4.

Then we define F on each of the segments A_t, B_t and C_t as represented in Fig. 13.4. That is:

- The segment A_t is sent to the segment joining v and $(0, 0)$ in the base of Z, sending v to $(0, 0)$ and the point $(t, \frac{2}{3})$ to v.
- The segment B_t is sent to the segment $\{t - 1\} \times [0, 1]$, sending the point $(t, \frac{2}{3})$ to v and the point $(t, \frac{1}{3})$ to the point $(t - 1, 0)$.
- The segment C_t is sent to the segment $[t + 1, t - 1] \times \{0\}$, sending the point $(t, \frac{1}{3})$ to the point $(t - 1, 0)$ and the point $(t, 0)$ to the point $(t + 1, 0)$.

In the spiral, the segment C_t is sent to an arc that completes one lap.

Note that F is a continuous function defined on $[1, \infty) \times [0, 1]$ and F does not have fixed points. Also observe that F can be continuously extended to the cone over the central circumference, by extending its definition to every segment going from v to a point on the base of S^1. Note that this extension is also free of fixed points.

In particular, we have defined F on the segment $\{1\} \times [0, 1]$. Its image is represented in the second part of Fig. 13.4, marked by the arrows $F(A) = F(B)$ and $F(C)$.

To finish the definition of F, we need to define it on the square $T = [0, 1]^2$. In order to do this, divide T using the arrows in Fig. 13.5. We are using two types of arrows.

The arrows of the first type are those arrows starting at the point $(1, \frac{2}{3})$, all of which are sent to the segment $\{0\} \times [0, 1]$. The arrows of the second type are the arrows of the form $\{t\} \times [0, \frac{1}{3}]$, all of which are sent to the segment $[0, 2] \times \{0\}$.

It is easy to see that F is continuous and does not have fixed points.

13.6 Exercises

Exercise 13.8 Prove that if a space X has the fpp, then X is connected.

Exercise 13.9 Let J be an arc in a continuum X. Suppose that $f : X \to X$ is a mapping such that $f(J)$ is an arc. Prove that if one of the following conditions holds: $J \subset f(J)$ or $f(J) \subset J$, then f has a fixed point in J. Conclude that $[0, 1]$ has the fpp.

Exercise 13.10 Let $f : [0, 1] \to [0, 1]$ be an onto mapping such that $f(0) = 0$ and there exists a point $x \in [0, 1]$ such that $f(x) < x$. Prove that f has a fixed point in the interval $(0, 1]$.

Exercise 13.11 Suppose that X and Y are subcontinua of a continuum Z and both have the fpp. Suppose also that $Z = X \cup Y$ and $X \cap Y = \{p\}$ for some $p \in Z$. Prove that Z has the fpp.

Exercise 13.12 Use Exercise 13.11 to prove that every tree has the fpp.

Exercise 13.13 Suppose that the space X has the fpp and $Y \subset X$ is a retract of X. Prove that Y has the fpp.

Exercise 13.14 Show that the continuum pictured in Fig. 13.6 has the fpp, but its cone does not have the fpp. This continuum is the union of the continuum X in Example 13.7 and the disk whose boundary is S^1. (Hint: use Exercise 13.13.)

Exercise 13.15 Let Y be the ray which is the graph of the mapping $\sin(\frac{1}{x})$ on the interval $(0, 1]$. Let $Z = Y \cup \{(0, 0)\}$. Prove that Z has the fpp (and Z is not compact).

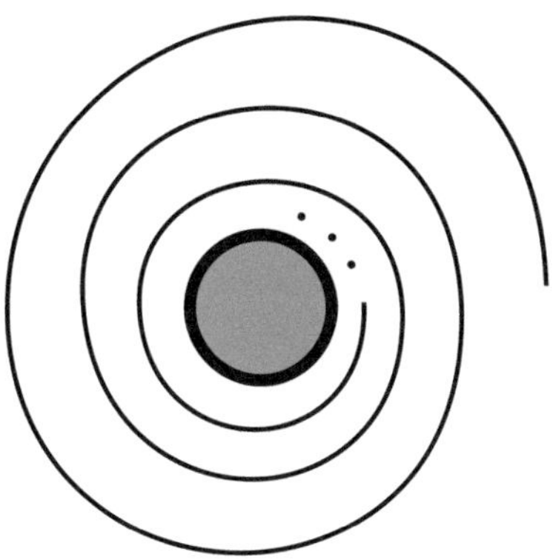

Fig. 13.6 Spiral with a disk

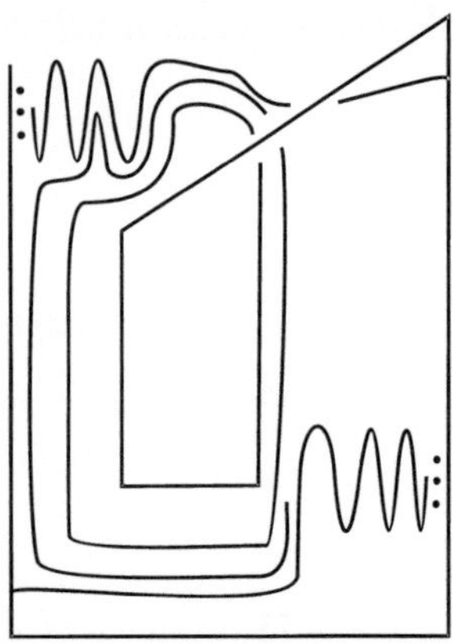

Fig. 13.7 Three rays

Exercise 13.16 Suppose that X is a continuum with metric d and that $f : X \to X$ is a mapping without fixed points. Prove that there exists an $\varepsilon > 0$ such that for every $p \in X$, $d(p, f(p)) > \varepsilon$.

Exercise 13.17 Prove that every chainable continuum has the fpp.

Exercise 13.18 A continuum X has the *surjective fixed point property* (*sfpp*) if each surjective mapping $f : X \to X$ has a fixed point. Prove that the properties fpp and sfpp are not equivalent.

Exercise 13.19 Using the dog-chasing-rabbit technique prove that every tree has the fpp.

Exercise 13.20 Using the dog-chasing-rabbit technique, show that the continuum pictured in Fig. 13.7 has the fpp, and the inner ray contains in its closure two copies of the continuum $\sin(\frac{1}{x})$.

Exercise 13.21 Prove that if X is a compactification of the ray $[0, \infty)$ such that its remainder has the fpp, then X has the fpp.

Exercise 13.22 Show that there exist compactifications of the ray $[0, \infty)$ without the fpp.

Exercise 13.23 The continuum pictured in Fig. 13.8 is called Bing's house. Show that it has the fpp. This continuum is a house with two rooms, the entrance to the lower room is by the roof and the entrance to the upper room is by the bottom. (Hint: use Exercise 13.13.)

Exercise 13.24 Let X be a continuum. Suppose that there exists a sequence of mappings $\{f_n\}_{n=1}^{\infty}$ from X to X uniformly converging to the identity mapping and satisfying that for each $n \in \mathbb{N}$, $f_n(X)$ has the fpp. Prove that X has the fpp.

Exercise 13.25 Prove that the Hilbert cube has the fpp.

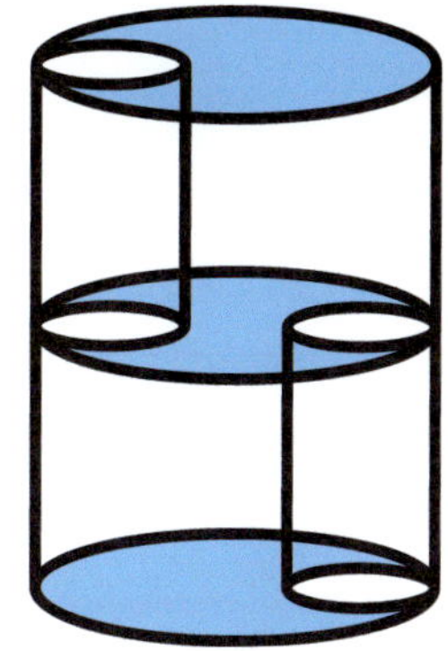

Fig. 13.8 Bing's house

Fig. 13.9 Bing's double tornado

Exercise 13.26 Prove that the Warsaw disk (the closure of the region bounded by the Warsaw circle) has the fpp.

Exercise 13.27 Prove that the unit n-sphere S^n does not have the fpp.

Exercise 13.28 The continuum X in Fig. 13.9 is called the "Bing's double tornado". This continuum is constructed by taking two cylinders of the form $S^1 \times [0, \infty)$ and spiraling each one on the base of the other, as shown in the figure. Prove that X is the intersection of a decreasing sequence of 3-cells. Give also a geometric argument to show that X does not have the fpp.

Chapter 14
Inverse Limits

Inverse limits have played a very important role in Continuum Theory. Their definition is technical and very precise. Consequently, their general properties are relatively easy to prove and, as we can see in this chapter, most of them can be left as exercises. Moreover, inverse limits are appropriate to construct continua with concrete and workable properties. Thus, they have been a powerful tool to create outstanding examples. However, there is a problem: it is extremely difficult to see the geometric behavior of the objects produced as inverse limits. In some cases, the Anderson–Choquet theorem (Theorem 14.7) helps to make a translation between geometric and technical ideas, but its extent is limited.

In this chapter we give a brief introduction to inverse limits. Fortunately, those who want deepen their knowledge on this topic can read the excellent books: [74], by W.T. Ingram and W.S. Mahavier, or [73], by W.T. Ingram.

Historical notes about inverse limits can be found in [22]. There is also a specific history on them on pages vii–ix of [74]

14.1 Definition and Examples

Definition 14.1 An *inverse sequence* is a sequence of ordered pairs

$$\{X_n, f_n\}_{n=1}^{\infty},$$

where for every $n \in \mathbb{N}$, X_n is a nonempty metric space and $f_n : X_{n+1} \to X_n$ is a mapping.

A. Illanes, *Continuum Theory*, Universitext,
https://doi.org/10.1007/978-3-031-91011-1_14

The usual representation of an inverse limit is as follows:

$$X_1 \underset{f_1}{\leftarrow} X_2 \underset{f_2}{\leftarrow} X_3 \leftarrow \cdots .$$

The mappings f_n are called *bonding mappings.*

We suppose that each X_n has a metric d_n bounded by the number 1 and the metric for the product

$$X = \prod_{n=1}^{\infty} X_n,$$

is given by

$$d_X((x_1, x_2, \ldots), (y_1, y_2, \ldots)) = \sum_{n=1}^{\infty} \frac{d_n(x_n, y_n)}{2^n}.$$

We denote by $\pi_m : X \to X_m$ the natural projection.

If $n < k$, define $f_n^k : X_k \to X_n$ as the composition

$$f_n^k = f_n \circ \cdots \circ f_{k-1}.$$

Given an inverse sequence $\{X_n, f_n\}_{n=1}^{\infty}$, define its *inverse limit* as:

$$\varprojlim\{X_n, f_n\}_{n=1}^{\infty} = \{\{x_n\}_{n=1}^{\infty} \in X : f_n(x_{n+1}) = x_n \text{ for every } n \in \mathbb{N}\}.$$

If it is not necessary to mention the inverse sequence, we simply denote the inverse limit $\lim_{\leftarrow}\{X_n, f_n\}_{n=1}^{\infty}$ by X_∞.

Theorem 14.2 *Let $\{X_n, f_n\}_{n=1}^{\infty}$ be an inverse sequence and A a compact subset of X_∞. Then A is homeomorphic to* $\lim_{\leftarrow}\{\pi_n(A), f_n|_{\pi_{n+1}(A)}\}_{n=1}^{\infty}$.

Proof Given $a \in A$ and $n \in \mathbb{N}$, by Exercise 14.13, $f_n(\pi_{n+1}(a)) = \pi_n(a) \in \pi_n(A)$. Then $f_n|_{\pi_{n+1}(A)} : \pi_{n+1}(A) \to \pi_n(A)$. So $\{\pi_n(A), f_n|_{\pi_{n+1}(A)}\}_{n=1}^{\infty}$ is an inverse sequence and then it is possible to define the space

$$A_\infty = \varprojlim\{\pi_n(A), f_n|_{\pi_{n+1}(A)}\}_{n=1}^{\infty}.$$

Observe that $A_\infty \subset X_\infty$.

Given $a = (a_1, a_2, \ldots) \in A$, since for each $n \in \mathbb{N}$, $f_n(a_{n+1}) = a_n$, we have that $f_n|_{\pi_{n+1}(A)}(a_{n+1}) = a_n$. Hence $a \in A_\infty$. Thus the mapping $\varphi : A \to A_\infty$ given by $\varphi(a) = a$ is continuous and one-to-one.

We check that φ is onto.

Take $x = (x_1, x_2, \ldots) \in A_\infty$. By definition, for each $n \in \mathbb{N}$, $f_n(x_{n+1}) = x_n$ and $x_n \in \pi_n(A)$.

Given $n \in \mathbb{N}$, choose $z^{(n)} \in A$ such that $\pi_n(z^{(n)}) = x_n$. Since A is compact, we can choose a convergent subsequence $\{z^{(n_k)}\}_{k=1}^{\infty}$ with limit a point $a \in A$. We claim that $a = x$.

Given $k \in \mathbb{N}$, since $\pi_{n_k}(z^{(n_k)}) = x_{n_k}$, we have that the n_k^{th}-coordinate of $z^{(n_k)}$ coincides with the n_k^{th}-coordinate of x. Since the points x and $z^{(n_k)}$ belong to $X_\infty = \lim_{\leftarrow}\{X_n, f_n\}_{n=1}^{\infty}$, we have that the first n_k coordinates of both points are calculated by applying compositions of the mappings $f_{n_k-1}, \ldots, f_1$. Thus the points x and z coincide in the first n_k coordinates. Hence $d_X(x, z^{(n_k)}) \leq \frac{1}{2^{n_k-1}}$.

This shows that $\lim_{k\to\infty} z^{(n_k)} = x$, so $x = z$. Hence $x \in A$. Therefore φ is onto.

This finishes the proof that φ is a homeomorphism. Therefore A is homeomorphic to A_∞. □

Example 14.3 As usual, we denote by S^1 the unit circle in the plane. Here, the plane is considered as the complex plane $\mathbb{C}$, with its algebraic structure.

Given $m \geq 2$, define $f : S^1 \to S^1$ as the mapping $f(z) = z^m$, and consider the inverse sequence $\{X_n, f_n\}_{n=1}^{\infty}$, where each X_n is equal to S^1 and each f_n is the mapping f. The space $\lim_{\leftarrow}\{X_n, f_n\}_{n=1}^{\infty}$ is known as the *m-solenoid* and is denoted by S_m, the 2-solenoid is also known as *dyadic solenoid.*

Definition 14.4 A *Knaster-type continuum* is a continuum that is the inverse limit of arcs with open, non-homeomorphic bonding mappings.

14.2 Indecomposability

Definition 14.5 An inverse sequence $\{X_n, f_n\}_{n=1}^{\infty}$ of continua is an *indecomposable inverse sequence* if for each $n \in \mathbb{N}$, the following condition holds: if A_{n+1} and B_{n+1} are subcontinua of X_{n+1} such that $X_{n+1} = A_{n+1} \cup B_{n+1}$, then $f_n(A_{n+1}) = X_n$ or $f_n(B_{n+1}) = X_n$.

Theorem 14.6 *Let* $\{X_n, f_n\}_{n=1}^{\infty}$ *be an indecomposable inverse sequence. Then its inverse limit is an indecomposable continuum.*

Proof By Exercise 14.17, X_∞ is a continuum. Let A and B be subcontinua of X_∞ such that $X_\infty = A \cup B$. We are going to show that either $A = X_\infty$ or $B = X_\infty$.

By Exercise 14.32, for each $n \in \mathbb{N}$, $X_{n+1} = \pi_{n+1}(A \cup B) = \pi_{n+1}(A) \cup \pi_{n+1}(B)$. Since the inverse sequence is indecomposable, we have that either $X_n = f_n(\pi_{n+1}(A)) = \pi_n(A)$ or $X_n = f_n(\pi_{n+1}(B)) = \pi_n(B)$.

We may assume then that the equality $X_n = \pi_n(A)$ holds for infinitely many positive integers $n_1 < n_2 < \cdots$.

We show that for every $n \in \mathbb{N}$, $X_n = \pi_n(A)$. Let $n \in \mathbb{N}$. Fix $k \in \mathbb{N}$ such that $n < n_k$. By Exercise 14.32,

$$\begin{aligned}\pi_n(A) &= f_n(\pi_{n+1}(A)) \\ &= (f_n \circ f_{n+1})(\pi_{n+2}(A)) = \cdots \\ &= (f_n \circ \cdots \circ f_{n_k-1})(\pi_{n_k}(A)) \\ &= (f_n \circ \cdots \circ f_{n_k-1})(X_{n_k}) \\ &= X_n.\end{aligned}$$

Hence for each $n \in \mathbb{N}$, $\pi_n(X_\infty) = X_n = \pi_n(A)$. Exercise 14.24 implies that $X_\infty = A$. We have shown that X_∞ is indecomposable. □

14.3 The Anderson–Choquet Theorem

Theorem 14.7 (Anderson–Choquet Theorem [4, Theorem I]) *Let Z be a compact metric space with metric d. Let $\{X_n, f_n\}_{n=1}^\infty$ be an inverse system, where each X_n is a nonempty compact subspace of Z and each mapping f_n is surjective. As before, if $n < k$, define $f_n^k : X_k \to X_n$ as the composition*

$$f_n^k = f_n \circ \cdots \circ f_{k-1}.$$

Consider the following properties.

(1) For each $\varepsilon > 0$, there exists an $m \in \mathbb{N}$ such that for every $p \in X_m$, $\text{diameter}(\bigcup\{(f_m^k)^{-1}(p) : k > m\}) < \varepsilon$.
(2) For each $n \in \mathbb{N}$ and each $\varepsilon > 0$, there exists a $\delta > 0$ such that if $m > n$ and $p, q \in X_m$ satisfy $d(p, q) < \delta$, then $d(f_n^m(p), f_n^m(q)) < \varepsilon$.

Then:

(a) if $\{X_n, f_n\}_{n=1}^\infty$ satisfies property (1), then for each $x = (x_1, x_2, \ldots) \in X_\infty = \lim_{\leftarrow}\{X_n, f_n\}_{n=1}^\infty$, the sequence $\{x_n\}_{n=1}^\infty$ converges to a point which we will call $h(x)$. Moreover, the function $h : X_\infty \to Z$ is continuous and its image is the set $Y = \bigcap_{n=1}^\infty(\text{cl}_Z(\bigcup_{m\geq n} X_m))$,
(b) if $\{X_n, f_n\}_{n=1}^\infty$ satisfies properties (1) and (2), the function h is also one-to-one and then X_∞ is homeomorphic to Y.

Proof

(a) Suppose property (1) holds.
Let $x = (x_1, x_2, \ldots) \in X_\infty$.
Given $\varepsilon > 0$, let $m \in \mathbb{N}$ be a positive integer as in property (1) for ε.

Given $k > m$, by the definition of inverse limit, $f_m^k(x_k) = x_m$. Thus, $x_k \in (f_m^k)^{-1}(x_m)$. By the choice of m, diameter$(\{x_{m+1}, x_{m+2}, \ldots\}) < \varepsilon$. Thus, for every $k, n > m$, $d(x_k, x_n) < \varepsilon$. This shows that the sequence $\{x_n\}_{n=1}^{\infty}$ is a Cauchy sequence. Since Z is compact, we obtain that there exists an element, which we denote by $h(x)$, such that the sequence $\{x_n\}_{n=1}^{\infty}$ converges to $h(x)$. Since $d(x_{m+1}, x_n) < \varepsilon$ for every $n > m$, we have that $d(x_{m+1}, h(x)) \leq \varepsilon$.

We check that h is continuous. Let $\varepsilon > 0$. Take $m \in \mathbb{N}$ as in (1) for $\frac{\varepsilon}{4}$. Let $x = (x_1, x_2, \ldots)$, $y = (y_1, y_2, \ldots) \in X_\infty$ be such that $d(x, y) < \frac{\varepsilon}{2^{m+3}}$. Since $\frac{d(x_{m+1}, y_{m+1})}{2^{m+1}} \leq d(x, y) < \frac{\varepsilon}{2^{m+3}}$, we have that $d(x_{m+1}, y_{m+1}) < \frac{\varepsilon}{2^2}$. As we showed before, the choice of m implies that $d(x_{m+1}, h(x)) \leq \frac{\varepsilon}{4}$ and $d(y_{m+1}, h(y)) \leq \frac{\varepsilon}{4}$. Thus $d(h(x), h(y)) < \varepsilon$. Therefore h is continuous.

Given $x = (x_1, x_2, \ldots) \in X_\infty$, by definition, for every $n \in \mathbb{N}$, $x_n \in X_n$. Then $\{x_n, x_{n+1}, \ldots\} \subset \text{cl}_Z(\bigcup_{m \geq n} X_m)$, and since this set is closed, we obtain that $h(x) \in Y$. Therefore $h(X_\infty) \subset Y$.

In order to prove the other inclusion, since $h(X_\infty)$ is compact (by Exercise 14.17) it is enough to show that $h(X_\infty)$ is dense in Y. To show this, take a point $w \in Y$, and let $\varepsilon > 0$. Let $m \in \mathbb{N}$ be as in property (1). As we showed before, if $x = (x_1, x_2, \ldots) \in X_\infty$, for every $k, n > m$, $d(x_k, x_n) < \varepsilon$. Since $w \in Y$, there exists a point $z \in \bigcup_{k \geq m+1} X_k$ such that $d(w, z) < \varepsilon$. Let $k > m$ be such that $z \in X_k$. Since the mappings f_i are surjective, it is possible to find points $z_{k+1}, z_{k+2}, \ldots$ such that $f_{k+1}(z_{k+1}) = z$, $f_{k+2}(z_{k+2}) = z_{k+1}$, etc. Hence the point $p = (f_1^k(z), \ldots, f_{k-1}^k(z), z, z_{k+1}, z_{k+2}, \ldots)$ belongs to X_∞. By the choice of m, for each $n > k$, $d(z, z_n) < \varepsilon$. Taking limits, we have that $d(z, h(p)) \leq \varepsilon$. Therefore, $d(w, h(p)) < 2\varepsilon$. This finishes the proof that $h(X_\infty)$ is dense in Y. Therefore $h(X_\infty) = Y$.

(b) Now suppose that properties (1) and (2) hold.

In order to show that h is one-to-one, let $x = (x_1, x_2, \ldots)$ and $y = (y_1, y_2, \ldots) \in X_\infty$ be such that $x \neq y$. Let $n \in \mathbb{N}$ be such that $x_n \neq y_n$. Set $\varepsilon = \frac{d(x_n, y_n)}{2}$ and take $\delta > 0$ as in property (2) for n and ε. Given $m > n$, if $d(x_m, y_m) < \delta$, we have that $d(x_n, y_n) = d(f_n^m(x_m), f_n^m(y_m)) < \varepsilon$, which is absurd. This shows that for every $m > n$, $d(x_m, y_m) \geq \delta$. Taking limits, we conclude that $d(h(x), h(y)) \geq \delta$. Thus $h(x) \neq h(y)$. Therefore h is one-to-one.

□

14.4 Chainable Continua as Inverse Limits

Definition 14.8 For $\varepsilon > 0$, an onto mapping between continua $f : X \to Y$ is an *ε-mapping* if for each $y \in Y$, diameter$(f^{-1}(y)) < \varepsilon$.

Given a family $\mathcal{P}$ of compact metric spaces, the continuum X is *$\mathcal{P}$-like* provided that for each $\varepsilon > 0$, there is an ε-mapping from X onto some member of $\mathcal{P}$. In the case when $\mathcal{P} = \{[0, 1]\}$, a $\mathcal{P}$-like continuum is called an *arc-like* continuum.

The aim of this section is to prove Theorem 14.16, which says that for a continuum X being chainable is equivalent to being arc-like and it is also equivalent to being an inverse limit of arcs. This theorem can be generalized to general families $\mathcal{P}$. The reader is referred to pages 731–733 of [22] for some historical remarks about this topic.

Definition 14.9 Let X be a continuum and let $\mathcal{C} = \{C_0, \ldots, C_n\}$ be a chain of open subsets of X, with $n \geq 1$. Then:

(a) $\mathcal{C}$ is a *proper covering chain* of X if $\mathcal{C}$ covers X and no proper subset of $\mathcal{C}$ covers X,
(b) $\mathcal{C}$ is a *taut chain* of X if $\mathcal{C}$ covers X, $\mathrm{cl}_X(C_i) \cap \mathrm{cl}_X(C_j) \neq \emptyset$ if and only if $|i - j| \leq 1$ and $C_0 \setminus \mathrm{cl}_X(C_1) \neq \emptyset \neq C_n \setminus \mathrm{cl}_X(C_{n-1})$. Observe that a taut chain is a proper covering chain.

Given a taut chain $\mathcal{C} = \{C_0, \ldots, C_m\}$ of X, a point $p \in C_0 \setminus \mathrm{cl}_X(C_1)$ and a point $q \in C_m \setminus \mathrm{cl}_X(C_{m-1})$, let $B_0 = \{p\}$, $B_{m+1} = \{q\}$ and

$$B_1 = \mathrm{cl}_X(C_0) \cap \mathrm{cl}_X(C_1), \ldots, B_m = \mathrm{cl}_X(C_{m-1}) \cap \mathrm{cl}_X(C_m).$$

For each $i \in \{0, \ldots, m\}$, the sets B_i and B_{i+1} are disjoint nonempty closed subsets of $\mathrm{cl}_X(C_i)$. By Urysohn's Lemma for metric spaces, there exists a mapping $g_i : \mathrm{cl}_X(C_i) \to [\frac{i}{m+1}, \frac{i+1}{m+1}]$ such that $g_i^{-1}(\frac{i}{m+1}) = B_i$ and $g_i^{-1}(\frac{i+1}{m+1}) = B_{i+1}$. Then, when $i < m$, $g_{i+1}^{-1}(\frac{i+1}{m+1}) = B_{i+1} = g_i^{-1}(\frac{i+1}{m+1})$. This implies that there exists a common continuous extension $g_{\mathcal{C}} : X \to [0, 1]$ of all the mappings g_i. Observe that for each $i \in \{0, \ldots, m\}$, $g_{\mathcal{C}}^{-1}([\frac{i}{m+1}, \frac{i+1}{m+1}]) = \mathrm{cl}_X(C_i)$.

We have shown that there exists a mapping $g_{\mathcal{C}} : X \to [0, 1]$ with the property that for each $i \in \{0, \ldots, m\}$, $g_{\mathcal{C}}^{-1}([\frac{i}{m+1}, \frac{i+1}{m+1}]) = \mathrm{cl}_X(C_i)$. We call any mapping with this property *a mapping induced by* $\mathcal{C}$. These mappings will be useful to prove properties of chainable continua.

Theorem 14.10 *Let X be a continuum and $\mathcal{C} = \{C_0, C_1, \ldots, C_n\}$ be a proper covering chain of X. Then there is a taut chain $\mathcal{D} = \{D_0, D_1, \ldots, D_n\}$ of X satisfying, for each $i \in \{0, 1, \ldots, n\}$, $\mathrm{cl}_X(D_i) \subset C_i$.*

Proof Since $X \setminus (C_1 \cup \cdots \cup C_n)$ and $X \setminus C_0$ are disjoint closed subsets of X, by the normality of X, there exist disjoint open subsets D_0 and U_0 of X such that $X \setminus (C_1 \cup \cdots \cup C_n) \subset D_0$ and $X \setminus C_0 \subset U_0$. Then $X = D_0 \cup C_1 \cup \cdots \cup C_n$, $\mathrm{cl}_X(D_0) \subset X \setminus U_0 \subset C_0$ and $\mathrm{cl}_X(D_0) \cap (C_2 \cup \cdots \cup C_n) = \emptyset$. Since $\mathcal{C}$ is a proper covering chain, $\emptyset \neq X \setminus (C_1 \cup \cdots \cup C_n)$, so $D_0 \neq \emptyset$. The connectedness of X implies that $D_0 \cap (C_1 \cup \cdots \cup C_n) \neq \emptyset$, so $D_0 \cap C_1 \neq \emptyset$. Thus the family $\{D_0, C_1, \ldots, C_n\}$ is a proper covering chain.

We repeat this step using C_1 instead of C_0. That is, observe that $X \setminus C_1$ and $X \setminus (D_0 \cup C_2 \cup \cdots \cup C_n)$ are closed disjoint subset of X. By the normality of X, there exist disjoint open subsets D_1 and U_1 of X such that $X \setminus C_1 \subset U_1$ and $X \setminus (D_0 \cup C_2 \cup \cdots \cup C_n) \subset D_1$. Proceeding as in the previous paragraph, we conclude that $X = D_0 \cup D_1 \cup C_2 \cup \cdots \cup C_n$, $\mathrm{cl}_X(D_1) \subset C_1$, $\mathrm{cl}_X(D_1) \cap (C_3 \cup \cdots \cup C_n) = \emptyset$, $D_1 \neq \emptyset$ and the family $\{D_0, D_1, C_2 \ldots, C_n\}$ is a proper covering chain.

Continuing in this way, substituting inductively each C_i for D_i, we can obtain a proper covering chain $\mathcal{D} = \{D_0, D_1, \ldots, D_n\}$ of X such that for each $i \in \{0, \ldots, n\}$, $\mathrm{cl}_X(D_i) \subset C_i$. Then $\mathrm{cl}_X(D_i) \cap \mathrm{cl}_X(D_j) \neq \emptyset$ if and only if $|i - j| \leq 1$. Since $\emptyset \neq C_0 \setminus C_1 \subset D_0 \setminus \mathrm{cl}_X(D_1)$, we have that $D_0 \setminus \mathrm{cl}_X(D_1) \neq \emptyset$. Similarly, $D_n \setminus \mathrm{cl}_X(D_{n-1}) \neq \emptyset$. Therefore, $\mathcal{D}$ is a taut chain. □

Lemma 14.11 *Suppose that X is a chainable continuum contained in the Hilbert cube $\mathbf{Q}$. Let $\varepsilon > 0$. Then there exists an arc L joining two points y_1 and y_2 in $\mathbf{Q}$, an open subset U of $\mathbf{Q}$ and a mapping $g : \mathrm{cl}_\mathbf{Q}(U) \to L$ such that $X \subset U \subset \mathrm{cl}_\mathbf{Q}(U) \subset N(X, \varepsilon)$, $X \subset N(L, \varepsilon)$, $L \subset N(X, \varepsilon)$, $g(X) = L$, $X \cap \mathrm{int}_\mathbf{Q}(g^{-1}(y_1)) \neq \emptyset \neq X \cap \mathrm{int}_\mathbf{Q}(g^{-1}(y_2))$ and $d(x, g(x)) < \varepsilon$ for all $x \in \mathrm{cl}_\mathbf{Q}(U)$.*

Proof By Lemma 14.10 there exists a taut chain $\mathcal{W} = \{W_1, \ldots, W_m\}$ of X such that $m \geq 3$ and $\mathrm{mesh}(\mathcal{W}) < \frac{\varepsilon}{2}$. The normality of $\mathbf{Q}$ implies that there are open subsets $U_1, \ldots, U_m$ of $\mathbf{Q}$ such that $U_1 \cap X \setminus \mathrm{cl}_\mathbf{Q}(U_2) \neq \emptyset \neq U_m \cap X \setminus \mathrm{cl}_X(U_{m-1})$, $\mathrm{diameter}(U_i) < \frac{\varepsilon}{2}$; $\mathrm{cl}_\mathbf{Q}(W_i) \subset U_i \subset \mathrm{cl}_\mathbf{Q}(U_i) \subset N(\varepsilon, X)$ for each $i \in \{1, \ldots, m\}$ and the families $\{U_1, \ldots, U_m\}$ and $\{\mathrm{cl}_\mathbf{Q}(U_1), \ldots, \mathrm{cl}_\mathbf{Q}(U_m)\}$ are chains.

Choose points $w_1 \in U_1 \cap X \setminus \mathrm{cl}_\mathbf{Q}(U_2)$ and $w_2 \in U_m \cap X \setminus \mathrm{cl}_\mathbf{Q}(U_{m-1})$, and open subsets R_1 and R_2 of $\mathbf{Q}$ such that $w_1 \in R_1 \subset \mathrm{cl}_\mathbf{Q}(R_1) \subset U_1 \setminus \mathrm{cl}_\mathbf{Q}(U_2)$ and $w_2 \in R_2 \subset \mathrm{cl}_\mathbf{Q}(R_2) \subset U_m \setminus \mathrm{cl}_\mathbf{Q}(U_{m-1})$.

By Exercise 14.48, there exist points $p_0, \ldots, p_m$ such that $p_i \in U_i \cap U_{i+1}$ for each $i \in \{1, \ldots, m-1\}$, $p_0 \in U_1 \setminus \mathrm{cl}_\mathbf{Q}(U_2)$, $p_m \in U_m \setminus \mathrm{cl}_\mathbf{Q}(U_{m-1})$ and the set $L = p_0p_1 \cup p_1p_2 \cup \ldots \cup p_{m-1}p_m$ is an arc joining p_0 and p_m, where $p_{i-1}p_i$ is the convex segment in $\mathbf{Q}$ joining p_{i-1} and p_i.

Set $U = U_1 \cup \ldots \cup U_m$.

For each $i \in \{1, \ldots, m\}$, by Urysohn's Lemma, there exists a mapping $g_i : \mathrm{cl}_\mathbf{Q}(U_i) \to p_{i-1}p_i$ such that

$$g_1(\mathrm{cl}_\mathbf{Q}(R_1)) = \{p_0\},$$
$$g_1(\mathrm{cl}_\mathbf{Q}(U_1) \cap \mathrm{cl}_\mathbf{Q}(U_2)) = \{p_1\};$$
$$g_m(\mathrm{cl}_\mathbf{Q}(U_{m-1}) \cap \mathrm{cl}_\mathbf{Q}(U_m)) = \{p_{m-1}\},$$
$$g_m(\mathrm{cl}_\mathbf{Q}(R_2)) = \{p_m\}$$

and if $0 < i < m$, then

$$g_i(\mathrm{cl}_\mathbf{Q}(U_{i-1}) \cap \mathrm{cl}_\mathbf{Q}(U_i)) = \{p_{i-1}\},$$
$$g_i(\mathrm{cl}_\mathbf{Q}(U_i) \cap \mathrm{cl}_\mathbf{Q}(U_{i-1})) = \{p_i\}.$$

Since the mappings g_i coincide on the corresponding intersections, there exists a continuous common extension $g : \mathrm{cl}_\mathbf{Q}(U) \to L$ of all mappings g_i.

Since $g(w_1) = p_0$ and $g(w_m) = p_m$, the continuity of g and the fact that X is connected imply that $g(X) = L$.

Given $i \in \{1, \ldots, m\}$, since $W_i \subset U_i$, we have that $U_i \cap X \neq \emptyset$. Since $p_{i-1}, p_i \in U_i$ and $\text{diameter}(U_i) < \frac{\varepsilon}{2}$, we have that $p_{i-1}, p_i \in N(X, \varepsilon)$. This implies that $p_{i-1}p_i \subset N(X, \varepsilon)$. Therefore $L \subset N(X, \varepsilon)$.

Given $x \in X$, there exists an $i \in \{1, \ldots, m\}$ such that $x \in U_i$, and since $p_i \in U_i$, we conclude that $x \in N(L, \varepsilon)$. Therefore $X \subset N(L, \varepsilon)$.

Observe that $w_1 \in X \cap \text{int}_{\mathbf{Q}}(g^{-1}(p_0))$ and $w_2 \in X \cap \text{int}_{\mathbf{Q}}(g^{-1}(p_m))$.

Given $x \in \text{cl}_{\mathbf{Q}}(U)$, there is an $i \in \{1, \ldots, m\}$ such that $x \in \text{cl}_{\mathbf{Q}}(U_i)$. Then $g(x) = g_i(x) \in p_{i-1}p_i$. Since $\max\{d(x, p_i), d(x, p_{i-1})\} < \text{diameter}(U_i) < \varepsilon$, we conclude that $d(x, g(x)) < \varepsilon$.

This concludes the proof of the lemma. □

Theorem 14.12 *Let X be a continuum. Then the following are equivalent.*

(a) X is chainable,
(b) X is an inverse limit of arcs, and
(c) X is arc-like.

Proof (a) $\Rightarrow$ (b). Suppose that X is a chainable continuum. By Theorem 1.1, we may suppose that X is a subspace of the Hilbert cube $\mathbf{Q}$, with the metric d defined in that theorem. Since X is chainable, by Theorem 14.10, it is possible to construct a sequence of taut chains $\{\mathcal{D}_n\}_{n=1}^{\infty}$ of X such that for each $n \in \mathbb{N}$, $\text{mesh}(\mathcal{D}_n) < \frac{1}{2^n}$. For each $n \in \mathbb{N}$, let $\mathcal{D}_n = \{D_0^{(n)}, \ldots, D_{m_n}^{(n)}\}$.

Now, we construct several sequences satisfying the following properties:

(A) $\{\varepsilon_n\}_{n=1}^{\infty}$, of positive integers,
(B) $\{L_n\}_{n=1}^{\infty}$, of arcs in $\mathbf{Q}$, each L_n joining points $y_1^{(n)}$ and $y_2^{(n)}$,
(C) $\{g_n\}_{n=1}^{\infty}$, of mappings, and
(D) $\{U_n\}_{n=1}^{\infty}$, of open subsets of $\mathbf{Q}$,

such that for every $n \in \mathbb{N}$:

$$L_{n+1} \subset U_n,$$

$$X \subset U_n,$$

$$\varepsilon_{n+1} < \frac{\varepsilon_n}{2} < \frac{1}{3 \cdot 2^{n+1}},$$

$$g_n : \text{cl}_{\mathbf{Q}}(U_n) \subset N(X, \varepsilon_n) \to L_n,$$

$$g_n(X) = L_n = g_n(L_{n+1}),$$

$$X \subset N(L_n, \varepsilon_n),$$

$$L_n \subset N(X, \varepsilon_n),$$

$$X \cap \mathrm{int}_{\mathbf{Q}}(g_n^{-1}(y_1^{(n)})) \neq \emptyset \neq X \cap \mathrm{int}_{\mathbf{Q}}(g_n^{-1}(y_2^{(n)})),$$
$$d(x, g_n(x)) < \varepsilon_n \text{ for all } x \in \mathrm{cl}_{\mathbf{Q}}(U_n), \text{ and}$$

if $m \leq n$, $x, y \in \mathrm{cl}_{\mathbf{Q}}(U_n)$ and $d(x, y) \leq 2\varepsilon_{n+1}$, then

$$(g_m^{n+1}(x), g_m^{n+1}(y)) < \frac{1}{3 \cdot 2^n},$$

where as usual, $g_m^{n+1} = g_m \circ \cdots \circ g_n$. Note that this composition will be well defined since for each $i \in \{m+1, \ldots, n\}$, $\mathrm{Im}\, g_i = g_i(\mathrm{cl}_{\mathbf{Q}}(U_i)) \subset L_i \subset U_{i-1}$ and U_{i-1} is contained in the domain of g_{i-1}.

We construct the sequences inductively. Choose $\varepsilon_1 > 0$ such that $\varepsilon_1 < \frac{1}{3 \cdot 2^2}$. Then take L_1, U_1 and g_1 as in Lemma 14.11 applied to ε_1. Note that, for $n = 1$, all properties are satisfied.

Now suppose that $\{\varepsilon_m\}_{m=1}^n$, $\{L_m\}_{m=1}^n$, $\{g_m\}_{m=1}^n$ and $\{U_m\}_{m=1}^n$ have been constructed and satisfy the required properties.

Fix points $w_1 \in X \cap \mathrm{int}_{\mathbf{Q}}(g_n^{-1}(y_1^{(n)}))$ and $w_2 \in X \cap \mathrm{int}_{\mathbf{Q}}(g_n^{-1}(y_2^{(n)}))$. Choose $\varepsilon_{n+1} > 0$ satisfying $\varepsilon_{n+1} < \min\{\frac{\varepsilon_n}{2}, \frac{1}{3 \cdot 2^{n+2}}\}$ and for each $i \in \{1, 2\}$, $B(w_i, \varepsilon_{n+1}) \subset g_n^{-1}(y_i^{(n)})$. By the uniform continuity of $g_1^{n+1}, \ldots, g_n^{n+1}$, it is possible to ask that if $m \leq n$, $x, y \in \mathrm{cl}_{\mathbf{Q}}(U_n)$ and $d(x, y) \leq 2\varepsilon_{n+1}$, then $d(g_m^{n+1}(x), g_m^{n+1}(y)) < \frac{1}{3 \cdot 2^n}$. Finally, we ask that ε_{n+1} also satisfies $N(X, \varepsilon_{n+1}) \subset U_n$.

Then we apply Lemma 14.11 to ε_{n+1} to obtain L_{n+1}, U_{n+1} and g_{n+1}. Note that we only need to verify that $L_{n+1} \subset U_n$ and $g_{n+1}(L_{n+1}) = L_n$. Given $y \in L_{n+1}$, since $g_{n+1}(X) = L_{n+1}$, there exists an $x \in X$ such that $y = g_{n+1}(x)$. We also know that $d(x, g_{n+1}(x)) < \varepsilon_{n+1}$. Thus $y \in N(X, \varepsilon_{n+1}) \subset U_n$. Therefore $L_{n+1} \subset U_n$.

Since $w_1 \in X \subset N(L_{n+1}, \varepsilon_{n+1})$, there exists a $y \in L_{n+1}$ such that $y \in B(w_1, \varepsilon_{n+1})$. Thus $g_n(y) = y_1^{(n)}$ and $y_1^{(n)} \in g_n(L_{n+1})$. Similarly, $y_2^{(n)} \in g_n(L_{n+1})$. Since $g_n(L_{n+1})$ is a connected subset of L_n, we conclude that $g_n(L_{n+1}) = L_n$. This completes the inductive construction.

Using the Anderson–Choquet Theorem (Theorem 14.7), we will prove that X is homeomorphic to $X_\infty = \lim_{\leftarrow}\{L_n, f_n\}_{n=1}^\infty$, where $f_n : L_{n+1} \to L_n$ is given by $f_n = g_n|_{L_{n+1}}$. Observe that $g_n|_{L_{n+1}}$ makes sense since $L_{n+1} \subset U_n$.

First, we check that Property (1) of Theorem 14.7 holds.

Let $\varepsilon > 0$, take $m \in \mathbb{N}$ such that $\frac{1}{3 \cdot 2^{m-1}} < \frac{\varepsilon}{3}$.

Take $p \in L_m$, $n > m$ and $q \in L_n$ such that $g_m^n(q) = p$. For each $i \in \{m, \ldots, n-2\}$, we have that $d(g_{i+1}^n(q), g_i^n(q)) = d(g_i(g_{i+1}^n(q), g_{i+1}^n(q))) < \varepsilon_i$. Then

$$\begin{aligned} d(q, p) &= d(q, g_m^n(q)) = d(q, (g_m \circ \cdots \circ g_{n-1})(q)) \\ &\leq d(q, g_{n-1}(q)) + d(g_{n-1}(q), (g_{n-2} \circ g_{n-1})(q))) + \cdots \\ &\quad + d((g_{m+1} \circ \cdots \circ g_{n-1})(q), (g_m \circ \cdots \circ g_{n-1})(q)) \\ &= d(q, g_{n-1}(q)) + d(g_{n-1}^n(q), g_{n-2}^n(q)) + \cdots \end{aligned}$$

$$
\begin{aligned}
&+ d(g^n_{m+1}(q), g^n_m(q)) \\
&< \varepsilon_{n-1} + \varepsilon_{n-2} + \ldots + \varepsilon_m \\
&< \frac{1}{3 \cdot 2^{n-1}} + \frac{1}{3 \cdot 2^{n-2}} + \ldots + \frac{1}{3 \cdot 2^m} \\
&< \frac{1}{3 \cdot 2^{m-1}} < \frac{\varepsilon}{3}
\end{aligned}
$$

Thus for each $q \in \bigcup\{(g^n_m)^{-1}(p) : m < n\}$, we have that $d(q, p) < \frac{\varepsilon}{3}$. So the diameter of this set is less than ε.

Therefore $\{L_n, f_n\}_{n=1}^{\infty}$ satisfies Property (1) of Theorem 14.7.

So, we have that for each $x = (x_1, x_2, \ldots) \in X_\infty$, the limit $\lim_{n\to\infty} x_n$ exists, and if we define $h(x) = \lim_{n\to\infty} x_n$, then h is well defined and continuous. Moreover, the image of h is the set $Y = \bigcap_{n=1}^{\infty}(\mathrm{cl}_Z(\bigcup_{m\geq n} L_m))$.

Now we check that $X = Y$.

Given $n \in \mathbb{N}$, $L_n \subset N(X, \varepsilon_n)$, and since for each $m > n$, $\varepsilon_m < \varepsilon_n$, we have that $\bigcup_{m\geq n} L_m \subset N(X, \varepsilon_n)$. Then $Y \subset \mathrm{cl}_Z(\bigcup_{m\geq n} L_m) \subset N(X, 2\varepsilon_n)$. Since $\lim_{n\to\infty} \varepsilon_n = 0$, we conclude that $Y \subset X$.

In order to see the other inclusion, take $x \in X$. Since for each $n \in \mathbb{N}$, $x \in N(L_n, \varepsilon_n)$, there exists an $x_n \in L_n$ such that $d(x, x_n) < \varepsilon_n$. Since $\lim_{n\to\infty} \varepsilon_n = 0$, we have that $\lim_{n\to\infty} x_n = x$.

Given $n \in \mathbb{N}$, since $x \in \mathrm{cl}_Z(\{x_m : m \geq n\}) \subset \mathrm{cl}_Z(\bigcup_{m\geq n} L_m)$, we conclude that $x \in Y$.

This shows that $X \subset Y$. Therefore $X = Y$.

Finally, we check that h is one-to-one. Take $x, y \in X_\infty$ such that $x \neq y$. Fix $m \in \mathbb{N}$ such that $x_m \neq y_m$.

Given $n > m + 1$, since $d(x_{n+1}, g_n(x_{n+1})) < \varepsilon_n$, we have $d(x_n, x_{n+1}) < \varepsilon_n$, $d(x_{n+1}, x_{n+2}) < \varepsilon_{n+1} < \frac{\varepsilon_n}{2}$, and in general, for every $k \in \mathbb{N}$, $d(x_{n+k}, x_{n+k+1}) < \varepsilon_{n+k} < \frac{\varepsilon_n}{2^k}$.

By the triangle inequality, $d(x_n, x_{n+k+1}) < \varepsilon_n + \frac{\varepsilon_n}{2} + \cdots + \frac{\varepsilon_n}{2^k} < 2\varepsilon_n$. Taking the limit as k tends to ∞, we have that $d(x_n, h(x)) \leq 2\varepsilon_n$.

Given $n > m$, $d(x_{n+1}, h(x)) \leq 2\varepsilon_{n+1}$. Since $x_{n+1} \in L_{n+1} \subset U_n$ and $h(x) \in Y = X \subset U_n$, we have that $d(g^{n+1}_m(x_n), g^{n+1}_m(h(x))) < \frac{1}{3\cdot 2^n}$. That is, $d(x_m, g^{n+1}_m(h(x))) \leq \frac{1}{3\cdot 2^n}$. Similarly, $d(y_m, g^{n+1}_m(h(y))) \leq \frac{1}{3\cdot 2^n}$.

Fix $n > m$ such that $d(x_m, y_m) > \frac{2}{3\cdot 2^n}$. By the triangle inequality, $g^{n+1}_m(h(x)) \neq g^{n+1}_m(h(y))$. Thus $h(x) \neq h(y)$. Therefore h is one-to-one.

This finishes the proof that h is a homeomorphism between X_∞ and X.

(b) $\Rightarrow$ (c). Suppose that $X = X_\infty = \lim_{\leftarrow}\{[0, 1]_n, f_n\}_{n=1}^{\infty} \subset \mathbf{Q}$. We consider $\mathbf{Q}$ with the metric d defined by Eq. (1.1). For each $n \in \mathbb{N}$, let $\pi_n : X_\infty \to [0, 1]_n$ be the nth-projection. Given $x = (x_1, x_2, \ldots)$, $y = (y_1, y_2, \ldots) \in X_\infty$ such that $x_n = \pi_n(x) = \pi_n(y) = y_n$, we have that $(x_1, \ldots, x_n) = (y_1, \ldots, y_n)$, so

$$
d(x, y) = \sum_{m=1}^{\infty} \frac{|x_m - y_m|}{2^m} = \sum_{m=n+1}^{\infty} \frac{|x_m - y_m|}{2^m} \leq \sum_{m=n+1}^{\infty} \frac{1}{2^m} = \frac{1}{2^n}.
$$

This implies that for each $t \in [0,1]_n$, diameter$(\pi_n^{-1}(t)) \leq \frac{1}{2^n}$. Since for the definition of X_∞, we ask that f_n is surjective, it follows that π_n is surjective. Therefore X satisfies (c).

(c) $\Rightarrow$ (a). Let $\varepsilon > 0$. We are supposing that there exists an onto mapping $f : X \to [0,1]$ such that for each $t \in [0,1]$, diameter$(f^{-1}(t)) < \varepsilon$. By Exercise 1.53, there exists a $\delta > 0$ such that if $J \subset [0,1]$ and diameter$(J) < \delta$, then diameter$(f^{-1}(J)) < \varepsilon$. Let $m \in \mathbb{N} \setminus \{1\}$ be such that $\frac{1}{m} < \frac{\delta}{2}$. For each $i \in \{1,\ldots,m\}$, let $J_i = (\frac{i-1}{m} - \frac{1}{3m}, \frac{i}{m} + \frac{1}{3m}) \cap [0,1]$ and $E_i = f^{-1}(J_i)$. Then E_i is open in X, diameter$(J_i) < \delta$ and diameter$(E_i) < \varepsilon$. Since $\mathrm{cl}_X(E_i) \subset \mathrm{cl}_X(f^{-1}(J_i))$, and $\mathcal{J} = \{J_1,\ldots,J_m\}$ is a taut chain of $[0,1]$, we conclude that $\mathcal{E} = \{E_1,\ldots,E_m\}$ is a taut chain of X. □

14.5 Generalized Inverse Limits

Let $\{X_n\}_{n=1}^\infty$ be a sequence of continua, $\{M_n\}_{n=1}^\infty$ a sequence of compact sets such that for each $n \in \mathbb{N}$, $M_n \subset X_{n+1} \times X_n$, and let $X = \prod_{n=1}^\infty X_n$. The *Generalized Inverse Limit* of the sequence $\{X_n, M_n\}_{n=1}^\infty$ is defined as

$$\varprojlim\{X_n, M_n\}_{n=1}^\infty = \{\{x_n\}_{n=1}^\infty \in (x_{n+1}, x_n) \in M_n \text{ for every } n \in \mathbb{N}\}.$$

Generalized inverse limits (GILs) were introduced by W.S. Mahavier in (2004) [101]. In Mahavier's presentation of these inverse limits in the Spring Topology and Dynamics Conference, someone in the audience said "Mahavier has just given us a new toy to play with". This metaphor turned out to be prophetic. Since 2004, a large amount of research and applications on generalized inverse limits has been developed.

There is a big difference between the variety of spaces we can get with inverse limits (ILs) and GILs. For example, in the specific case where all the continua X_n are arcs, we know that ILs are chainable continua (Theorem 14.12), so they are hereditarily unicoherent (Exercise 12.29) and they have the fpp. On the other hand, in the case where all continua X_n are arcs, a very large variety of compact sets can be obtained as GILs. A very small sample of this is given in Exercise 14.47. In fact, for some time, it was an open problem to determine if all continua can be obtained as GILs with each continuum X_n being an arc. This problem was solved by S. Greenwood and R. Suabedissen [43], who proved that the only connected 2-manifold that can be obtained as a GIL of arcs is the torus. The problem of determining what continua can be obtained when each X_n is an arc and all M_n are the same set M has also been widely studied. For example, recently, I. Banič, E. Goran and J.A. Kennedy [7] showed that the Lelek fan can be obtained as this kind of GIL by taking a particular arc M.

Another important line of research in this area is to establish what results on ILs can be extended to GILs. The many results that cannot be extended have open opportunities for study and discovery. In this direction, there is a long and interesting list of problems at the end of the book [73].

Very recently a group of experts on GILs (I. Banič, G. Erceg, J.A. Kennedy, C. Mouron, V.C. Nall) have joined efforts and have announced how to use GILs to study the dynamics of mappings on some continua, particularly on fans.

Even though GILs have been widely developed over the last 20 years, they still offer many opportunities for further research. The book by W.T. Ingram [73] offers a quick introduction and many open problems on this subject (see also [73]).

14.6 Exercises

In these exercises we use the notation introduced in this chapter.

Exercise 14.13 Show that for each $m \in \mathbb{N}$, $\pi_m = f_m \circ \pi_{m+1}$.

Exercise 14.14 Show that a base for the topology on X_∞ is given by the family

$$\mathcal{B} = \{\pi_m^{-1}(U_m) \subset X_\infty : m \in \mathbb{N} \text{ and } U_m \text{ is open in } X_m\}.$$

Exercise 14.15 Show that every metric space can be expressed as an inverse limit.

Exercise 14.16 Given an inverse sequence $\{X_n, f_n\}_{n=1}^\infty$ and given $m \in \mathbb{N}$, define $Z_m = \{\{x_n\}_{n=1}^\infty \in \prod_{n=1}^\infty X_n : f_n(x_{n+1}) = x_n \text{ for every } n \leq m\}$. Show that the following properties hold.

(a) For each $m \in \mathbb{N}$, $Z_{m+1} \subset Z_m$,
(b) For each $m \in \mathbb{N}$, Z_m is homeomorphic to $X_{m+1} \times X_{m+2} \times \cdots$, and
(c) $\lim_{\leftarrow}\{X_n, f_n\}_{n=1}^\infty = \bigcap_{m=1}^\infty Z_m$.

Exercise 14.17 Prove that an inverse limit of nonempty compact spaces is nonempty and compact. Also show that the inverse limit of continua is a continuum.

Exercise 14.18 $\{X_n, f_n\}_{n=1}^\infty$ and $\{Y_n, g_n\}_{n=1}^\infty$ be inverse sequences. Suppose that for each $n \in \mathbb{N}$, there exists a mapping $\varphi_n : X_n \to Y_n$ such that each diagram of the form:

$$\begin{array}{ccc} X_n & \xleftarrow[f_n]{} & X_{n+1} \\ \downarrow{\scriptstyle \varphi_n} & \circlearrowleft & \downarrow{\scriptstyle \varphi_{n+1}} \\ Y_n & \xleftarrow[g_n]{} & Y_{n+1} \end{array}$$

is commutative.

Define $\varphi : X_\infty \to Y_\infty$ by $\varphi((x_1, x_2, \ldots)) = (\varphi_1(x_1), \varphi_2(x_2), \ldots)$.
Prove that:

(a) φ is well defined and continuous,
(b) if each φ_n is one-to-one, then φ is one-to-one,
(c) if each φ_n is surjective and each X_n is compact, then φ is surjective, and
(d) if each φ_n is a homeomorphism, then φ is a homeomorphism.

Exercise 14.19 Suppose that a sequence of metric spaces $\{X_n\}_{n=1}^{\infty}$ satisfies $X_1 \supset X_2 \supset X_3 \supset \cdots$, then $\bigcap_{n=1}^{\infty} X_n$ is homeomorphic to $\lim_{\leftarrow}\{X_n, f_n\}_{n=1}^{\infty}$, where each f_n is the natural embedding.

Exercise 14.20 Let $\{X_n\}_{n=1}^{\infty}$ be a sequence of compact metric spaces. Find a way to express $X = \prod_{n=1}^{\infty} X_n$ as an inverse limit of the spaces $X_1 \times \cdots \times X_n$.

Exercise 14.21 Use Exercise 14.15 to express the continuum $\sin(\frac{1}{x})$ as an inverse limit in a very easy way. Also use the Anderson–Choquet Theorem to express the continuum $\sin(\frac{1}{x})$ as an inverse limit of arcs.

Exercise 14.22 Show that an inverse limit can be empty, but if each bonding mapping is surjective, then the inverse limit is nonempty.

Exercise 14.23 Let $\{X_n, f_n\}_{n=1}^{\infty}$ be an inverse sequence and $n_1 < n_2 < \cdots$ an increasing sequence in $\mathbb{N}$. Prove that $\lim_{\leftarrow}\{X_n, f_n\}_{n=1}^{\infty}$ is homeomorphic to $\lim_{\leftarrow}\{X_{n_k}, f_{n_k}^{n_{k+1}}\}_{k=1}^{\infty}$.

Exercise 14.24 Prove that if A, B are closed subsets of X_∞, then $A \subset B$ if and only if for each $n \in \mathbb{N}$, $\pi_n(A) \subset \pi_n(B)$. Also prove that if A is a closed proper subset of X_∞, then there exists an $n \in \mathbb{N}$ such that $\pi_n(A)$ is a proper subset of X_n.

Exercise 14.25 Using the Anderson–Choquet Theorem and Exercise 14.19, prove that the dyadic solenoid defined in Example 14.3 is homeomorphic to the continuum constructed in Example 1.11.

In Exercises 14.26–14.29, m denotes a positive integer with $m \geq 2$ and S_m denotes the m-solenoid as defined in Example 14.3.

Exercise 14.26 Prove that S_m is a topological group. That means, it is possible to define a binary group operation continuous from $S_m \times S_m$ into S_m such that the function of taking the inverse from S_m into S_m is also continuous.

Exercise 14.27 Prove that every non-degenerate proper subcontinuum of the m-solenoid S_m is an arc. Given $z_1 \in S^1$, prove that the set $\{(x_1, x_2, \ldots) \in S_m : x_1 = z_1\}$ is homeomorphic to $\prod_{n=1}^{\infty}\{1, 2, \ldots, m\}_n$, where each space of the form $\{1, 2, \ldots, m\}_n$ is considered with the discrete topology. By Exercise 8.22, this space is homeomorphic to the Cantor set.

Exercise 14.28 Prove that S_m has a base of open subsets such that each of its elements is homeomorphic to the space $C \times (0, 1)$, where C denotes the Cantor set.

Exercise 14.29 Prove that there exists a one-to-one mapping $f : \mathbb{R} \to S_m$ such that the sets $f((-\infty, 0])$ and $f([0, \infty))$ are dense in S_m.

Exercise 14.30 Prove that an open mapping between continua is surjective. Suppose that if $f : [0, 1] \to [0, 1]$ is an open mapping, then there exists a partition $P : 0 = t_1 < t_2 < \cdots < t_k = 1$ of $[0, 1]$ such that for each $i \in \{1, \ldots, k\}$, $f|_{[t_{i-1}, t_i]} : [t_{i-1}, t_i] \to [0, 1]$ is a homeomorphism.

Exercise 14.31 The *tent mapping* is the mapping $f : [0, 1] \to [0, 1]$ defined by:

$$f(t) = \begin{cases} 2t, & \text{if } 0 \leq t \leq \frac{1}{2}, \\ 2 - 2t, & \text{if } \frac{1}{2} \leq t \leq 1. \end{cases}$$

Using the Anderson–Choquet theorem, give a geometric argument to show that the continuum $X_\infty = \lim_{\leftarrow}\{X_n, f_n\}_{n=1}^{\infty}$ is homeomorphic to the buckethandle continuum defined in Example 1.8.

Exercise 14.32 Prove that if $\{X_n, f_n\}_{n=1}^{\infty}$ is an indecomposable inverse sequence, then each f_n is surjective and each projection $\pi_n : X_\infty \to X_n$ is also surjective.

Exercise 14.33 Show that the inverse sequences that define the m-solenoids and the Knaster-type continua are indecomposable.

Exercise 14.34 Prove that an inverse limit of chainable continua is chainable.

Exercise 14.35 Let A and B be compact subsets of X_∞, and let $C = A \cap B$. For each $n \in \mathbb{N}$, let $C_n = \pi_n(A) \cap \pi_n(B)$. Prove that $C = \lim_{\leftarrow}\{C_n, f_n|C_{n+1}\}_{n=1}^{\infty}$.

Exercise 14.36 Prove that an inverse limit of unicoherent continua with surjective bonding mappings is a unicoherent continuum.

Exercise 14.37 Prove that an inverse limit of indecomposable continua with surjective bonding mappings is an indecomposable continuum.

Exercise 14.38 Let $\{X_n, f_n\}_{n=1}^{\infty}$ be an inverse sequence, where each X_n is a continuum and each f_n is surjective. Show that each f_n is monotone if and only if each projection from X_∞ onto X_n is monotone.

Exercise 14.39 Prove that an inverse limit of Peano continua with monotone bonding mappings is a Peano continuum.

Exercise 14.40 Prove that an inverse limit of dendrites with monotone bonding mappings is a dendrite.

Exercise 14.41 Show that an inverse limit of arcs with onto monotone bonding mappings is an arc. (Hint: use Exercise 5.16.)

Exercise 14.42 Show that an inverse limit of simple closed curves with monotone bonding mappings is a simple closed curve. (Hint: Use Exercise 5.8.)

Exercise 14.43 Prove that solenoids are not chainable. (Hint: use Exercises 12.29 and 14.25.)

Exercise 14.44 Prove that the product of two inverse limits is the inverse limit of the products.

Exercise 14.45 Let $\{X_n, f_n\}_{n=1}^{\infty}$ be an inverse sequence, where each X_n is a continuum. Prove that each f_n is an open mapping if and only if each projection from X_∞ onto X_n is an open mapping.

Exercise 14.46 Let Y be a continuum and $\{X_n, f_n\}$ be an inverse sequence of continua. Suppose that for each $n \in \mathbb{N}$ there is a mapping $g_n : Y \to X_n$ such that $g_n = f_n \circ g_{n+1}$. Then it is possible to define a natural mapping from Y in $X = \lim_{\leftarrow}\{X_n, f_n\}$. Prove that this mapping is one-to-one if some g_n is one-to-one.

Exercise 14.47 Suppose that M is a compact subset of $[0, 1] \times [0, 1]$. For each $n \in \mathbb{N}$, let $X_n = [0, 1]$ and $M_n = M$. Denote the generalized inverse limit $\lim_{\leftarrow}\{X_n, M_n\}_{n=1}^{\infty}$ by M_∞. Prove the following:

(a) if $M = [0, 1] \times \{\frac{1}{2}\}$, then $M_\infty = \{(\frac{1}{2}, \frac{1}{2}, \ldots)\}$,
(b) if $M = \{(x, x) : x \in [0, 1]\}$, then M_∞ is an arc,
(c) if $M = [0, 1] \times \{0, 1\}$, then M_∞ is the Cantor set,
(d) if $M = [0, 1] \times [0, 1]$, then M_∞ is the Hilbert cube, and
(e) if $M = \{(x, x) : x \in [0, 1]\} \cup \{(x, 1 - x) : x \in [0, 1]\}$, then M_∞ is the cone over the Cantor set.

Exercise 14.48 As usual we denote by $\mathbf{Q}$ the Hilbert cube and by d its metric, as defined by Eq. (1.1). Let $W_1, \ldots, W_m$ be nonempty open disjoint subsets of $\mathbf{Q}$. Then there exist points $p_1, \ldots, p_m$ such that for each $i \in \{1, \ldots, m\}$, $p_i \in W_i$ and the union of the convex arcs $p_1p_2, p_2p_3, \ldots, p_{m-1}p_m$ is an arc joining p_1 and p_m.

Chapter 15
Homogeneity of the Hilbert Cube

15.1 Introduction

Definition 15.1 A topological space X is *homogeneous* if for every $p, q \in X$, there exists a homeomorphism $h : X \to X$ such that $h(p) = q$.

When we observe the n-cell $D = [0, 1]^n \subset \mathbb{R}^n$, we intuit that there are two kinds of points: those in its boundary $\operatorname{Fr}(D)$ and those in the interior $\operatorname{int}(D)$. It is relatively easy to see that if $p, q \in \operatorname{Fr}(D)$, then there exists a homeomorphism $h : D \to D$ such that $h(p) = q$. The same happens if we take $p, q \in \operatorname{int}(D)$. However, the Invariance of Domain Theorem [50, Theorem VI 9, §6, Chapter 6, p. 95] implies that if $p \in \operatorname{Fr}(D)$ and $q \in \operatorname{int}(D)$, then there is no such homeomorphism. Thus D is not homogeneous. We know that the points in $\operatorname{Fr}(D)$ are those points in D with some coordinate in $\{0, 1\}$. So we can define the pseudo-boundary of the Hilbert cube $\mathbf{Q} = [0, 1]^\infty$ as the set of those points in $\mathbf{Q}$ with some coordinate in $\{0, 1\}$, and the question is if we can find a homeomorphism from $\mathbf{Q}$ onto $\mathbf{Q}$ that takes a point in the pseudo-boundary to one not in the pseudo-boundary. The surprising fact is that, indeed, it is possible to find such a homeomorphism, and then we can prove that $\mathbf{Q}$ is homogeneous. This result was first proved by O.H. Keller in 1931 [79]. In this chapter we present the standard nice proof of the homogeneity of $\mathbf{Q}$.

15.2 The Proof

Definition 15.2 The *pseudo-interior* $\mathbf{P}$ of $\mathbf{Q}$ is defined by

$$\mathbf{P} = (0, 1)^\infty.$$

A. Illanes, *Continuum Theory*, Universitext,
https://doi.org/10.1007/978-3-031-91011-1_15

The *pseudo-boundary* $\mathbf{B}$ of $\mathbf{Q}$ is

$$\mathbf{B} = \mathbf{Q} \setminus \mathbf{P}.$$

Given a compact metric space X, the space of homeomorphisms of X is denoted by $\mathcal{H}(X)$. Given $f, g \in \mathcal{H}(X)$, let

$$\rho(f, g) = \sup\{d(f(x), g(x)) : x \in X\}, \text{ and } D(f, g) = \rho(f, g) + \rho(f^{-1}, g^{-1}).$$

Theorem 15.3 *The metrics ρ and D are equivalent for the space $\mathcal{H}(X)$.*

Proof By Exercise 15.12, D is a metric for the space $\mathcal{H}(X)$. Since for every $f, g \in \mathcal{H}(X)$, $\rho(f, g) \leq D(f, g)$, we have that the identity function

$$id : (\mathcal{H}(X), D) \to (\mathcal{H}(X), \rho)$$

is continuous.

Now, suppose that $\{f_n\}_{n=1}^{\infty}$ is a sequence in $\mathcal{H}(X)$ converging to an element $f \in \mathcal{H}(X)$, in the metric ρ. Since $\rho(k, g) = \rho(k \circ h, g \circ h)$ for every $k, g, h \in \mathcal{H}(X)$ (Exercise 15.12), we have that the sequence $\{f_n \circ f^{-1}\}_{n=1}^{\infty}$ converges, in the metric ρ, to the identity function $id : X \to X$.

By Exercise 15.12, the inequality $\rho(h, id) < \varepsilon$ implies that $\rho(h^{-1}, id) < \varepsilon$, for every $h \in \mathcal{H}(X)$. Therefore the sequence $\{f \circ f_n^{-1}\}_{n=1}^{\infty}$ also converges to the identity function in the metric ρ.

We are going to show that $\{f_n^{-1}\}_{n=1}^{\infty}$ converges to f^{-1} in the metric ρ. In this way we will have that $\{f_n\}_{n=1}^{\infty}$ converges to f in the metric D.

Let $\varepsilon > 0$. Since f^{-1} is uniformly continuous, there exists a $\delta > 0$ such that if $p, q \in X$ and $d(p, q) < \delta$, then $d(f^{-1}(p), f^{-1}(q)) < \frac{\varepsilon}{2}$. Since $f \circ f_n^{-1}$ converges to the identity function in the metric ρ, there exists an $N \in \mathbb{N}$ such that $\rho(f \circ f_n^{-1}, id) < \delta$ for every $n \geq N$.

Given $n \geq N$, we have that for each $p \in X$, $d((f \circ f_n^{-1})(p), p) < \delta$, then $d(f_n^{-1}(p), f^{-1}(p)) = d(f^{-1}((f \circ f_n^{-1})(p)), f^{-1}(p)) < \frac{\varepsilon}{2}$. Therefore $\rho(f_n^{-1}, f^{-1}) < \varepsilon$ for every $n \geq N$.

This proves that $\{f_n^{-1}\}_{n=1}^{\infty}$ converges to f^{-1} in the metric ρ.

We have shown that the identity function $id : (\mathcal{H}(X), \rho) \to (\mathcal{H}(X), D)$ is continuous. Therefore the identity function is a homeomorphism, and the topologies induced in $\mathcal{H}(X)$ by ρ and D coincide. □

Theorem 15.4 *The metric D is complete.*

Proof Let $\{f_n\}_{n=1}^{\infty}$ be a Cauchy sequence in $\mathcal{H}(X)$ with the metric D.

From the definition of D, we have that $\{f_n\}_{n=1}^{\infty}$ and $\{f_n^{-1}\}_{n=1}^{\infty}$ are Cauchy sequences with the metric ρ. Then for each $p \in X$, we have that $\{f_n(p)\}_{n=1}^{\infty}$ and $\{f_n^{-1}(p)\}_{n=1}^{\infty}$ are Cauchy sequences in X. Since X is compact, for each $p \in X$, we can define $f(p) = \lim_{n\to\infty} f_n(p)$ and $F(p) = \lim_{n\to\infty} f_n^{-1}(p)$.

We check that f is continuous. Let $p \in X$ and $\varepsilon > 0$. Since $\{f_n\}_{n=1}^{\infty}$ is a Cauchy sequence with ρ, there exists an $N \in \mathbb{N}$ such that $\rho(f_n, f_m) < \frac{\varepsilon}{4}$ for every $n, m \geq N$. Since $f(p) = \lim_{n\to\infty} f_n(p)$, we may suppose that $d(f(p), f_n(p)) < \frac{\varepsilon}{4}$ for every $n \geq N$. Then for every $n, m \geq N$, $\rho(f_n, f_m) < \frac{\varepsilon}{4}$ and $d(f(p), f_n(p)) < \frac{\varepsilon}{4}$. By the continuity of f_N, there exists a $\delta > 0$ such that if $d(p, q) < \delta$, then $d(f_N(p), f_N(q)) < \frac{\varepsilon}{4}$.

Thus, if $n \geq N$ and $d(p, q) < \delta$, we have that

$$d(f(p), f_n(q)) \leq d(f(p), f_N(p)) + d(f_N(p), f_N(q)) + d(f_N(q), f_n(q)) \leq \frac{3\varepsilon}{4}.$$

Taking the limit as n tends to ∞, we have that if $d(p, q) < \delta$, then $d(f(p), f(q)) \leq \frac{3\varepsilon}{4} < \varepsilon$. Therefore f is continuous.

In a similar way, it can be shown that F is continuous.

Now, we check that $f \circ F = F \circ f = id$. Let $p \in X$ and $\varepsilon > 0$. Since $\{f_n\}_{n=1}^{\infty}$ is a Cauchy sequence with the metric ρ, there exists an $N_1 \in \mathbb{N}$ such that $\rho(f_n, f_m) < \frac{\varepsilon}{4}$ for every $n, m \geq N_1$. Since $f(F(p)) = \lim_{n\to\infty} f_n(F(p))$, we may suppose that $d(f_n(F(p)), f(F(p))) < \frac{\varepsilon}{4}$ for every $n \geq N_1$. Since f_{N_1} is continuous, there exists a $\delta > 0$ such that if $q \in X$ and $d(p, q) < \delta$, then $d(f_{N_1}(p), f_{N_1}(q)) < \frac{\varepsilon}{4}$. Since $F(p) = \lim_{n\to\infty} f_n^{-1}(p)$, there exists an $N \geq N_1$ such that for every $n \geq N$, $d(f_n^{-1}(p), F(p)) < \delta$. Thus, if $n \geq N$, we have that $d(f_n(f_n^{-1}(p)), f_{N_1}(f_n^{-1}(p))) \leq \rho(f_n, f_{N_1}) < \frac{\varepsilon}{4}$, $d(f_{N_1}(f_n^{-1}(p)), f_{N_1}(F(p))) < \frac{\varepsilon}{4}$ and $d(f_{N_1}(F(p)), f(F(p))) < \frac{\varepsilon}{4}$. Then for every $n \geq N$, we have that

$$\begin{aligned} d(p, f(F(p))) &\leq d(f_n(f_n^{-1}(p)), f_{N_1}(f_n^{-1}(p))) + d(f_{N_1}(f_n^{-1}(p)), f_{N_1}(F(p))) \\ &\quad + d(f_{N_1}(F(p)), f(F(p))) \\ &< \varepsilon. \end{aligned}$$

Since ε is arbitrary, we conclude that $p = f(F(p))$. Similarly, $p = F(f(p))$. Therefore $f, F \in \mathcal{H}(X)$.

Finally, we show that $\lim_{n\to\infty} \rho(f_n, f) = 0 = \lim_{n\to\infty} \rho(f_n^{-1}, F)$.

Let $\varepsilon > 0$. Then there exists an $M \in \mathbb{N}$ such that $\rho(f_M, f_n) < \frac{\varepsilon}{2}$ for every $n \geq M$. Thus for every $p \in X$ and $n \geq M$, we have that $d(f_M(p), f(p)) \leq \frac{\varepsilon}{2}$ and $d(f_M(p), f_n(p)) < \frac{\varepsilon}{2}$. This implies that $\rho(f, f_n) \leq \varepsilon$ for every $n \geq M$. Therefore $\lim_{n\to\infty} \rho(f_n, f) = 0$.

Analogously, it can be shown that $\lim_{n\to\infty} \rho(f_n^{-1}, F) = 0$. This shows that $\lim_{n\to\infty} D(f_n, f) = 0$. Therefore D is a complete metric space. □

Lemma 15.5 *Let $f \in \mathcal{H}(X)$ and $\varepsilon > 0$. Then there exists a $\delta > 0$ such that if $g \in \mathcal{H}(X)$ and $\rho(g, id) < \delta$, then $D(g \circ f, f) < \varepsilon$.*

Proof Since f^{-1} is uniformly continuous, there exists a $\gamma > 0$ such that if $d(p, q) < \gamma$, then $d(f^{-1}(p), f^{-1}(q)) < \frac{\varepsilon}{2}$. Let $\delta < \min\{\gamma, \frac{\varepsilon}{2}\}$. If $g \in \mathcal{H}(X)$ and $\rho(g, id) < \delta$, by Exercise 15.12, we have that $\rho(g^{-1}, id) < \delta$. So for each

$p \in X$, we have that $d(f^{-1}(g^{-1}(p)), f^{-1}(p)) < \frac{\varepsilon}{2}$. Therefore

$$D(g \circ f, f) = \rho(g \circ f, f) + \rho(f^{-1} \circ g^{-1}, f^{-1}) \leq \rho(g, id) + \frac{\varepsilon}{2} < \varepsilon.$$

□

Theorem 15.6 *Let $a = (a_1, a_2, \ldots) \in \mathbf{Q}$, $\delta > 0$ and $n \in \mathbb{N}$. Then there is an $h = (h_1, h_2, \ldots) \in \mathcal{H}(\mathbf{Q})$ such that $\rho(h, id) < \delta$, for each $i \in \{1, \ldots, n-1\}$, h_i coincides with the ith-projection from $\mathbf{Q}$ onto $[0, 1]$ and $h_n(a) \in (0, 1)$.*

Proof If $a_n \in (0, 1)$, then we simply take h as the identity in $\mathbf{Q}$.

We show how to find h when $a_n = 1$. The case $a_n = 0$ is similar.

Fix a number $k > n$ such that $\frac{1}{2^k} < \frac{\delta}{2}$. By Exercise 15.14, there exists a homeomorphism $f = (f_1, f_2) : [0, 1]^2 \to [0, 1]^2$ such that:

(1) $f|_{[0,1-\frac{\delta}{2}]\times[0,1]}$ is the identity, and
(2) $f(\{1\} \times [0, 1]) \subset (1 - \frac{\delta}{2}, 1) \times \{1\}$.

For each $m \in \mathbb{N}$, let $\pi_m : \mathbf{Q} \to [0, 1]$ be the mth-projection. Define

$$h = (h_1, h_2, \ldots) : \mathbf{Q} \to \mathbf{Q}$$

by

(a) $h_m = \pi_m$ for every $m \in \mathbb{N} \setminus \{n, k\}$,
(b) $h_n(x_1, x_2, \ldots) = f_1(x_n, x_k)$ and $h_k(x_1, x_2, \ldots) = f_2(x_n, x_k)$.

It is easy to see that h is a homeomorphism.

Since $a_n = 1$, we have that $(a_n, a_k) \in \{1\}\times[0, 1]$, so $f(a_n, a_k) \in (1-\frac{\delta}{2}, 1)\times\{1\}$ and $h_n(a) = f_1(a_n, a_k) \in (1 - \frac{\delta}{2}, 1)$.

Finally, we check that $\rho(h, id) < \delta$. Let $x = (x_1, x_2, \ldots) \in \mathbf{Q}$ and $y = (y_1, y_2, \ldots) = h(x)$. Since f satisfies (1), we have, for $x_n \leq 1 - \frac{\delta}{2}$, that $f(x_n, x_k) = (x_n, x_k)$. Thus $y_n = h_n(x) = f_1(x_n, x_k) = x_n$; and if $x_n > 1 - \frac{\delta}{2}$, $y_n = f_1(x_n, x_k) > 1 - \frac{\delta}{2}$ (since f is one-to-one). Hence $x_n, y_n \in (1 - \frac{\delta}{2}, 1]$. Therefore

$$d((x_1, x_2, \ldots), (y_1, y_2, \ldots)) = \sum_{i=1}^{\infty} \frac{|x_i - y_i|}{2^i} = \frac{|x_n - y_n|}{2^n} + \frac{|x_k - y_k|}{2^k} < \frac{\delta}{2} + \frac{1}{2^k} < \delta.$$

We have shown that $\rho(h, id) < \delta$. □

Theorem 15.7 *If $p \in \mathbf{Q}$, then there exists a homeomorphism $h : \mathbf{Q} \to \mathbf{Q}$ such that $h(p) \in \mathbf{P}$.*

Proof We will define, inductively, a sequence of homeomorphisms $h_1, h_2, \ldots$, from $\mathbf{Q}$ onto $\mathbf{Q}$, satisfying the following properties for each $n \in \mathbb{N}$.

(a) the first n coordinates of $(h_n \circ h_{n-1} \circ \cdots \circ h_1)(p)$ belong to $(0, 1)$,
(b) for every $q \in \mathbf{Q}$, the points q and $h_n(q)$ coincide in the first $n - 1$ coordinates.
(c) $D(h_n \circ \cdots \circ h_1, h_{n-1} \circ \cdots \circ h_1) < \frac{1}{2^n}$.

Construction of h_1.

By Lemma 15.5, there is a $\delta > 0$ such that if $g \in \mathcal{H}(\mathbf{Q})$ and $\rho(g, id) < \delta$, then $D(g, id) < \varepsilon$. By Theorem 15.6, applied to p, δ and $n = 1$, there exists an $h_1 \in \mathcal{H}(\mathbf{Q})$ such that the first coordinate of $h_1(p)$ belongs to $(0, 1)$.

Suppose, inductively, that we have defined homeomorphisms $h_1, h_2, \ldots, h_k$ satisfying (a), (b) and (c).

In order to construct h_{k+1}, take $\delta > 0$ as in Lemma 15.5, for the homeomorphism $h_k \circ \cdots \circ h_1$ and $\varepsilon = \frac{1}{2^{k+1}}$. Then we have the following property: if $f \in \mathcal{H}(\mathbf{Q})$ and $\rho(f, id) < \delta$, then $D(f \circ h_k \circ \cdots \circ h_1, h_k \circ \cdots \circ h_1) < \frac{1}{2^{k+1}}$. Let $x = (h_k \circ \cdots \circ h_1)(p)$. By Theorem 15.6, applied to the point x, δ and the positive integer $k+1$, there exists an $h_{k+1} \in \mathcal{H}(\mathbf{Q})$ such that $\rho(h_{k+1}, id) < \delta$ (and then $D(h_{k+1} \circ h_k \circ \cdots \circ h_1, h_k \circ \cdots \circ h_1) < \frac{1}{2^{k+1}}$), the $(k + 1)^{\text{th}}$-coordinate of $h_{k+1}(x)$ belongs to the interval $(0, 1)$ and for every $q \in \mathbf{Q}$, q and $h_{k+1}(q)$ coincide in the first k coordinates.

By the induction hypothesis, the first k coordinates of $x = (h_k \circ \cdots \circ h_1)(p)$ belong to $(0, 1)$ and since x and $h_{k+1}(x)$ coincide in the first k coordinates, we have that the first k coordinates of $h_{k+1}((h_k \circ \cdots \circ h_1)(p))$ belong to $(0, 1)$. Moreover, the $(k + 1)^{\text{th}}$-coordinate of $h_{k+1}(x)$ belongs to $(0, 1)$. Thus the first $k + 1$ coordinates of $(h_{k+1} \circ h_k \circ \cdots \circ h_1)(p)$ belong to $(0, 1)$.

We have shown that properties (a), (b) and (c) hold for $k + 1$. This completes the inductive construction.

By (c), the sequence $\{h_n \circ \cdots \circ h_1\}_{n=1}^{\infty}$ is a Cauchy sequence in $\mathcal{H}(\mathbf{Q})$, with the metric D, and since D is complete (Theorem 15.4), there exists the limit

$$h = \lim_{n \to \infty} (h_n \circ \cdots \circ h_1),$$

and h is a homeomorphism from $\mathbf{Q}$ onto $\mathbf{Q}$.

Let $w = (w_1, w_2, \ldots) = h(p)$. We claim that $w \in \mathbf{P}$. Given $n \in \mathbb{N}$, put $v = (v_1, v_2, \ldots) = (h_n \circ \cdots \circ h_1)(p)$. By (a), $v_n \in (0, 1)$. By (b), for each $k > n$, $(h_k \circ \cdots \circ h_1)(p)$ and $(h_n \circ \cdots \circ h_1)(p)$ coincide in the nth-coordinate. Thus for each $k > n$, the nth-coordinate of $(h_k \circ \cdots \circ h_1)(p)$ is equal to v_n. Hence $w_n = v_n \in (0, 1)$. Therefore $h(p) \in \mathbf{P}$. This finishes the proof of the theorem. □

Combining Theorem 15.7 and Exercise 15.11, we obtain the following theorem.

Theorem 15.8 *The Hilbert cube is homogeneous.*

15.3 Exercises

Exercise 15.9 Prove that the spaces S^1, (0, 1) and the solenoids are homogeneous, while [0, 1] and $[0, 1]^2$ are not homogeneous. (Hint: use that $[0, 1]^2 \setminus \{(\frac{1}{2}, \frac{1}{2})\}$ is not unicoherent.)

Exercise 15.10 Prove that products of homogeneous continua are homogeneous.

Exercise 15.11 Prove that if $p, q \in \mathbf{P}$, then there exists a homeomorphism $h : \mathbf{Q} \to \mathbf{Q}$ such that $h(p) = q$.

Exercise 15.12 Prove that D is a metric for $\mathcal{H}(X)$ and that for every $f, g, h \in \mathcal{H}(X)$, we have that $\rho(f \circ h, g \circ h) = \rho(f, g)$ and that $\rho(h, id) = \rho(h^{-1}, id)$.

Exercise 15.13 Prove that the space $(\mathcal{H}(\mathbf{Q}), \rho)$ is not complete. (Hint: first prove that the space of homeomorphisms from [0, 1] onto [0, 1] with the supremum metric is not complete.)

Exercise 15.14 For each $\delta > 0$, prove that there exists a homeomorphism $f : [0, 1]^2 \to [0, 1]^2$ such that

(1) $f|_{[0,1-\frac{\delta}{2}]\times[0,1]}$ is the identity, and
(2) $f(\{1\} \times [0, 1]) \subset (1 - \frac{\delta}{2}, 1) \times \{1\}$.

Chapter 16
Absolute Retracts

16.1 General Theory

Definition 16.1 Let X be a continuum. We say that:

(a) X is an *absolute retract* (*AR*) if X is a retract of every continuum containing X. This means that the following implication holds: if Z is a continuum and $h : X \to Z$ is an embedding, then $h(X)$ is a retract of Z.
(b) X is an *absolute extensor* (*AE*) if the following implication holds: if Z is a continuum, A is a closed subset of Z and $f : A \to X$ is a mapping, then there exists a mapping $F : Z \to X$ such that $F|_A = f$.
(c) X is an *absolute neighborhood retract* (*ANR*) if X is a retract of a neighborhood in each continuum containing X. This means that the following implication holds: if Z is a continuum and $h : X \to Z$ is an embedding, then there exists an open subset W of Z such that $h(X) \subset W$ and $h(X)$ is a retract of W.
(d) X is an *absolute neighborhood extensor* (*ANE*) if the following implication holds: if Z is a continuum, A is a closed subset of Z and $f : A \to X$ is a mapping, then there exists an open subset W of Z and a mapping $F : W \to X$ such that $A \subset W$ and $F|_A = f$.

Note that the Tietze Extension Theorem implies that the interval $[0, 1]$ is an AE. So we have the following.

Theorem 16.2 (Tietze Extension Theorem) $[0, 1]$ *is an AE.*

In the early 1930s K. Borsuk defined the concepts of absolute retract and absolute neighborhood retract ([14] and [15]). AR continua have strong topological properties, namely, they are locally contractible, locally connected, arcwise connected, unicoherent, contractible and have the fixed point property (Theorem 16.7). AR continua have played an important role in continuum theory, especially in the theory of hyperspaces. In 1938–39, M. Wojdysławski proved that a continuum X is locally connected if and only if 2^X and $C(X)$ are absolute retracts ([124] and [125]). He

A. Illanes, *Continuum Theory*, Universitext,
https://doi.org/10.1007/978-3-031-91011-1_16

mentioned that this reinforced the possibility of answering the following question in the positive: if X is any locally connected continuum, then is 2^X homeomorphic to the Hilbert cube? [124, p. 248]. For some time, this was one of the most important problems in hyperspaces and its solution produced one of the most significant results in this area. In 1974, R.M. Schori and J.E. West [29, 30] proved that a continuum X is locally connected if and only if 2^X is homeomorphic to the Hilbert cube, and they also showed that if X is a locally connected continuum such that each arc in X has empty interior, then $C(X)$ is homeomorphic to the Hilbert cube.

Theorem 16.3 *Every product of at most countably many absolute extensors is an absolute extensor.*

Proof Let $\{X_\alpha : \alpha \in J\}$ be a family of AEs, where J is at most countable. Let Z be a continuum, A be a closed subset of Z and $f : A \to \Pi_{\alpha\in J} X_\alpha$ be a mapping.

For each $\beta \in J$, let $\pi_\beta : \Pi_{\alpha\in J} X_\alpha \to X_\beta$ be the natural projection. Since $f_\beta = \pi_\beta \circ f : A \to X_\beta$ is a mapping and X_β is an AE, we have that there exists a mapping $F_\beta : Z \to X_\beta$ such that $F_\beta|_A = f_\beta$.

Define $F : Z \to \Pi_{\alpha\in J} X_\alpha$ by:

$$F(x) = \{F_\alpha(x)\}_{\alpha\in J}.$$

Note that F is continuous since each of its coordinate functions is continuous.

Given $a \in A$, we have that

$$F(a) = \{F_\alpha(a)\}_{\alpha\in J} = \{f_\alpha(a)\}_{\alpha\in J} = \{\pi_\alpha \circ f(a)\}_{\alpha\in J} = f(a).$$

Then $F|_A = f$. Therefore $\Pi_{\alpha\in J} X_\alpha$ is an AE. □

Corollary 16.4 *The Hilbert cube is an AE and for each $n \in \mathbb{N}$, each n-cell is an AE.*

Theorem 16.5 *Let X be a continuum. Then X is an AR if and only if X is an AE.*

Proof (Necessity.) Suppose that X is an AR. By Theorem 1.1, we may suppose that X is contained in the Hilbert cube, and X is a retract of $\mathbf{Q}$. By Exercise 16.14, X is an AE.

(Sufficiency.) Let $h : X \to Z$ be an embedding, where Z is a continuum. Consider the mapping $h^{-1} : h(X) \to X$. Since X is an AE, and $h(X)$ is closed in Z, there exists a mapping $g : Z \to X$ such that $g|_{h(X)} = h^{-1}$. Given $y \in h(X)$, there exists an $x \in X$ such that $y = h(x)$. Thus $g(y) = h^{-1}(h(x)) = x$. Then $h(g(y)) = h(x) = y$. Therefore $h \circ g$ is a retraction from Z onto $h(X)$. □

Definition 16.6 A continuum X is *locally contractible* at a point $p \in X$ if for each open subset U of X with $p \in U$, there exists an open subset V of X such that $p \in V \subset U$, there exists a mapping $G : V \times [0, 1] \to U$ and there exists a point $u \in U$ such that for every $v \in V$, $G(v, 0) = v$ and $G(v, 1) = u$. The continuum X is *locally contractible* if it is locally contractible at each of its points.

The following theorem is left as Exercise 16.18.

Theorem 16.7 *If a continuum X is an AR (equivalently, an AE), then X is locally contractible, locally connected, arcwise connected, unicoherent, contractible and has the fixed point property.*

Example 16.8 The *Hawaiian earring* is the continuum Y defined as the union of a sequence of simple closed curves $\{C_n\}_{n=1}^{\infty}$ that contains a special point p with the property that $C_n \cap C_m = \{p\}$ whenever $n \neq m$ and $\lim_{n\to\infty} \text{diameter}(C_n) = 0$. In Exercise 16.24, it is asked to show that the cone over the Hawaiian earring is a locally connected, arcwise connected, unicoherent, contractible continuum with the fixed point property, but it is not an AR.

Theorem 16.9 *Let $\{X_n : n \in \mathbb{N}\}$ be a family of continua. Then the space $X = \prod_{n=1}^{\infty} X_n$ is an ANR if and only if each X_n is an ANR and there exists an $m \in \mathbb{N}$ such that X_n is an AR for all $n \geq m$.*

Proof (Necessity) Suppose that X is an ANR. Given $m \in \mathbb{N}$, since X_m is homeomorphic to a slice of X and this slice is a retract of X, by Exercise 16.14, we conclude that X_m is an ANR.

By Theorem 1.1, we may assume that each X_n is contained in a Hilbert cube $\mathbf{Q}_n$. Then $X = \prod_{n=1}^{\infty} X_n$ is a compact subset of $\mathbf{Q}_\infty = \prod_{n=1}^{\infty} \mathbf{Q}_n$. Observe that $\mathbf{Q}_\infty$ is a Hilbert cube, so it is an AR. For each $n \in \mathbb{N}$, let $\pi_n : \mathbf{Q}_\infty \to \mathbf{Q}_n$ be the nth-projection.

Fix a point $p = (p_1, p_2, \ldots) \in X$. Since X is an ANR, there exists an open subset U of $\mathbf{Q}_\infty$ such that $X \subset U$ and there exists a retraction $r : U \to X$. Since $p \in U$, there exists a basic neighborhood

$$M = U_1 \times \cdots \times U_m \times \mathbf{Q}_{m+1} \times \mathbf{Q}_{m+2} \times \cdots$$

of $\mathbf{Q}_\infty$ such that $p \in M \subset U$.

Consider the continua $A = \{p_1\} \times \cdots \times \{p_m\} \times \mathbf{Q}_{m+1} \times \mathbf{Q}_{m+2} \times \cdots$, and $B = \{p_1\} \times \cdots \times \{p_m\} \times X_{m+1} \times X_{m+2} \times \cdots$. Observe that $B \subset A \subset M \subset U$ and $B \subset X$.

Define $f : A \to B$ by

$$f(a) = (p_1, \ldots, p_m, \pi_{m+1}(r(a)), \pi_{m+2}(r(a)), \ldots).$$

Given $b \in B$,

$$\begin{aligned} f(b) &= (p_1, \ldots, p_m, \pi_{m+1}(r(b)), \pi_{m+2}(r(b)), \ldots) \\ &= (p_1, \ldots, p_m, \pi_{m+1}(b), \pi_{m+2}(b), \ldots) = b. \end{aligned}$$

Hence f is a retraction.

By Theorems 16.3 and 16.5, A is an AR. Then Exercise 16.14, implies that B is an AR. Since each continuum X_n ($n \geq m+1$) is homeomorphic to a slice of B, and slices of B are retracts of B, Exercise 16.14 implies that $X_{m+1}, X_{m+2}, \ldots$ are ARs.

(Sufficiency) Let $m \in \mathbb{N}$ be such that X_n is an AR for every $n \geq m$.

By Exercise 16.16, we only have to prove that X is an ANE.

Let Z be a continuum, A a closed subset of Z and $f : A \to X$ a mapping.

For each $k \in \mathbb{N}$, let $\pi_k : X \to X_k$ be the kth-projection. Since the function $\pi_k \circ f : A \to X_k$ is continuous, for each $k \leq m$, we can choose an open subset U_k of Z and a mapping $f_k : U_k \to X_k$ such that $A \subset U_k$ and f_k is an extension of $\pi_k \circ f$. For each $k > m$, we can choose a mapping $f_k : Z \to X_k$ that extends $\pi_k \circ f$.

Let $U = U_1 \cap \cdots \cap U_m$. Then U is open in Z, $A \subset U$ and f_k is defined on U for every $k \in \mathbb{N}$.

Define $F : U \to X$ by

$$F(u) = \{f_n(u)\}_{n=1}^{\infty}.$$

Then F is continuous since each of its coordinate functions is continuous, and if $a \in A$,

$$F(a) = \{f_n(a)\}_{n=1}^{\infty} = \{\pi_n(f(a))\}_{n=1}^{\infty} = f(a).$$

Hence F extends f. Therefore X is an ANR. □

Theorem 16.10 *Let X be a continuum, and let X_1 and X_2 be subcontinua of X such that $X = X_1 \cup X_2$. Then*

(a) if $X_1 \cap X_2$, X_1 and X_2 are ARs, then X is an AR,
(b) if X and $X_1 \cap X_2$ are ARs, then X_1 and X_2 are ARs,
(c) if $X_1 \cap X_2$, X_1 and X_2 are ANRs, then X is an ANR,
(d) if X and $X_1 \cap X_2$ are ANRs, then X_1 and X_2 are ANRs.

Proof We only prove (a) and (b), the proofs for (c) and (d) are similar.

Suppose that $X_1 \cap X_2$, X_1 and X_2 are ARs. We prove that X is an AR.

Let Z be a continuum. Suppose that X is a subcontinuum of Z. We check that X is a retract of Z. Let d be a metric for Z.

Define

$$Z_1 = \{p \in Z : \text{dist}(p, X_1) \leq \text{dist}(p, X_2)\}, \text{ and}$$
$$Z_2 = \{p \in Z : \text{dist}(p, X_1) \geq \text{dist}(p, X_2)\}.$$

Then

$$Z_1 \cap Z_2 = \{p \in Z : \text{dist}(p, X_1) = \text{dist}(p, X_2)\}.$$

Note that Z_1, Z_2 and $Z_1 \cap Z_2$ are closed in Z, $X_1 \subset Z_1$, $X_2 \subset Z_2$ and $Z = Z_1 \cup Z_2$.

Since $X_1 \cap X_2$ is an AR, there exists a retraction $r_0 : Z \to X_1 \cap X_2$.

Given $i \in \{1, 2\}$, define $f_i : (Z_1 \cap Z_2) \cup X_i \to X_i$ by

$$f_i(p) = \begin{cases} r_0(p), & \text{if } p \in Z_1 \cap Z_2, \\ p, & \text{if } p \in X_i. \end{cases}$$

Given $p \in (Z_1 \cap Z_2) \cap X_i$, we have that $\text{dist}(p, X_1) = \text{dist}(p, X_2)$ and $\text{dist}(p, X_i) = 0$. Then $p \in X_1 \cap X_2$. Thus $r_0(p) = p$. This shows that f_i is a well-defined continuous function.

By Theorem 16.5, X_i is an AE. Therefore there exists a continuous extension $r_i : Z \to X_i$ of the mapping f_i.

Define $r : Z \to X$ by

$$r(p) = \begin{cases} r_1(p), & \text{if } p \in Z_1, \\ r_2(p), & \text{if } p \in Z_2. \end{cases}$$

Given $p \in Z_1 \cap Z_2$, $r_1(p) = f_1(p) = r_0(p) = f_2(p) = r_2(p)$. This shows that r is well defined and continuous.

Given $p \in X$, there exists an $i \in \{1, 2\}$ such that $p \in X_i$. Then $r(p) = r_i(p) = f_i(p) = p$. Thus r is a retraction from Z onto X. This completes the proof that X is an AR.

In order to prove (b), suppose that X and $X_1 \cap X_2$ are ARs. We prove that X_1 is an AR.

Since $X_1 \cap X_2$ is an AR and a subspace of X_2, we have that $X_1 \cap X_2$ is a retract of X_2. Then there exists a retraction $f : X_2 \to X_1 \cap X_2$. Define $r : X \to X_1$ by

$$r(p) = \begin{cases} f(p), & \text{if } p \in X_2, \\ p, & \text{if } p \in X_1. \end{cases}$$

Given $p \in X_1 \cap X_2$, $f(p) = p$. Thus r is well defined and continuous.

We have shown that X_1 is a retract of X and since X is an AR, we conclude that X_1 is an AR (see Exercise 16.14). □

Theorem 16.11 *Let X be a continuum. Then X is an AR if and only if X is a contractible ANR.*

Proof The necessity follows from Theorem 16.7.

Now, suppose that X is a contractible ANR.

Let $G : X \times [0, 1] \to X$ and $p_0 \in X$ be such that G is continuous and for each $x \in X$, $G(x, 0) = x$ and $G(x, 1) = p_0$.

Let Z be a continuum such that $X \subset Z$. Then there exists an open subset U of Z such that $X \subset U$ and there exists a retraction $r : U \to X$.

By Urysohn's Lemma, there exists a mapping $f : Z \to [0, 1]$ such that $f(X) \subset \{0\}$ and $f(Z \setminus U) \subset \{1\}$.

Define $R : Z \to X$ by

$$R(p) = \begin{cases} G(r(z), 2f(z)), & \text{if } 0 \leq f(z) \leq \frac{1}{2}, \\ p_0, & \text{if } \frac{1}{2} \leq f(z) \leq 1. \end{cases}$$

If $f(z) \leq \frac{1}{2}$, then $z \in U$, so we can apply r to z and $r(z) \in X$. Then we can apply G to the pair $(r(z), 2f(z))$. If $f(z) = \frac{1}{2}$, then $G(r(z), 2f(z)) = G(r(z), 1) = p_0$. This shows that R is well defined and it is continuous.

If $z \in X$, then $R(z) = G(r(z), 0) = r(z) = z$. Therefore R is a retraction from Z onto X. □

16.2 A Characterization

By Theorem 16.3 and Exercise 16.14 each retract of an n-cell (or of the Hilbert cube) is an AR. Since each continuum is embeddable in the Hilbert cube (Theorem 1.1), we can say that a continuum is an AR if and only if it is a retract of the Hilbert cube. In 1951, J. Dugundji [32, Corollary 4.2] gave a more general way to obtain AR continua by proving that any convex set in a linear space L is an AR. Dugundji's proof can be used in other structures. Generalizing the convex structures in linear spaces, in 1985, D.W. Curtis [28] introduced a more general definition of a convex structure (see Definition 16.12) and M. Lynch [97] (see also [69, Theorem 66.4]) used it to prove that each space of the form $\{A \in C(X) : p \in \mathcal{A}\}$, where X is a continuum, $p \in X$ and $\mathcal{A}$ is a Whitney level for $C(X)$, is an AR. The author has used Lynch's procedure to prove that the space of Whitney levels $WL(X)$ of a continuum X is an AR (metric) [52], as a previous step to show that $WL(X)$ is homeomorphic to the Hilbert space l_2. He also used Lynch's ideas to prove that the hyperspace $F_3(X)$ has the fixed point property, when X is a chainable continuum [62].

Definition 16.12 For each $n \in \mathbb{N}$, let

$$\Delta_n = \{(t_1, \ldots, t_n) \in [0, 1]^n : t_1 + \cdots + t_n = 1\}.$$

Given a continuum X, with a metric d, bounded by 1, let $\mathcal{M}_n = X^n \times \Delta_n$ and

$$\mathcal{M} = \bigcup\{\mathcal{M}_n : n \in \mathbb{N}\}.$$

Fixing a point $x_0 \in X$, we identify X^n with $X^n \times \{x_0\} \times \{x_0\} \times \cdots$ and $[0, 1]^n$ with $[0, 1]^n \times \{0\} \times \{0\} \times \cdots$. In this way we have a topology defined on $\mathcal{M}$, when we consider this set as a subset of the topological space

$$(X \times X \times \cdots) \times ([0, 1] \times [0, 1] \times \cdots).$$

This product is endowed with the metric

$$d_{\mathcal{M}}(((x_1, x_2, \ldots), (t_1, t_2, \ldots)), ((y_1, y_2, \ldots), (s_1, s_2, \ldots)))$$
$$= \sum_{n=1}^{\infty} \frac{d(x_n, y_n)}{2^{n+1}} + \frac{|t_n - s_n|}{2^{n+1}}.$$

A *convex structure* [28, p. 748] in X is a uniformly continuous function $C : \mathcal{M} \to X$ such that for every $n \in \mathbb{N}$, $x \in X$, $(x_1, \ldots, x_n) \in X^n$, $i \in \{1, \ldots, n\}$ and $(t_1, \ldots, t_n) \in \Delta_n$:

(a) $C((x, \ldots, x), (t_1, \ldots, t_n)) = x$, and
(b) $C((x_1, \ldots, x_n), (t_1, \ldots, t_{i-1}, 0, t_{i+1}, \ldots, t_n)) =$
$C((x_1, \ldots, x_{i-1}, x_{i+1}, \ldots, x_n), (t_1, \ldots, t_{i-1}, t_{i+1}, \ldots, t_n))$.

Theorem 16.13 *A continuum X is an AR if and only if X admits a convex structure.*

Proof By Exercise 16.23, we only need to prove the sufficiency.

Let d be a metric for X, upper bounded by 1.

Let (Z, ρ) be a continuum, A be a nonempty closed subset of Z and $f : A \to X$ be a mapping.

Given $z \in Z \setminus A$, we consider the ball $D_z = B(z, \frac{1}{2} \operatorname{dist}(z, A))$.

Since the space $Z \setminus A$ is metric, it is also paracompact. Thus there exists an open locally finite refinement $\mathcal{U} = \{U_\alpha : \alpha \in J\}$ of its cover $\{D_z : z \in Z \setminus A\}$. Let $\mathcal{P} = \{\varphi_\alpha : \alpha \in J\}$ be a partition of unity subordinated to $\mathcal{U}$. We suppose that J is endowed with a well-ordering $\prec$.

For each $\alpha \in J$, we fix a point $p_\alpha \in U_\alpha$ and we choose a point $a_\alpha \in A$ such that $\rho(p_\alpha, a_\alpha) < 2 \operatorname{dist}(p_\alpha, A)$.

Define $F : Z \to X$ as follows.

If $z \in A$, define $F(z) = f(z)$.

If $z \in Z \setminus A$, let $\alpha_1, \ldots, \alpha_n$ be the elements of J such that $\varphi_{\alpha_i}(z) > 0$. We suppose that $\alpha_1 \prec \cdots \prec \alpha_n$.

Then define

$$F(z) = C((f(a_{\alpha_1}), \ldots, f(a_{\alpha_n})), (\varphi_{\alpha_1}(z), \ldots, \varphi_{\alpha_n}(z))).$$

Note that by Property (b) of Definition 16.12, the definition of F does not depend on the order in which the elements $\alpha_1, \ldots, \alpha_n$ is taken. Thus F is well defined.

Let $z \in \operatorname{Fr}_Z(A)$. In order to show that F is continuous at z, let $\varepsilon > 0$.

Since C is uniformly continuous, there is a $\lambda > 0$ such that if $(p, s), (q, t) \in \mathcal{M}$ and its distance is less than λ, then $d(C(p, s), C(q, t)) < \varepsilon$.

Since f is uniformly continuous, there exists a $\delta > 0$ such that if $a, b \in A$ and $\rho(a, b) < 9\delta$, then $d(f(a), f(b)) < \lambda$.

Let $w \in Z \setminus A$ be such that $\rho(z, w) < \delta$.

Let $\alpha_1, \ldots, \alpha_n$ be the elements of J satisfying the inequality $\varphi_{\alpha_i}(w) > 0$, where $\alpha_1 \prec \cdots \prec \alpha_n$.

Given $i \in \{1, \ldots, n\}$, since $\varphi_{\alpha_i}(w) > 0$, we have that $w \in U_{\alpha_i}$. Let $v_i \in Z \setminus A$ be such that $U_{\alpha_i} \subset D_{v_i} = B(v_i, \frac{1}{2}\operatorname{dist}(v_i, A))$. Thus $\rho(v_i, w) < \frac{1}{2}\operatorname{dist}(v_i, A)$.

Since $\operatorname{dist}(v_i, A) \leq \rho(v_i, z) \leq \rho(v_i, w) + \rho(w, z) < \frac{1}{2}\operatorname{dist}(v_i, A) + \delta$, we have $\frac{1}{2}\operatorname{dist}(v_i, A) < \delta$. Therefore $\rho(v_i, w) < \delta$.

Since $p_{\alpha_i} \in U_{\alpha_i}$, we have $\rho(p_{\alpha_i}, v_i) < \frac{1}{2}\operatorname{dist}(v_i, A) < \delta$. Therefore $\rho(p_{\alpha_i}, A) \leq \rho(p_{\alpha_i}, z) \leq \rho(p_{\alpha_i}, v_i) + \rho(v_i, w) + \rho(w, z) < 3\delta$.

By the choice of a_{α_i}, we have $\rho(p_{\alpha_i}, a_{\alpha_i}) < 2\rho(p_{\alpha_i}, A)$. So, $\rho(p_{\alpha_i}, a_{\alpha_i}) < 6\delta$.

Hence $\rho(z, a_{\alpha_i}) \leq \rho(z, w) + \rho(w, v_i) + \rho(v_i, p_{\alpha_i}) + \rho(p_{\alpha_i}, a_{\alpha_i}) < 9\delta$.

Therefore $\rho(z, a_{\alpha_i}) < 9\delta$ and $d(f(z), f(a_{\alpha_i})) < \lambda$.

Then the distance from the point $((f(z), \ldots, f(z)), (\varphi_{\alpha_1}(w), \ldots, \varphi_{\alpha_n}(w)))$ to the point $((f(a_{\alpha_1}), \ldots, f(a_{\alpha_n})), (\varphi_{\alpha_1}(w), \ldots, \varphi_{\alpha_n}(w)))$ is less than λ.

By Property (a) of Definition 16.12 and the choice of λ, we have $d(f(z), F(w)) < \varepsilon$.

This finishes the proof that F is continuous at the points of $\operatorname{Fr}_X(A)$.

Now, take a point $z \in Z \setminus A$. Since $\mathcal{U}$ is a locally finite cover, there exists an open subset V of Z such that $z \in V$ and V only intersects a finite number $U_{\alpha_1}, \ldots, U_{\alpha_n}$ of elements of $\mathcal{U}$, where $\alpha_1 \prec \cdots \prec \alpha_n$.

Let $w \in V$ and $\alpha \in J$ be such that $\varphi_\alpha(w) > 0$. Then $w \in U_\alpha$, so $\alpha = \alpha_i$ for some $i \in \{1, \ldots, n\}$. Thus the set L of indices α for which $\varphi_\alpha(w) > 0$ is contained in $\{\alpha_1, \ldots, \alpha_n\}$. By Property (b) of Definition 16.12, this shows that for every $w \in V$:

$$F(w) = C((f(a_{\alpha_1}), \ldots, f(a_{\alpha_n})), (\varphi_{\alpha_1}(w), \ldots, \varphi_{\alpha_n}(w))).$$

Since C and the functions $\varphi_{\alpha_1}, \ldots, \varphi_{\alpha_n}$ are continuous, we conclude that the function F is continuous at V. Hence F is continuous in $Z \setminus A$.

Finally, since f is continuous, we have that F is also continuous at the points in the interior of A. Therefore F is a continuous extension of the mapping f.

We have shown that X is an AE, and by Theorem 16.5, X is also an AR. □

16.3 Exercises

Exercise 16.14 Show that a retract of an ANR (respectively, AR, AE, ANE) is also an ANR (respectively AR, AE, ANE).

Exercise 16.15 Let $n \in \mathbb{N} \setminus \{1\}$. Show that the nth-unit sphere S^n, contained in $\mathbb{R}^{n+1}$ is an ANR. (Hint: use Exercise 16.14.)

Exercise 16.16 Prove that a continuum is an ANR if and only if it is an ANE.

Exercise 16.17 Let $\{X_n : n \in \mathbb{N}\}$ be a family of continua. Prove that $\prod_{n=1}^{\infty} X_n$ is an AR if and only if each X_n is an AR.

Exercise 16.18 Prove Theorem 16.7.

Exercise 16.19 Prove Properties (c) and (d) in Theorem 16.10.

Exercise 16.20 Prove, using Theorem 16.10, that each tree is an AR.

Exercise 16.21 Prove that every finite graph is an ANR.

Exercise 16.22 Prove that the function C defined for the Hilbert cube by

$$C((x_1, \ldots, x_n), (t_1, \ldots, t_n)) = t_1x_1 + \cdots + t_nx_n$$

is a convex structure.

Exercise 16.23 Prove that each AR continuum admits a convex structure.

Exercise 16.24 Suppose that Y is the Hawaiian earring described in Example 16.8. Prove that the cone over Y is a locally connected, arcwise connected, unicoherent, contractible continuum with the fixed point property, but it is not an AR.

Chapter 17
Stronger Properties of the Pseudo-Arc

In the first volumes of *Fundamenta Mathematicae* two questions were posed that have had great influence on the development of Continuum Theory.

Question 17.1 (B. Knaster and K. Kuratowski, 1920, [84]). *Un continu (borné) plan, topologiquement homogène, est-il nécessairement homéomorphe à une circonférence ? (Must a homogeneous plane continuum necessarily be a simple closed curve?)*

Question 17.2 (S. Mazurkiewicz, 1921, [106]) *Un continu dans l'espace à m dimensions qui est homéomorphe à tout continu qu'il contient, est il nécessairement un arc simple ? (Must a continuum which is homeomorphic to each of its non-degenerate subcontinua necessarily be an arc?)*

Question 17.2 was answered by E.E. Moise in 1948 [108], who constructed a plane indecomposable continuum which is homeomorphic to each of its non-degenerate subcontinua. He called this continuum the "pseudo-arc". The question whether there are continua, different from an arc or the pseudo-arc, with this property remains open. Recently this problem was solved for plane continua; L.C. Hoehn and L.G. Oversteegen [48] have proven that if a *plane* continuum X is homeomorphic to each of its non-degenerate subcontinua, then X is either an arc or a pseudo-arc.

Question 17.1 was answered the same year (1948) by RH Bing [9] who generalized Moise's construction to obtain a homogeneous hereditarily indecomposable plane continuum. In 1951 Bing [10] gave a characterization of the pseudo-arc and showed that his continuum, Moise's continuum and the continuum constructed by Knaster in his thesis [82] are homeomorphic.

Certainly Bing's proof of the homogeneity of the pseudo-arc is not easy to read, but it is correct. However it is interesting to note that it was not accepted by everyone. The following paragraph was written by J.J. Charatonik [22, p. 734].

"A few years later Isaac Kapuano claimed [78] that the pseudo-arc is not homogeneous. However, an error was discovered in his work, so he published an

A. Illanes, *Continuum Theory*, Universitext,
https://doi.org/10.1007/978-3-031-91011-1_17

attempt to correct it [77]. Mathematicians seemed more inclined to accept the results of Bing and Moise than those of Kapuano, but A.S. Esenin-Vol'pin, a reviewer of *Referativnyĭ Zhurnal*, wrote in 1955 that "in the light of this, the problem of Knaster and Kuratowski remains open" [37]. It is not surprising that the discussion greatly interested B. Knaster, the discoverer of the pseudo-arc, who asked in 1955 two of his students, Andrzej Lelek and Marek Rochowski, to verify Bing and Kapuano's arguments and clarify the situation. They did this hard work, the results of which were presented to Knaster in the form of a handwritten 60-page paper (in Polish; never published), and which concluded that Bing was right."

According to C.O. Christenson and W.L. Voxman, "The pseudo-arc has probably been the most scrutinized of all the indecomposable continua, and its properties are for the most part as spectacular as they are difficult to prove." [25, Ch. 9, pp. 247–248].

The pseudo-arc has played an important role in many aspects of Continuum Theory. In particular, L.C. Hoehn and L.G. Oversteegen [47] have shown that there exist exactly three homogeneous plane continua: the simple closed curve, the pseudo-arc and the circle of pseudo-arcs.

The interested reader can find a lot of information about the pseudo-arc in the papers by W. Lewis [94] and [95].

This chapter is devoted to proving four of the more important properties of the pseudo-arc, namely: being homogeneous, being homeomorphic to all of its proper non-degenerate subcontinua, being a hereditarily indecomposable chainable continuum, and being the unique chainable homogeneous continuum. The proofs are mainly based on Bing's arguments, but with some improvement of the ideas due to W. Lewis, contained in [96], and A. Illanes, P. Minc and F. Sturm in [71]. Another interesting work that contains simplifications of Bing's and Moise's arguments for proving these properties of the pseudo-arc was written by L.G. Oversteegen and E.D. Tymchatyn [115].

17.1 Chains

Definition 17.1 Given a continuum X, a point $p \in X$ is *a final point of* X if for every pair of subcontinua A and B of X such that $p \in A \cap B$, we have that $A \subset B$ or $B \subset A$.

A point $p \in X$ is *a terminal point of* X if for each $\varepsilon > 0$, there exists a proper covering chain $C = \{C_0, C_1, \ldots, C_n\}$ such that $\text{mesh}(C) < \varepsilon$ and $p \in C_0 \setminus C_1$.

Lemma 17.2 *Let X be a continuum.*

(a) *If* $C = \{0, 1, \ldots, n\}$ *is a taut chain of* X *and* $g_C : X \to [0, 1]$ *is a mapping induced by* C *(see Sect. 14.4), then for each* $t \in [0, 1]$, $\text{diameter}(g_C^{-1}(t)) \leq \text{mesh}(C)$.

(b) *If X is chainable, p is a terminal point of X and $\varepsilon > 0$, then there exists a $K \in \mathbb{N}$ such that for each $m \geq K$, there exists a taut chain $\mathcal{E} = \{E_0, E_1, \ldots, E_m\}$ such that* $\mathrm{mesh}(\mathcal{E}) < \varepsilon$ *and* $p \in E_0 \setminus \mathrm{cl}_X(E_1)$.

Proof

(a) Given $t \in [0, 1]$, there exists an $i \in \{0, 1, \ldots, n\}$, such that $t \in [\frac{i}{n+1}, \frac{i+1}{n+1}]$. Then $g_{\mathcal{C}}^{-1}(t) \subset \mathrm{cl}_X(C_i)$. Therefore $\mathrm{diameter}(g_{\mathcal{C}}^{-1}(t)) \leq \mathrm{mesh}(\mathcal{C})$.

(b) By definition, there is a proper covering chain $\mathcal{C} = \{C_0, C_1, \ldots, C_n\}$ such that $\mathrm{mesh}(\mathcal{C}) < \varepsilon$ and $p \in C_0 \setminus C_1$. By Theorem 14.10, there is a taut chain $\mathcal{D} = \{D_0, D_1, \ldots, D_n\}$ of X satisfying that for each $i \in \{0, 1, \ldots, n\}$, $\mathrm{cl}_X(D_i) \subset C_i$. This implies that $p \notin \mathrm{cl}_X(D_1) \cup \cdots \cup \mathrm{cl}_X(D_n)$. Thus $p \in D_0 \setminus \mathrm{cl}_X(D_1)$. Let $g_{\mathcal{D}} : X \to [0, 1]$ be a mapping induced by $\mathcal{D}$, with $g_{\mathcal{D}}(p) = 0$. By (a), for each $t \in [0, 1]$, $\mathrm{diameter}(g_{\mathcal{C}}^{-1}(t)) \leq \mathrm{mesh}(\mathcal{C}) < \varepsilon$. By Exercise 1.53, there exists a $\delta > 0$ such that if $J \subset [0, 1]$ and $\mathrm{diameter}(J) < \delta$, then $\mathrm{diameter}(g_{\mathcal{C}}^{-1}(J)) < \varepsilon$. Let $K \in \mathbb{N} \setminus \{1\}$ be such that $\frac{1}{K} < \frac{\delta}{2}$. If $m \geq K$, then $\frac{1}{m} < \frac{\delta}{2}$. For each $i \in \{0, 1, \ldots, m\}$, let $J_i = (\frac{i}{m+1} - \frac{1}{3(m+1)}, \frac{i+1}{m+1} + \frac{1}{3(m+1)}) \cap [0, 1]$ and $E_i = g_{\mathcal{D}}^{-1}(J_i)$. Then E_i is open in X, $\mathrm{diameter}(J_i) < \delta$ and $\mathrm{diameter}(E_i) < \varepsilon$. Since $\mathrm{cl}_X(E_i) \subset \mathrm{cl}_X(g_{\mathcal{D}}^{-1}(J_i))$, and $\mathcal{J} = \{J_0, J_1 \ldots, J_m\}$ is a taut chain of $[0, 1]$, we conclude that $\mathcal{E} = \{E_0, E_1, \ldots, E_m\}$ is a taut chain of X. Since $g_{\mathcal{D}}(p) = 0$, we have that $p \in E_0 \setminus \mathrm{cl}_X(E_1)$.

□

17.2 Terminal and Final Points

One important step in the proof of the homogeneity of the pseudo-arc is the result by Bing [10, Theorem 13] that says that in a chainable continuum X a point is final if and only if it is terminal. The sufficiency is easy to prove (Lemma 17.3), however the necessity requires a complicated argument and its development is the main result of this section.

Given $n \in \mathbb{N}$, define $\langle n \rangle = \{0, 1, \ldots, n\}$.

Lemma 17.3 *Let X be a chainable continuum and $p \in X$. If p is a terminal point of X, then p is a final point.*

Proof Suppose that p is a terminal point of X. To prove that p is a final point of X, suppose to the contrary that there exist subcontinua A and B of X such that $p \in A \cap B$, $A \nsubseteq B$ and $B \nsubseteq A$. Choose points $a \in A \setminus B$ and $b \in B \setminus A$. Let $\varepsilon > 0$ be such that $B(a, \varepsilon) \cap B = \emptyset$ and $B(b, \varepsilon) \cap A = \emptyset$. By hypothesis there exists a proper covering chain $\mathcal{C} = \{C_0, C_1, \ldots, C_n\}$ such that $p \in C_0$ and $\mathrm{mesh}(\mathcal{C}) < \varepsilon$. Let $i, j \in \langle n \rangle$ be such that $a \in C_i$ and $b \in C_j$. We may assume that $i \leq j$. Since B intersects the links C_0 and C_j, we have that B intersects the link C_i. Thus there exists a $q \in B \cap C_i$. This implies that $q \in B \cap B(a, \varepsilon)$, which contradicts the choice of ε. This finishes the proof that p is a final point of X. □

Definition 17.4 Let X be a continuum, let $\mathcal{C} = \{C_0, C_1, \ldots, C_n\}$ be a proper covering chain of X and let Y be a non-degenerate subcontinuum of X. Given $j, k \in \langle n \rangle$ with $j \leq k$, the subchain $\mathcal{C}(j, k) = \{C_j, \ldots, C_k\}$ of $\mathcal{C}$ *exactly covers* Y if $Y \subset C_j \cup \cdots \cup C_k$ and $Y \cap C_i \neq \emptyset$ for each $j \leq i \leq k$.

Let i_1 be the maximum element $i \in \langle n \rangle$ such that $Y \subset \mathcal{C}(i, n)$. Let j_1 be the minimum element $j \in \langle n \rangle$ such that $Y \subset \mathcal{C}(0, j)$. Let i_2 (respectively, j_2) be the minimum (respectively, maximum) element $i \in \langle n \rangle$ such that $Y \cap C_i \neq \emptyset$. Define $[Y] = \{i_1, \ldots, j_1\}$ and $\langle Y \rangle = \{i_2, \ldots, j_2\}$. Note that $[Y] \subset \langle Y \rangle$ and the inclusion can be proper.

If Z is a subcontinuum of X, the pair (Y, Z) is a *step* if Y is non-degenerate, $Y \subset Z$ and one of the following holds:

(a) there exist $j, k, l \in \langle n \rangle$ such that $0 \leq j < k \leq l \leq n$, the subchain $\mathcal{C}(k, l)$ exactly covers Y, the subchain $\mathcal{C}(j, l)$ exactly covers Z and $Z \cap C_j \setminus Y \neq \emptyset$,
(b) there exist $j, k, l \in \langle n \rangle$ such that $0 \leq l \leq k < j \leq n$, the subchain $\mathcal{C}(l, k)$ exactly covers Y, the subchain $\mathcal{C}(l, j)$ exactly covers Z and $Z \cap C_j \setminus Y \neq \emptyset$.

If (a) holds, then (Y, Z) is a *left step*, and if (b) holds, then (Y, Z) is a *right step*. In both cases the step (Y, Z) is said to *start* at l and to *finish* at j.

Lemma 17.5 *Let X be a chainable continuum and let $\mathcal{C} = \{C_0, C_1, \ldots, C_n\}$ be a proper covering chain of X. Let Y, Z be subcontinua of X such that (Y, Z) is a step starting at l and finishing at j. Let $p \in Y$ be such that p is a final point of X. Suppose that there exists a chain $\mathcal{D} = \{D_0, D_1, \ldots, D_m\}$ of open subsets of X satisfying the following properties:*

(a) *$\mathcal{D}$ exactly covers Y,*
(b) *$\mathcal{D}$ refines $\mathcal{C}$,*
(c) *$p \in D_0 \setminus D_1$,*
(d) *$D_m \subset C_l$.*

Then there exists a chain $\mathcal{E} = \{E_0, E_1, \ldots, E_r\}$ of open subsets of X satisfying the following properties

(a') *$\mathcal{E}$ exactly covers Z,*
(b') *$\mathcal{E}$ refines $\mathcal{C}$,*
(c') *$p \in E_0 \setminus E_1$,*
(d') *$E_r \subset C_j$.*

Proof We suppose that (Y, Z) is a left step. The other case is similar. So the numbers $j, k, l \in \langle n \rangle$ satisfy the condition (a) in Definition 17.4 ((Y, Z) starts at l and ends at j). Then we can choose a point $z_0 \in Z \cap C_j \setminus Y$.

Since $\mathcal{D}$ exactly covers Y, we have that $Y \cap D_m$ is a nonempty (and therefore non-degenerate) open subset of the continuum Y. Then we can choose a point $w \in Y \cap D_m \setminus \{p\}$. Choose an open subset W of X such that $w \in W$, $W \subset D_m$, $z_0 \notin W$ and $p \notin W$.

Let F be the component of $X \setminus W$ containing p. Then F and Y are subcontinua of X and both contain p. Since p is a final point of X, F and Y are comparable by inclusion. Since $w \in Y \setminus F$, we conclude that $F \subset Y$. In particular, $z_0 \notin F$.

Let $R = \bigcup \mathcal{D}$. Then R is an open subset of X such that $F \subset Y \subset R$ and $W \subset R$.

Since F is a component of $X \setminus W$, it is possible to apply the Cut Wire Theorem (Theorem 3.3) to the space $X \setminus W$ and its closed subsets F and $(X \setminus R) \cup \{z_0\}$. Then there exist closed subsets L and M of $X \setminus W$ (and then closed in X) such that $F \subset L$, $(X \setminus R) \cup \{z_0\} \subset M$, $L \cap M = \emptyset$ and $X \setminus W = L \cup M$.

By the normality of X, there exist disjoint open subsets U and V of X such that $L \subset U$ and $M \subset V$. Since $L \subset X \setminus M \subset R$, we may suppose that $U \subset R$. Since $z_0 \notin L$, we may also assume that $z_0 \notin U$.

We are ready to define the required chain $\mathcal{E} = \{E_0, E_1, \ldots, E_r\}$.

Set $r = m + l - j$.

Define:

$$E_0 = D_0 \cap U,\ E_1 = D_1 \cap U,\ \ldots,\ E_{m-1} = D_{m-1} \cap U;$$

$$E_m = D_m \cup (C_l \cap V);$$

$$E_{m+1} = C_{l-1} \cap V,\ E_{m+2} = C_{l-2} \cap V,\ \ldots,\ E_{m+(l-j)} = C_j \cap V.$$

We show that $\mathcal{E}$ satisfies properties (a')–(d').

Since $D_m \subset C_l$ and $\mathcal{D}$ refines C, we have that $\mathcal{E}$ refines C.

Take two non-consecutive links $E_{i'}$ and $E_{j'}$ of $\mathcal{E}$. We check that they do no intersect. Since $\mathcal{D}$ is a chain, if $E_{i'}$ and $E_{j'}$ are in the first line in the definition of $\mathcal{E}$, then they do not intersect. Since C is a chain, the same happens if they belong to the third line. If one of the links is in the first line and the other is in the third one, they are disjoint since $U \cap V = \emptyset$. If $i' = m$ and $j' < m-1$, then $E_{i'} \cap E_{j'} = \emptyset$ because $\mathcal{D}$ is a chain and $U \cap V = \emptyset$, and if $i' = m$ and $m+1 < j'$, then $E_{i'} \cap E_{j'} = \emptyset$ because C is a chain and $D_m \subset C_l$. This ends the proof that, in any case, $E_{i'} \cap E_{j'} = \emptyset$.

Now, we check that $\mathcal{E}$ covers Z. Recall that $Z \subset C(j, l)$ and $X = L \cup W \cup M = U \cup W \cup V$. Take $z \in Z$. Let $i \in \{j, \ldots, l\}$ be such that $z \in C_i$.

If $z \in V$, then depending on i, z belongs to E_m (if $i = l$) or to some E_s in the third line of the definition of $\mathcal{E}$ (if $i < l$).

If $z \in U \subset R$, there exists a $t \in \langle m \rangle$ such that $z \in D_t$. Depending on t, we have that z belongs to E_m (if $t = m$) or to some E_s in the first line (if $i < m$).

If $z \in W$, then $z \in D_m \subset E_m$.

This completes the proof that $\mathcal{E}$ covers Z.

Since $p \in D_0 \setminus D_1$ and $p \in F \subset L \subset U$, we have that $p \in E_0 \setminus E_1$. Since $z_0 \notin W \cup U$, we have that $z_0 \in V$, and since $z_0 \in C_j$, we conclude that $z_0 \in E_r$. By Exercise 17.20, we conclude that $\mathcal{E}$ is a chain exactly covering Z.

Since, clearly $E_r \subset C_j$, we obtain that $\mathcal{E}$ satisfies the required properties. □

Lemma 17.6 *Let X be a continuum, let $p \in X$ and let $C = \{C_0, C_1, \ldots, C_n\}$ be a proper covering chain of X with $n \geq 4$. Then there exists a finite sequence of continua $\{p\} \subsetneq A_0 \subsetneq A_1 \subsetneq \cdots \subsetneq A_m = X$ such that for each $i \in \{1, \ldots, m\}$, the* *pair (A_{i-1}, A_i) is a step and the sequence satisfies the following property: if the pair (A_{i-1}, A_i) satisfies one of the conditions (a) or (b), then the pair (A_i, A_{i+1}) satisfies the other one. Moreover, the pair (A_i, A_{i+1}) starts where the pair (A_{i-1}, A_i) ends.*

Proof Let $i_0 \in \langle n \rangle$ be such that $p \in C_{i_0}$.

Fix a non-degenerate subcontinuum A_0 of X such that $p \in A_0 \subset C_{i_0}$. We remark that we will not need more properties of A_0.

Note that there exist $k, l \in \langle n \rangle$ such that $k \leq l$ and $\langle A_0 \rangle = \{k, \ldots, l\}$. Observe that $C(k, l)$ exactly covers A_0 and $i_0 - 1 \leq k \leq i_0 \leq l \leq i_0 + 1$.

Since $n \geq 4$, we have that the set $G = \langle n \rangle \setminus \langle A_0 \rangle$ is nonempty.

Let

$$s_M = \max\{i \in \langle n \rangle : \text{ there exists a subcontinuum } Z \text{ of } X \text{ such that } A_0 \subset Z \text{ and } C(k, i) \text{ exactly covers } Z\}, \text{ and}$$

$$s_m = \min\{i \in \langle n \rangle : \text{ there exists a subcontinuum } Z \text{ of } X \text{ such that } A_0 \subset Z \text{ and } C(i, l) \text{ exactly covers } Z\}.$$

Since $Z = A_0$ is exactly covered by $C(k, l)$, we have that s_M and s_m are well defined and $s_m \leq k$ y $l \leq s_M$.

Claim $s_m < k$ or $l < s_M$.

We prove this claim. By Corollary 9.18, there exists an order arc $\alpha : [0, 1] \to C(X)$ such that $\alpha(0) = A_0$ and $\alpha(1) = X$.

Let $U = \bigcup\{C_i : i \in \langle A_0 \rangle\}$ and $V = \bigcup\{C_i : i \in G\}$. Then U and V are nonempty open subsets of X such that $X = U \cup V$ and $A_0 \subset U$.

Let $P = \{t \in [0, 1] : \alpha(t) \subset U\}$ and $Q = \{t \in [0, 1] : \alpha(t) \cap V \neq \emptyset\}$. Then P and Q are open subsets of $[0, 1]$ (see Exercise 9.31) such that $[0, 1] = P \cup Q$. Since $\alpha(0) = A_0 \subset U$ and $\alpha(1) = X$, we have that $0 \in P$ and $1 \in Q$. The connectedness of $[0, 1]$ implies that there exists a $t_0 \in [0, 1]$ such that $\alpha(t_0) \in P \cap Q$.

Let $B = \alpha(t_0)$. Then $B \subset C_k \cup \cdots \cup C_l$ and there exists a $j \in \langle n \rangle \setminus \{k, \ldots, l\}$ such that $B \cap C_j \neq \emptyset$. Thus $j = k - 1$ or $j = l + 1$.

Since $A_0 \subset B$, we have that B intersects each of the links $C_k, \ldots, C_l$.

We analyze the case $j = l + 1$. The case $j = k - 1$ is similar. Note that it is possible that B intersects C_{k-1} and C_{l+1}, but this is not important in what follows.

Since $B \subset C_k \cup \cdots \cup C_l \subset C_k \cup \cdots \cup C_l \cup C_{l+1}$, and B intersects all these sets, we have that $C(k, l + 1)$ exactly covers B. Therefore $l + 1 \leq s_M$ and $l < s_M$.

This finishes the proof of the claim.

In order to construct A_1, we suppose that $l < s_M$ (the case $s_m < k$ is similar). From the definition of s_M, we have that there exists a subcontinuum A_1 of X such that $A_0 \subset A_1$ and $C(k, s_M)$ exactly covers A_1. Since $\langle A_0 \rangle = \{k, \ldots, l\}$ and $l < s_M$, we have that $A_0 \cap C_{s_M} = \emptyset$ and $A_1 \cap C_{s_M} \neq \emptyset$. Thus $A_1 \cap C_{s_M} \setminus A_0 \neq \emptyset$. Therefore the pair (A_0, A_1) is a right step.

In order to continue the inductive procedure, we suppose that continua $A_0, \ldots, A_r$ have been constructed, with $r \geq 1$, and they satisfy that there exist

$k, l \in \langle n \rangle$ such that $0 \le k \le l \le n$, $C(k, l)$ exactly covers A_{r-1}, and one of the following conditions holds.

(1) (A_{r-1}, A_r) is a right step, $l < s_M$ and A_r is exactly covered by $C(k, s_M)$, where $s_M = \max\{i \in \langle n \rangle$: there exists a subcontinuum Z of X such that $A_{r-1} \subset Z$ and $C(k, i)$ exactly covers $Z\}$, or
(2) (A_{r-1}, A_r) is a left step, $s_m < k$ and A_r is exactly covered by $C(s_m, l)$, where $s_m = \min\{i \in \langle n \rangle$: there exists a subcontinuum Z of X such that $A_{r-1} \subset Z$ and $C(i, l)$ exactly covers $Z\}$.

We prove that if $\langle A_r \rangle \neq \langle n \rangle$, then we can construct a subcontinuum A_{r+1} of X satisfying the required properties and with the additional property that $[A_r] \subsetneq [A_{r+1}]$. We suppose that condition (1) holds (that is, (A_{r-1}, A_r) is a right step). The case in which condition (2) holds is similar. In particular, we are assuming that (A_{r-1}, A_r) finishes at s_M.

In the case when $s_M = n$, set $A_{r+1} = X$. Since $\langle A_r \rangle \neq \langle n \rangle$ and $\{k, \dots, s_M\} \subset \langle A_r \rangle$, we have that $0 < k \le s_M = n$. Thus the subchain $C(k, n)$ exactly covers A_r, the subchain $C(0, n)$ exactly covers X and since C is a proper covering chain of X, we conclude that $\emptyset \neq X \cap C_0 \setminus (C_1 \cup \dots \cup C_n) \subset A_{r+1} \cap C_0 \setminus A_r$. Hence (A_r, A_{r+1}) is a left step starting at $n = s_M$ and (A_{r-1}, A_r) finishes at n. Thus we are done in this case. So, we suppose that $s_M < n$.

In the case when $k = 0$, we have that the subchain $C(0, n)$ exactly covers X. The maximality of s_M implies that $n \le s_M$, which contradicts our assumption. Therefore $0 < k$.

Define $L = X \setminus (C_{s_M+1} \cup \dots \cup C_n)$. Then L is closed in X.

If $A_r \cap C_{s_M+1} \neq \emptyset$, since $C(k, s_M)$ exactly covers A_r, we have that $C(k, s_M + 1)$ also exactly covers A_r, and since $A_{r-1} \subset A_r$, we obtain a contradiction with the maximality of s_M. Since $A_r \subset C_k \cup \dots \cup C_{s_M}$, we conclude that $A_r \subset L$.

Let $\beta : [0, 1] \to C(X)$ be an order arc such that $\beta(0) = A_r$ and $\beta(1) = X$.

Since $\beta(0) \subset L$, we can define $t_0 = \max\{t \in [0, 1] : \beta(t) \subset L\}$. Let $Q = \beta(t_0)$. Then $A_r \subset Q \subset L$.

We prove that $Q \cap C_{k-1} \setminus C_k \neq \emptyset$. Suppose the contrary. Since $A_r \subset Q$, we have that $C_k \cap Q \neq \emptyset$. This implies that $Q \subset C_k \cup \dots \cup C_{s_M}$. Since $C_k \cup \dots \cup C_{s_M}$ is open in X, there exists a $t_1 \in (t_0, 1)$ such that $\beta(t_1) \subset C_k \cup \dots \cup C_{s_M}$. By the maximality of t_0, we have that $\beta(t_1) \not\subset L$, so $\beta(t_1) \cap (C_{s_M+1} \cup \dots \cup C_n) \neq \emptyset$. Since $\beta(t_1) \subset C_k \cup \dots \cup C_{s_M}$, we have that $A_{r-1} \subset \beta(t_1) \subset C_k \cup \dots \cup C_{s_M+1}$ and $\beta(t_1)$ intersects to all these links. Thus $C(k, s_M + 1)$ exactly covers $\beta(t_1)$. This contradicts the maximality of s_M and finishes the proof that $Q \cap C_{k-1} \setminus C_k \neq \emptyset$.

Define $t_m = \min\{i \in \langle n \rangle$: there exists a subcontinuum Z of X such that $A_r \subset Z$ and $C(i, s_M)$ exactly covers $Z\}$.

Let $l_0 = \min\{i \in \langle n \rangle : Q \cap C_i \neq \emptyset\}$. Then $l_0 \le k - 1$. Since $Q \subset L$, we have that $Q \subset C_{l_0} \cup \dots \cup C_{s_M}$. Since $A_r \subset Q$ and $A_r \cap C_{s_M} \neq \emptyset$, we have that the subchain $C(l_0, s_M)$ exactly covers Q. Hence $t_m \le l_0 \le k - 1$.

Choose a subcontinuum A'_{r+1} of X such that $A_r \subset A'_{r+1}$ and $C(t_m, s_M)$ exactly covers A'_{r+1}. Define $A_{r+1} = A'_{r+1} \cup Q$. Then A_{r+1} is a subcontinuum of X, $A_r \subset$

$A_{r+1} \subset C_{t_m} \cup \cdots \cup C_{s_M}$ and A_{r+1} intersects each of these links. Therefore $C(t_m, s_M)$ exactly covers A_{r+1}. Moreover $A_{r+1} \cap C_{k-1} \setminus C_k \neq \emptyset$.

Then $0 \leq t_m < k \leq s_M \leq n$, the subchain $C(k, s_M)$ exactly covers A_r and the subchain $C(t_m, s_M)$ exactly covers A_{r+1}. In order to see that (A_r, A_{r+1}) is a left step, it only remains to check that $A_{r+1} \cap C_{t_m} \setminus A_r \neq \emptyset$.

If $t_m = k - 1$, then we know that $A_{r+1} \cap C_{k-1} \setminus C_k \neq \emptyset$, and since $A_r \subset C_k \cup \cdots \cup C_{s_M}$, we conclude that $A_{r+1} \cap C_{k-1} \setminus A_r \neq \emptyset$.

If $t_m < k - 1$, then we know that $A_{r+1} \cap C_{t_m} \neq \emptyset$. In this case, $A_{r+1} \cap C_{t_m} \subset X \setminus (C_k \cup \cdots \cup C_{s_M})$, so $A_{r+1} \cap C_{t_m} \setminus A_r \neq \emptyset$.

This finishes the proof that (A_r, A_{r+1}) is a left step.

Note that (A_r, A_{r+1}) starts at s_M and the pair (A_{r-1}, A_r) finishes at s_M.

Since $C(k, s_M)$ exactly covers A_r, we have that $[A_r] \subset \{k, \ldots, s_M\}$. Since $A_r \subset A_{r+1}$ and $A_{r+1} \cap C_{k-1} \setminus C_k \neq \emptyset$, we have that $k - 1 \in [A_{r+1}]$, so $[A_r] \subsetneq [A_{r+1}]$. This finishes the inductive step.

We have shown that while $\langle A_r \rangle \neq \langle n \rangle$, it is possible to continue constructing more continua A_i. Since $[A_0] \subsetneq [A_1] \subsetneq \cdots$, and all sets $[A_r]$ are subsets of $\langle n \rangle$, this procedure must end. Therefore there exists an $r \geq 0$ such that $\langle A_{r+1} \rangle = \langle n \rangle$.

To finish the proof of the lemma we need to see that it is possible to finish the sequence of sets A_i at the continuum X. If $A_{r+1} = X$, then we are done. So, we suppose that $A_{r+1} \neq X$.

We return to the inductive construction of A_{r+1}. As before, we consider the case when (A_{r-1}, A_r) satisfies condition (1) ((A_{r-1}, A_r) is a right step). We are supposing that $\langle A_{r+1} \rangle = \langle n \rangle$. In particular, $A_{r+1} \cap C_n \neq \emptyset \neq A_{r+1} \cap C_0$. We showed that if $s_M = n$, then we can finish by making $A_{r+1} = X$. Thus we assume that $s_M < n$. As before, we define $t_m = \min\{i \in \langle n \rangle :$ there exists a subcontinuum Z of X such that $A_r \subset Z$ and $C(i, s_M)$ exactly covers $Z\}$ and we construct A_{r+1} such that the subchain $C(t_m, s_M)$ exactly covers A_{r+1}. Since $A_{r+1} \cap C_{s_M} \neq \emptyset$ and $A_{r+1} \subset C_{t_m} \cup \cdots \cup C_{s_M}$, we have that $(C_{t_m} \cup \cdots \cup C_{s_M}) \cap C_n \neq \emptyset$, so $s_M = n - 1$. Thus the subchain $C(0, s_M)$ exactly covers A_{r+1}. The minimality of t_m implies that $t_m = 0$.

Since C is a proper covering chain of X, we have that $\emptyset \neq X \cap C_n \setminus (C_0 \cup \cdots \cup C_{s_M}) \subset X \cap C_n \setminus A_{r+1}$. Thus the subchain $C(0, s_M)$ exactly covers A_{r+1}, the subchain $C(0, n)$ exactly covers X and $X \cap C_n \setminus A_{r+1} \neq \emptyset$. This shows that the pair (A_{r+1}, X) is a right step. Then if we define $A_{r+2} = X$, we obtain the required sequence. □

Theorem 17.7 *Let X be a chainable continuum and let $p \in X$. Then p is a terminal point if and only if p is a final point.*

Proof By Lemma 17.3, we only need to prove the sufficiency.

Suppose that p is a final point. To prove that p is a terminal point, take $\varepsilon > 0$. Let $\mathcal{F} = \{F_0, F_1, \ldots, F_k\}$ be a proper covering chain of X such that $\text{mesh}(\mathcal{F}) < \min\{\frac{\varepsilon}{3}, \frac{\text{diameter}(X)}{7}\}$. Then $k \geq 7$.

Choose $j_0 \in \langle k \rangle$ such that $p \in F_{j_0}$.

If $j_0 \leq 2$, take the proper covering chain $\mathcal{F}_1 = \{F_0 \cup F_1 \cup F_2, F_3, \ldots, F_k\}$ of X which has mesh less than ε and has p in its first link, and we are done. If $k-2 \leq j_0$, we can group the last three links and take the chain with inverted indices and then obtain the required chain. Thus we may suppose that $3 \leq j_0 \leq k-3$.

Consider the following chain.

$$C = \{F_0, F_1, \ldots, F_{j_0-2}, F_{j_0-1} \cup F_{j_0} \cup F_{j_0+1}, F_{j_0+2}, \ldots, F_k\}.$$

Set $C = \{C_0, C_1, \ldots, C_n\}$ and $C_{i_0} = F_{j_0-1} \cup F_{j_0} \cup F_{j_0+1}$. Then $n \geq 4$, C is a proper covering chain of X, $\text{mesh}(C) < \varepsilon$ and $p \in F_{j_0} \subset C_{i_0} \setminus (C_1 \cup \cdots \cup C_{i_0-1} \cup C_{i_0+1} \cup \cdots \cup C_n)$.

Let $\{p\} \subsetneq A_0 \subsetneq A_1 \subsetneq \cdots \subsetneq A_m = X$ be a finite sequence satisfying the properties of Lemma 17.6. As we observed in that lemma, we can take any small enough A_0, so we may assume that $\{p\} \subsetneq A_0 \subset F_{i_0}$. Thus $A_0 \cap C_i = \emptyset$ for each $i \neq i_0$.

We suppose that (A_0, A_1) is a left step (the case in which this pair is a right step can be treated in a similar way). Inductively, we check that each A_i can be covered by an open (in X) chain that refines C and has the point p in the first link. Since $X = A_m$, we will conclude that p is a terminal point.

The first step in the induction is as follows.

Since (A_0, A_1) is a left step, there exist $j_1, k, l \in \langle n \rangle$ such that $0 \leq j_1 < k \leq l \leq n$, the subchain $C(k, l)$ exactly covers A_0, the subchain $C(j_1, l)$ exactly covers A_1 and $A_1 \cap C_{j_1} \setminus A_0 \neq \emptyset$. Since $C(k, l)$ exactly covers A_0, we have that A_0 intersects each link $C_k, \ldots, C_l$. By the choice of A_0, we have that $k = i_0 = l$. For this step we consider the chain $\mathcal{D}_1 = \{C_{i_0}, \ldots, C_{j_1}\}$ (here, we take descending indices). Note that $\mathcal{D}_1$ has the following properties: $\mathcal{D}_1$ exactly covers A_1, $\mathcal{D}_1$ refines C, p only belongs to the first link of $\mathcal{D}_1$, the last link of $\mathcal{D}_1$ is contained in C_{j_1}. Note that the pair (A_0, A_1) finishes at j_1.

Since (A_1, A_2) is a right step and, by hypothesis, starts at j_1, there exist $k, j_2 \in \langle n \rangle$ such that $0 \leq j_1 \leq k < j_2 \leq n$, the subchain $C(j_1, k)$ exactly covers A_1, the subchain $C(j_1, j_2)$ exactly covers A_2 and $A_2 \cap C_{j_2} \setminus A_1 \neq \emptyset$. Then the step (A_1, A_2) starts at j_1 and finishes at j_2. Since the last link of $\mathcal{D}_1$ is contained in C_{j_1}, we can apply Lemma 17.5 to obtain a chain $\mathcal{D}_2$ of open subsets of X such that $\mathcal{D}_2$ exactly covers A_2, $\mathcal{D}_2$ refines C, p only belongs to the first link of $\mathcal{D}_2$ and the last link of $\mathcal{D}_2$ is contained in C_{j_2}.

The argument of the last paragraph can be repeated for each of the pairs (A_2, A_3), (A_3, A_4), etcetera. Therefore we can conclude that there exists a chain $\mathcal{D}_m$ of open subsets of X such that $\mathcal{D}_m$ exactly covers X, $\mathcal{D}_m$ refines C and p belongs only to the first link of $\mathcal{D}_m$.

This ends the proof that p is a terminal point of X. □

17.3 An Auxiliary Result

Lemma 17.8 *Let X be a hereditarily indecomposable continuum, let A and B be nonempty disjoint closed subsets of X, and let U_0 and V_0 be open subsets of X such that $A \subset U_0$ and $B \subset V_0$. Then there exist closed subsets F_0, F_1 and F_2 of X such that $X = F_0 \cup F_1 \cup F_2$, $F_0 \cap F_2 = \emptyset$, $A \subset F_0$, $B \subset F_2$, $F_0 \cap F_1 \subset V_0$ and $F_1 \cap F_2 \subset U_0$.*

Proof Let U and V be open subsets of X such that $A \subset U \subset \mathrm{cl}_X(U) \subset U_0$, $B \subset V \subset \mathrm{cl}_X(V) \subset V_0$ and $\mathrm{cl}_X(U) \cap \mathrm{cl}_X(V) = \emptyset$.

Let

$$A_0 = \bigcup\{D \in C(X) : D \text{ is component of } X \setminus V \text{ and } D \cap A \neq \emptyset\}, \text{ and}$$

$$B_0 = \bigcup\{D \in C(X) : D \text{ is component of } X \setminus U \text{ and } D \cap B \neq \emptyset\}.$$

By Exercise 17.21, A_0 and B_0 are closed in X. We claim that $A_0 \cap B_0 = \emptyset$. Suppose to the contrary that there exists a point $p \in A_0 \cap B_0$. Then there exist components D of $X \setminus V$ and E of $X \setminus U$ such that $D \cap A \neq \emptyset \neq E \cap B$ and $p \in D \cap E$. Since X is hereditarily indecomposable, and $D \cap E \neq \emptyset$, we have that either $D \subset E$ or $E \subset D$. Without loss of generality, suppose that $D \subset E$. Then $E \subset X \setminus U$ and $\emptyset \neq A \cap D \subset U \cap E$. This contradiction shows that $A_0 \cap B_0 = \emptyset$.

Then the sets A_0 and $B_0 \setminus V$ are closed, disjoint and they are contained in $X \setminus V$. Since A_0 is a union of components of $X \setminus V$, we have that if a component D of $X \setminus V$ intersects A_0, then $D \subset A_0$. This proves that no component of $X \setminus V$ intersects both sets A_0 and $B_0 \setminus V$. Thus we can apply the Cut Wire Theorem (Theorem 3.3) to the space $X \setminus V$ and to the closed sets A_0 and $B_0 \setminus V$ to obtain disjoint closed subsets D_1 and D_2 of $X \setminus V$ (and then closed in X) such that $X \setminus V = D_1 \cup D_2$, $A_0 \subset D_1$ and $B_0 \setminus V \subset D_2$.

Since B_0 is a union of components of $X \setminus U$, we have that if a component D of $X \setminus U$ intersects B_0, then $D \subset B_0$. Since $D_1 \subset X \setminus V$, $D_1 \cap D_2 = \emptyset$ and $B_0 \subset D_2 \cup V$, we have that $D_1 \cap B_0 = \emptyset$. So we can apply again the Cut Wire Theorem (Theorem 3.3) to the space $X \setminus U$ and closed subsets $D_1 \setminus U$ and B_0 to obtain that there exist disjoint closed subsets E_1 and E_2 of $X \setminus U$ (and therefore closed in X) such that $X \setminus U = E_1 \cup E_2$, $B_0 \subset E_1$ and $D_1 \setminus U \subset E_2$.

We are ready to define the required sets.

Set

$$F_0 = D_1 \cup (\mathrm{cl}_X(V) \cap E_2),$$

$$F_1 = (D_2 \cap E_2) \cup (\mathrm{cl}_X(U) \cap D_2) \cup (\mathrm{cl}_X(V) \cap E_2), \text{ and}$$

$$F_2 = E_1 \cup (\mathrm{cl}_X(U) \cap D_2).$$

Note that the sets F_0, F_1 and F_2 are closed in X.

Since $A \subset A_0 \subset D_1 \subset F_0$ and $B \subset B_0 \subset E_1 \subset F_2$, we have that $A \subset F_0$ and $B \subset F_2$.

Since $U \subset X \setminus V = D_1 \cup D_2$, we have that $U \subset D_1 \cup (U \cap D_2)$. Moreover $V \subset X \setminus U = E_1 \cup E_2$, so $V \subset E_1 \cup (V \cap E_2)$. Then $U \cup V \subset D_1 \cup (U \cap D_2) \cup E_1 \cup (V \cap E_2) \subset F_0 \cup F_2$. Therefore $U \cup V \subset F_0 \cup F_2$.

We check that $X = F_0 \cup F_1 \cup F_2$. Let $p \in X$. If $p \in U \cup V$, by the previous paragraph, $p \in F_0 \cup F_2$. Suppose then that $p \in X \setminus (U \cup V) = (X \setminus U) \cap (X \setminus V) = (E_1 \cup E_2) \cap (D_1 \cup D_2)$. Since $D_1 \subset F_0$ and $E_1 \subset F_2$, we may suppose that $p \in D_2 \cap E_2 \subset F_1$. This finishes the proof that $X = F_0 \cup F_1 \cup F_2$.

Given $p \in F_0 \cap F_1$, if $p \notin \mathrm{cl}_X(V)$, by the definition of F_0 and F_1, we have that $p \in D_1 \cap D_2$, which is impossible. Hence $F_0 \cap F_1 \subset \mathrm{cl}_X(V) \subset V_0$.

Given $p \in F_1 \cap F_2$, if $p \notin \mathrm{cl}_X(U)$, by the definition of F_1 and F_2, we have that $p \in E_1 \cap E_2$, which is impossible. Thus $F_1 \cap F_2 \subset \mathrm{cl}_X(U) \subset U_0$.

Now we show that $F_0 \cap F_2 = \emptyset$. Suppose to the contrary that there exists a point $p \in F_0 \cap F_2$. Since $D_1 \cap D_2 = \emptyset$, $E_1 \cap E_2 = \emptyset$ and $\mathrm{cl}_X(V) \cap \mathrm{cl}_X(U) = \emptyset$, by the definition of F_0 and F_2, we conclude that $p \in D_1 \cap E_1$. Since $E_1 \subset X \setminus U$, we have that $p \in (D_1 \setminus U) \cap E_1 \subset E_2 \cap E_1 = \emptyset$, which is impossible. Therefore $F_0 \cap F_2 = \emptyset$. □

17.4 Patterns

Definition 17.9 A *pattern* is a function $\lambda : \langle m \rangle \to \langle n \rangle$, where $n, m \in \mathbb{N}$ and $|\lambda(i) - \lambda(i-1)| \leq 1$ for each $i \in \{1, \ldots, m\}$. The pattern λ is *strict* if $\lambda(0) = 0$, $\lambda(m) = n$ and $|\lambda(i) - \lambda(i-1)| = 1$ for each $i \in \{1, \ldots, m\}$. A strict pattern λ is *non-trivial* if $m > n$. A *wiggle* for a pattern λ is a quadruple of integers (i, j, k, l) satisfying the following:

(a) $0 \leq i < j < k < l \leq m$,
(b) $\lambda(i) = \lambda(k) \neq \lambda(j) = \lambda(l)$, and
(c) $\min\{\lambda(i), \lambda(j)\} \leq \lambda(r) \leq \max\{\lambda(i), \lambda(j)\}$ for every $r \in \{i, \ldots, l\}$.

The wiggle (i, j, k, l) is *straight* if λ restricted to each of the sets $\{i, \ldots, j\}$, $\{j, \ldots, k\}$ and $\{k, \ldots, l\}$ is strictly monotone. Observe that in this case we have that $l - k = k - j = j - i$, so $j - l = j - k + k - l = 2(i - j)$.

Given chains $C = \{C_0, C_1, \ldots, C_n\}$ and $\mathcal{D} = \{D_0, D_1, \ldots, D_m\}$, and a pattern $\lambda : \langle m \rangle \to \langle n \rangle$, the chain $\mathcal{D}$ *follows the pattern* λ *in* C if for each $i \in \langle m \rangle$, $D_i \subset C_{\lambda(i)}$.

Lemma 17.10 *If* $\lambda : \langle m \rangle \to \langle n \rangle$ *is a strict, non-trivial pattern, then* λ *has a straight wiggle.*

Proof We start by proving the following claim.

Claim If $s, t \in \langle m \rangle$ satisfy $s < t$, $\min\{\lambda(s), \lambda(t)\} \leq \lambda(r) \leq \max\{\lambda(s), \lambda(t)\}$ for every $r \in \{s, \ldots, t\}$, and λ is not monotone in $\{s, \ldots, t\}$, then λ has a wiggle in $\{s, \ldots, t\}$.

In order to prove this claim, first note that since λ is strict, $\lambda(s+1) = \lambda(s) + 1$ or $\lambda(s+1) = \lambda(s) - 1$. We suppose that $\lambda(s+1) = \lambda(s) + 1$, the other case being similar. Then we have that $\lambda(s) < \lambda(s+1)$, so $\lambda(s) \leq \lambda(r) \leq \lambda(t)$ for every $r \in \{s, \ldots, t\}$.

Let $j = \max\{r \in \{s, \ldots, t\} : \lambda \text{ is monotone in } \{s, \ldots, r\}\}$. This set is nonempty because it contains the element $s+1$. Since $\lambda(s) < \lambda(s+1)$ and λ is strict, we have that λ is strictly increasing in $\{s, \ldots, j\}$.

Note that $\lambda(s) < \lambda(j)$. By our assumption, $s + 1 \leq j < t$. By the maximality of j, we have that $\lambda(j+1) = \lambda(j) - 1$.

Since λ is strict, the values of λ move along one by one. Since $\lambda(j+1) < \lambda(j) \leq \lambda(t)$ and the values of λ move along one by one, we have that there exists an $l \in \{j+1, \ldots, t\}$ such that $\lambda(l) = \lambda(j)$. Suppose that l is minimal with this property. Given $r \in \{j+1, \ldots, l\}$, the minimality of l and the fact that the values of λ move along one by one imply that $\lambda(r) \leq \lambda(l) = \lambda(j)$.

Let $j \leq k \leq l$ be such that $\lambda(k)$ is the minimum of the set $\lambda(\{j, \ldots, l\})$.

Since $\lambda(j) = \lambda(l)$ and $\lambda(j+1) < \lambda(j)$, we have that $\lambda(k) \leq \lambda(j+1)$. This implies that $j < k < l$. The minimality of $\lambda(k)$ implies that $\lambda(k) \leq \lambda(r) \leq \lambda(j)$ for every $r \in \{j, \ldots, l\}$.

Since $\lambda(s) \leq \lambda(k) < \lambda(j)$ and the values of λ move along one by one, we have that there exists an $i \in \{s, \ldots, j\}$ such that $\lambda(i) = \lambda(k)$. Since λ is strictly increasing in the set $\{s, \ldots, j\}$, we have that $\lambda(i) \leq \lambda(r) \leq \lambda(j)$ for every $r \in \{i, \ldots, j\}$.

Therefore the quadruple (i, j, k, l) satisfies the conditions for being a wiggle for λ. This finishes the proof of the claim.

Since the pattern is non-trivial, $m > n$ and λ is not one-to-one. Then λ is not strictly monotone. Since λ is strict, we have that λ cannot be monotone. Therefore λ satisfies the hypothesis of the claim for $s = 0$ and $t = m$. Thus λ has a wiggle in $\{0, \ldots, m\}$. Then we can take a wiggle (i, j, k, l) for λ with the property that $l - i$ is minimal.

We check that (i, j, k, l) is a straight wiggle for λ.

If λ is not monotone in $\{i, \ldots, j\}$, then we can apply the claim and conclude that λ has a wiggle (i', j', k', l') in $\{i, \ldots, j\}$. Then $l' - i' \leq j - i < l - i$, contradicting the minimality of $l - i$. Thus λ is monotone in $\{i, \ldots, j\}$ and since λ is strict, we have that λ is strictly monotone in $\{i, \ldots, j\}$.

In a similar way it can be shown that λ is strictly monotone in each of the sets $\{j, \ldots, k\}$ and $\{k, \ldots, l\}$. Therefore λ is a straight wiggle. □

Theorem 17.11 *Suppose that X is a hereditarily indecomposable continuum. Let $\mathcal{C} = \{C_0, C_1, \ldots, C_n\}$ be a proper covering chain of X. Let A and B be nonempty closed subsets of X such that $A \subset C_0 \setminus \mathrm{cl}_X(C_1)$ and $B \subset C_n \setminus \mathrm{cl}_X(C_{n-1})$. Let $\lambda : \langle m \rangle \to \langle n \rangle$ be a pattern such that $\lambda(0) = 0$ and $\lambda(m) = n$ (then λ is onto and $m \geq n$). Then there is a taut chain $\mathcal{D} = \{D_0, D_1, \ldots, D_m\}$ of X that follows the pattern λ in $\mathcal{C}$ and satisfies $A \subset D_0 \setminus \mathrm{cl}_X(D_1)$ and $B \subset D_m \setminus \mathrm{cl}_X(D_{m-1})$.*

Proof We divide the proof into two steps.

First Step. λ is a strict pattern.

We start constructing a proper covering chain $\mathcal{E} = \{E_0, E_1, \dots, E_m\}$ satisfying the required conditions with the possible exemption that $\mathcal{E}$ is taut. The construction of $\mathcal{E}$ will be done for each $m \geq n$, by induction on m.

If $m = n$, then since λ is strict, we have that $\lambda(0) = 0$ and $\lambda(n) = n$. Moreover, since the values of λ move along one by one, we have that λ is surjective and also one-to-one. This implies that λ is the identity. In this case we put $\mathcal{E} = C$ and we are done.

Now, suppose that $m > n$ and it is possible to construct $\mathcal{E}$ for the numbers $n, n+1, \dots, m-1$. We construct the chain for m.

Since λ is a strict pattern, by Lemma 17.10, there exists a straight wiggle (i, j, k, l) for λ.

To use the induction hypothesis, we define a pattern $\mu : \langle m + j - 1 \rangle \to \langle n \rangle$ as follows:

$$\mu(r) = \begin{cases} \lambda(r), & \text{if } 0 \leq r \leq j, \\ \lambda(r + l - j), & \text{if } j \leq r \leq m + j - l. \end{cases}$$

Note that μ and λ coincide from 0 to j, and from j to $m + j - l$, we have: $\mu(j) = \lambda(l) = \lambda(j)$, $\mu(j+1) = \lambda(l+1), \dots, \mu(m+j-l) = \lambda(m)$. Then μ is well defined and μ "avoids the wiggle". Since $0 \leq i < j < l \leq m$, we have that $1 < m + j - l < m$. Clearly, for each $r \in \{0, 1, \dots, m + j - l - 1\}$, $|\mu(r) - \mu(r+1)| = 1$. Moreover $\mu(0) = \lambda(0) = 0$ and $\mu(m+j-l) = \lambda(m) = n$. Therefore μ is a strict pattern. This implies that μ is surjective and $n \leq m + j - l$.

Now we can apply the induction hypothesis. Thus there exists a proper covering chain $\mathcal{G} = \{G_0, G_1, \dots, G_{m+j-l}\}$ of X that follows the pattern μ in C, $A \subset G_0 \setminus \operatorname{cl}_X(G_1)$ and $B \subset G_{m+j-l} \setminus \operatorname{cl}_X(G_{m+j-l-1})$.

We need to recover the wiggle that we eliminated when we defined μ. In order to do this, we start defining some sets.

Let

$$U = G_0 \cup \dots \cup G_i,$$

$$V = G_j \cup \dots \cup G_{m+j-l},$$

$$A_0 = X \setminus (G_{i+1} \cup \dots \cup G_{m+j-l}), \text{ and}$$

$$B_0 = X \setminus (G_0 \cup \dots \cup G_{j-1}).$$

Note that A_0 and B_0 are closed. Since $i \leq j-1$, $G_0 \cup \dots \cup G_i \subset G_0 \cup \dots \cup G_{j-1}$, so $G_0 \cup \dots \cup G_{j-1} \cup G_{i+1} \cup \dots \cup G_{m+j-l} = X$. This implies that $A_0 \cap B_0 = \emptyset$ and $A_0 \subset U$. Since $j - 1 < m + j - 1$, we have that $B \subset B_0$.

Since $B_0 \subset V$, $\emptyset \neq A \subset A_0$ and $\emptyset \neq B \subset B_0$, we can apply Lemma 17.8 to the continuum X, the closed sets A_0 and B_0, and the open sets U and V. Then there exist closed subsets F_0, F_1 and F_2 of X such that

$$X = F_0 \cup F_1 \cup F_2,\ F_0 \cap F_2 = \emptyset,\ A_0 \subset F_0,\ B_0 \subset F_2,\ F_0 \cap F_1 \subset V \text{ and } F_1 \cap F_2 \subset U.$$

Define $\mathcal{E} = \{E_0, E_1, \ldots, E_m\}$ as follows:

$$\begin{aligned}
E_0 &= G_0 \setminus (F_1 \cup F_2),\ E_1 = G_1 \setminus (F_1 \cup F_2), \ldots, \\
E_{j-1} &= G_{j-1} \setminus (F_1 \cup F_2); \\
E_j &= G_j \setminus F_2; \\
E_{j+1} &= G_{j-1} \setminus (F_0 \cup F_2),\ E_{j+2} = G_{j-2} \setminus (F_0 \cup F_2), \ldots, \\
E_{j+(j-i-1)} &= G_{j-(j-i-1)} \setminus (F_0 \cup F_2) = G_{i+1} \setminus (F_0 \cup F_2); \\
E_{2j-i} &= G_i \setminus F_0; \\
E_{2j-i+1} &= G_{i+1} \setminus (F_0 \cup F_1),\ E_{2j-i+2} = G_{i+2} \setminus (F_0 \cup F_1), \ldots, \\
E_{2j-i+(m-2j+i)} &= E_m = G_{i+m-2j+i} \setminus (F_0 \cup F_1) = G_{m-2j+2i} \setminus (F_0 \cup F_1).
\end{aligned}$$

Clearly, for each $i \in \{0, 1, \ldots, m\}$, E_i is an open subset of X.

We check that $\mathcal{E}$ covers X. Taking $p \in X$, we analyze five cases.

Case 1. $p \in G_0 \cup \cdots \cup G_{i-1} \subset U$.

Since $\mathcal{G}$ is a chain and $i < j$, we have that $p \in A_0 \subset F_0$ and $p \notin V$. Thus $p \notin F_2$ and since $F_0 \cap F_1 \subset V$, we have that $p \notin F_1$. Then $p \notin F_1 \cup F_2$. This implies that $p \in E_0 \cup \cdots \cup E_{i-1}$.

Case 2. $p \in G_i \subset U$.

If $p \notin F_0$, then $p \in E_{2j-i}$. Suppose then that $p \in F_0$. Hence $p \notin F_2$. If $p \notin F_1$, then $p \in E_i$. Suppose then that $p \in F_1$. In this case, $p \in F_0 \cap F_1 \subset V$, so $p \in U \cap V$. Since $\mathcal{G}$ is a chain, $p \in G_i \cap G_j$. Therefore $p \in E_j$.

Case 3. $p \in G_{i+1} \cup \cdots \cup G_{j-1}$.

If $p \notin F_0 \cup F_2$, we have that $p \in E_{j+1} \cup \cdots \cup E_{2j-i-1}$. Then we assume that $p \in F_0 \cup F_2$.

Suppose first that $p \in F_0$. Then $p \notin F_2$. If $p \notin F_1$, we have that $p \notin F_1 \cup F_2$, so $p \in E_{i+1} \cup \cdots \cup E_{j-1}$. If $p \in F_1$, then $p \in F_0 \cap F_1 \subset V$. Thus $p \in G_{j-1} \cap G_j$. Hence $p \in E_j$.

Now suppose that $p \in F_2$. Thus $p \notin F_0$. If $p \notin F_1$, then $p \notin F_0 \cup F_1$, so $p \in E_{2j-i+1} \cup \cdots \cup E_m$. If $p \in F_1$, then $p \in F_1 \cap F_2 \subset U$. Thus $p \in G_i \cap G_{i+1}$. Therefore $p \in E_{2j-i}$.

Case 4. $p \in G_j \subset V$.

If $p \notin F_2$, we have $p \in E_j$. Suppose then that $p \in F_2$. Thus $p \notin F_0$. If $p \notin F_1$, then $p \in E_{2j-i+(j-i)}$. If $p \in F_1$, we have $p \in F_1 \cap F_2 \subset U$. Thus $p \in U \cap V$. Hence $p \in G_i \cap G_j$. Therefore $p \in E_{2j-i}$.

Case 5. $p \in G_{j+1} \cup \cdots \cup G_{m+j-l}$.

Since $\mathcal{G}$ is a chain and $i < j$, we have that $p \in B_0 \subset F_2$ and $p \notin U$. Thus $p \notin F_0$, and since $F_1 \cap F_2 \subset U$, we have that $p \notin F_1$. Then $p \notin F_0 \cup F_1$. This implies that $p \in E_{2j-i+(j-i+1)} \cup \cdots \cup E_m$ (since (i, j, k, l) is a straight wiggle, we have that $j - l = 2(i - j)$, so $G_{m+j-l} = G_{m-2j+2i}$).

Since the five cases cover all the possibilities, we conclude that $\mathcal{E}$ covers X.

We check that if $E_r \cap E_s \neq \emptyset$, then $|r - s| \leq 1$. That is, the sets E_r can only intersect if they have consecutive indices.

The sets $E_0, \ldots, E_m$ are divided into three (non-disjoint) groups, namely, $E_0, \ldots, E_j$, $E_j, \ldots, E_{2j-i}$ and $E_{2j-i}, \ldots, E_m$. For each group, the definition of the sets E_r is given in terms of the elements of the chain $\mathcal{G}$. Thus, for each group the sets E_r can only intersect if they are consecutive. Given a point p in $E_0 \cup \cdots \cup E_{j-1}$, we have that $p \notin F_1 \cup F_2$. If $p \in E_{j+1} \cup \cdots \cup E_m$, then $p \notin F_0$. This contradicts the fact that $X = F_0 \cup F_1 \cup F_2$, and proves that $(E_0 \cup \cdots \cup E_{j-1}) \cap (E_{j+1} \cup \cdots \cup E_m) = \emptyset$.

Similarly, $(E_j \cup \cdots \cup E_{2j-i-1}) \cap (E_{2j-i+1} \cup \cdots \cup E_m) = \emptyset$. Therefore the sets E_r can only intersect if they have consecutive indices.

Given $p \in A \subset A_0 \subset F_0$, since $p \in G_0 \setminus \mathrm{cl}_X(G_1)$, we have that $p \notin V$. Then $p \notin F_2$ and since $F_0 \cap F_1 \subset V$, we have that $p \notin F_1$. Thus $p \in E_0$. This shows that $A \subset E_0$ and since $\mathrm{cl}_X(E_1) \subset \mathrm{cl}_X(G_1)$, we conclude that $A \subset E_0 \setminus \mathrm{cl}_X(E_1)$. Similarly, $B \subset E_m \setminus \mathrm{cl}_X(E_{m-1})$.

By Exercise 17.20, we obtain that for each $r \in \{1, \ldots, m\}$, $E_{r-1} \cap E_r \neq \emptyset$. Hence $\mathcal{E}$ is a chain of open sets covering X. Since $\emptyset \neq A \subset E_0 \setminus \mathrm{cl}_X(E_1)$ and $\emptyset \neq B \subset E_m \setminus \mathrm{cl}_X(E_{m-1})$, we have that $\mathcal{E}$ is a proper covering chain of X.

We show that $\mathcal{E}$ follows the pattern λ in C. Let $r \in \langle m \rangle$. We analyze three cases.

Case 1. $0 \leq r \leq j$.

Since $\mathcal{G}$ follows the pattern μ in C, we have that $G_r \subset C_{\mu(r)}$, and since $\mu(r) = \lambda(r)$, the definition of E_r implies that $E_r \subset G_r \subset C_{\mu(r)} = C_{\lambda(r)}$.

Case 2. $j + 1 \leq r \leq 2j - i$.

By the definition of E_r, we have $E_r \subset G_{j-(r-j)}$. Note that $2j - (j + 1) \geq 2j - r \geq 2j - (2j - i)$, so $i \leq 2j - r \leq j - 1$. Since $\mathcal{G}$ follows the pattern μ in C, we have that $G_{2j-r} \subset C_{\mu(2j-r)} = C_{\lambda(2j-r)}$. Recall that (i, j, k, l) is a straight wiggle for λ. This implies that λ is strictly monotone in $\{i, \ldots, j\}$ and in $\{j, \ldots, k\}$. If $\lambda(i) < \lambda(j)$, then λ takes the values, in increasing order, one by one in the segment $i, \ldots, j$ and then takes the values in decreasing order, one by one, in the set $\{j, \ldots, k\}$. Then $\lambda(j+1) = \lambda(j-1)$, $\lambda(j+2) = \lambda(j-2)$, etc. Similarly, these equalities also hold in the case when $\lambda(i) > \lambda(j)$. Thus $\lambda(2j-r) = \lambda(j+j-r) = \lambda(j - (j - r)) = \lambda(r)$. Hence $E_r \subset G_{2j-r} \subset C_{\lambda(2j-r)} = C_{\lambda(r)}$.

Case 3. $2j - i + 1 \leq r \leq m$.

By the definition of E_r, $E_r \subset G_{r-2j+2i} \subset C_{\mu(r-2j+2i)}$. Since $2j - i + 1 \leq r$, we have that $i + 1 \leq r - 2j + 2i$.

If $r - 2j + 2i < j$, by the definition of μ, then we have that $\mu(r - 2j + 2i) = \lambda(r - 2j + 2i)$. Moreover, $2j - i + 1 \leq r < 2j - i + (j - i)$, so there exists $1 \leq s < j - i$ such that $r = 2j - i + s$. Since (i, j, k, l) is a straight wiggle for λ,

$\lambda(i+s) = \lambda(k+s) = \lambda(2j-i+s)$. Thus $\mu(r-2j+2i) = \lambda(r-2j+2i) = \lambda(i+s) = \lambda(2j-i+s) = \lambda(r)$. Therefore $\mu(r-2j+2i) = \lambda(r)$.

If $j \leq r-2j+2i$, then by the definition of μ, we have that $\mu(r-2j+2i) = \lambda(r-2j+2i+l-j) = \lambda(r+2i+l-3j)$. Since (i, j, k, l) is a straight wiggle for λ, we have that $j-l = 2(i-j)$. Thus $r+2i+l-3j = r+2i-2j+(l-j) = r$. Therefore $\mu(r-2j+2i) = \lambda(r)$.

We have shown that in both cases $\mu(r-2j+2i) = \lambda(r)$. Therefore $E_r \subset C_{\lambda(r)}$.

This completes the proof that $\mathcal{E}$ follows the pattern λ in C.

By Theorem 14.10, there exists a taut chain $\mathcal{D} = \{D_0, D_1, \ldots, D_m\}$ of X satisfying that for each $r \in \{0, 1, \ldots, m\}$, $\mathrm{cl}_X(D_r) \subset E_r$, $A \subset D_0 \setminus \mathrm{cl}_X(D_1) \subset E_0 \setminus \mathrm{cl}_X(E_1)$ and $B \subset D_m \setminus \mathrm{cl}_X(D_{m-1}) \subset E_m \setminus \mathrm{cl}_X(E_{m-1})$.

This completes the first step.

Second Step. λ is not necessarily a strict pattern.

We start constructing a division of the set $\{0, 1, \ldots, m\}$ in blocks to obtain an appropriate strict pattern.

Let j_0 be the maximum element in $\{0, 1, \ldots, m\}$ such that λ is constant in the set $\{0, 1, \ldots, j_0\}$. Since $\lambda(0) = 0$ and $\lambda(m) = n \geq 1$, we have that $j_0 < m$. Observe that $\lambda(j_0) \neq \lambda(j_0+1)$.

Let j_1 be the maximum element in $\{j_0+1, \ldots, m\}$ such that λ is constant in the set $\{j_0+1, \ldots, j_1\}$. If $j_1 < m$ we have that $\lambda(j_1) \neq \lambda(j_1+1)$.

If $j_1 < m$, then we can define j_2 as the maximum element in $\{j_1+1, \ldots, m\}$ such that λ is constant in the set $\{j_1+1, \ldots, j_2\}$.

Continuing in this way it is possible to define a sequence

$$0 \leq j_0 < j_1 < \cdots < j_r = m$$

such that λ is constant in the blocks

$$\{0, \ldots, j_0\},\ \{j_0+1, \ldots, j_1\},\ \{j_1+1, \ldots, j_2\},\ \ldots,\ \{j_{r-1}+1, \ldots, j_r\}.$$

Moreover, $1 \leq r$ and $0 = \lambda(0) = \lambda(j_0) \neq \lambda(j_0+1) = \lambda(j_1) \neq \lambda(j_1+1) = \lambda(j_2) \neq \cdots \neq \lambda(j_{r-2}+1) = \lambda(j_{r-1}) \neq \lambda(j_{r-1}+1) = \lambda(j_r) = \lambda(m) = n$.

Since λ is a pattern, we have that

$$1 = |\lambda(j_0) - \lambda(j_1)| = |\lambda(j_1) - \lambda(j_2)| = \cdots = |\lambda(j_{r-1}) - \lambda(j_r)|.$$

Define $\gamma : \langle r \rangle \to \langle n \rangle$ by $\gamma(i) = \lambda(j_i)$. Then γ is a strict pattern such that $\gamma(0) = \lambda(j_0) = 0$ and $\gamma(r) = \lambda(j_r) = n$.

By the first step, there exists a taut chain $\mathcal{F} = \{F_0, F_1, \ldots, F_r\}$ of X that follows the pattern γ in C and satisfies $A \subset F_0 \setminus \mathrm{cl}_X(F_1)$ and $B \subset F_r \setminus \mathrm{cl}_X(F_{r-1})$.

Given $i \in \{0, 1, \ldots, r\}$, let $t_{j_i} = \frac{i+1}{r+1}$. Apply Urysohn's Lemma for metric spaces (Exercise 1.55) to obtain a mapping $g_i : \mathrm{cl}_X(F_i) \to [\frac{i}{r+1}, \frac{i+1}{r+1}]$ satisfying:

(a) $\mathrm{cl}_X(F_{i-1}) \cap \mathrm{cl}_X(F_i) = g_i^{-1}(\{\frac{i}{r+1}\})$, $\mathrm{cl}_X(F_i) \cap \mathrm{cl}_X(F_{i+1}) = g_i^{-1}(\{\frac{i+1}{r+1}\})$, if $0 < i < r$,
(b) $A = g_0^{-1}(\{0\})$ and $\mathrm{cl}_X(F_0) \cap \mathrm{cl}_X(F_1) = g_0^{-1}(\{\frac{1}{r+1}\})$, and
(c) $\mathrm{cl}_X(F_{r-1}) \cap \mathrm{cl}_X(F_r) = g_0^{-1}(\{\frac{r}{r+1}\})$ and $B = g_0^{-1}(\{1\})$.

Set $t_{j_{-1}} = 0$ and fix a partition

$$T_i : \frac{i}{r+1} = t_{j_{i-1}} < t_{j_{i-1}+1} < \cdots < t_{j_i} = \frac{i+1}{r+1}$$

of the interval $[\frac{i}{r+1}, \frac{i+1}{r+1}]$.

In this way, for each $k \in \{-1, 0, 1, \ldots, m\}$, we have defined a number t_k, and taking the union of the partitions T_i, we obtain a partition

$$T : 0 = t_{-1} < t_0 < \cdots < t_{j_r-1} < t_{j_r} = t_m = 1$$

of the interval $[0, 1]$.

Let $g : X \to [0, 1]$ be the mapping that extends all the mappings g_i. Choose $\delta > 0$ such that $3\delta < \min\{t_k - t_{k-1} : 0 \leq k \leq m\}$. Given $i \in \{0, 1, \ldots, r\}$, $g^{-1}([t_{j_{i-1}}, t_{j_i}]) = g^{-1}([\frac{i}{r+1}, \frac{i+1}{r+1}]) \subset \mathrm{cl}_X(F_i) \subset C_{\gamma(i)}$, so we can ask that $g^{-1}([t_{j_{i-1}} - \delta, t_{j_i} + \delta]) \subset C_{\gamma(i)}$ (Exercise 1.54).

Given $k \in \{0, 1, \ldots, m\}$, define $D_k = g^{-1}((t_{k-1} - \delta, t_k + \delta))$. Observe that the family $\mathcal{D} = \{D_0, D_1, \ldots, D_m\}$ is a taut chain of X and satisfies $A \subset D_0 \setminus \mathrm{cl}_X(D_1)$ and $B \subset D_m \setminus \mathrm{cl}_X(D_{m-1})$.

Finally, we check that $\mathcal{D}$ follows the pattern λ in C. Let $k \in \{0, 1, \ldots, m\}$. Then there exists an $i \in \{0, 1, \ldots, r\}$ such that $[t_{k-1}, t_k] \subset [\frac{i}{r+1}, \frac{i+1}{r+1}] = [t_{j_{i-1}}, t_{j_i}]$, $t_{j_{i-1}} \leq t_{k-1} < t_k \leq t_{j_i}$ and $j_{i-1} \leq k - 1 < k \leq j_i$. So, $\lambda(k) = \lambda(j_i) = \gamma(i)$ Thus $D_k = g^{-1}((t_{k-1} - \delta, t_{k-1} + \delta)) \subset g^{-1}([t_{j_{i-1}} - \delta, t_{j_i} + \delta]) \subset C_{\gamma(i)} = C_{\lambda(k)}$. Therefore $\mathcal{D}$ follows the pattern λ in C. □

17.5 Stronger Properties of the Pseudo-Arc

Theorem 17.12 *Let X and Y be hereditarily indecomposable chainable continua, $p \in X$ and $q \in Y$. Then there exists a homeomorphism $h : X \to Y$ such that $h(p) = q$.*

Proof First, we will show, inductively, that it is possible to find two sequences of taut chains $\{\mathcal{C}_n\}_{n=1}^{\infty}$ and $\{\mathcal{D}_n\}_{n=1}^{\infty}$ of X and Y, respectively, such that for each $n \in \mathbb{N}$ they satisfy the following:

(a) $\mathcal{C}_n = \{C_0^{(n)}, \ldots, C_{m_n}^{(n)}\}$ and $\mathcal{D}_n = \{D_0^{(n)}, \ldots, D_{m_n}^{(n)}\}$,
(b) $\text{mesh}(\mathcal{C}_n)$ and $\text{mesh}(\mathcal{D}_n)$ are less than $\frac{1}{2^n}$,
(c) there exists a pattern $\lambda_n : \langle m_{n+1}\rangle \to \langle m_n\rangle$ such that $\mathcal{C}_{n+1}$ (respectively, $\mathcal{D}_{n+1}$) follows the pattern λ_n in $\mathcal{C}_n$ (respectively, in $\mathcal{D}_n$).
(d) $p \in C_0^{(n)} \setminus \text{cl}_X(C_1^{(n)})$ and $q \in D_0^{(n)} \setminus \text{cl}_Y(D_1^{(n)})$.

Since X and Y are hereditarily indecomposable, p is a final (and therefore terminal) point of X and q is a terminal point of Y. By Lemma 17.2 (b), there exist two taut chains $\mathcal{C}_1 = \{C_0^{(1)}, \ldots, C_{m_1}^{(1)}\}$ of X and $\mathcal{D}_1 = \{D_0^{(1)}, \ldots, D_{m_1}^{(1)}\}$ of Y, both with mesh less than 1 and satisfying $p \in C_0^{(1)} \setminus \text{cl}_X(C_1^{(1)})$ and $q \in D_0^{(1)} \setminus \text{cl}_Y(D_1^{(1)})$. This ends the first step of the induction.

The inductive step will be consequence of the following claim.

Claim Let $\mathcal{C} = \{C_0, C_1, \ldots, C_m\}$ and $\mathcal{D} = \{D_0, D_1, \ldots, D_m\}$ be taut chains of X and Y, respectively. Let $\varepsilon > 0$, $x \in C_0 \setminus \text{cl}_X(C_1)$ and $y \in D_0 \setminus \text{cl}_Y(D_1)$. Then there exist taut chains of $\mathcal{E} = \{E_0, E_1, \ldots, E_r\}$ and $\mathcal{F} = \{F_0, F_1, \ldots, F_r\}$ of X and Y, respectively, and there exists a pattern $\lambda : \langle r\rangle \to \langle m\rangle$ such that $\mathcal{E}$ follows the pattern λ in $\mathcal{C}$, $\mathcal{F}$ follows the pattern λ in $\mathcal{D}$, $x \in E_0 \setminus \text{cl}_X(E_1)$, $y \in F_0 \setminus \text{cl}_Y(F_1)$ and the mesh of $\mathcal{F}$ is less that ε.

We prove this claim. Fix a point $z \in D_m \setminus \text{cl}_Y(D_{m-1})$. Let $\delta > 0$ be such that $\delta < \varepsilon$ and δ is a Lebesgue number for the family $\mathcal{D}$ (see Exercise 1.26). Since Y is chainable and y is a terminal point, by Lemma 17.2 (b), there exists a taut chain $\mathcal{F} = \{F_0, F_1, \ldots, F_r\}$ of Y such that $y \in F_0 \setminus \text{cl}_Y(F_1)$, $r \geq m$ and its mesh is less than ε. Since δ is a Lebesgue number for $\mathcal{D}$, we have that $\mathcal{F}$ refines $\mathcal{D}$.

For each $i \in \langle r\rangle$, we can choose a number $\lambda(i) \in \langle m\rangle$ such that F_i is contained in $D_{\lambda(i)}$. Given $i \in \{1, 2, \ldots, r\}$, we have that $\emptyset \neq F_{i-1} \cap F_i \subset D_{\lambda(i-1)} \cap D_{\lambda(i)}$, so $|\lambda(i-1) - \lambda(i)| \leq 1$. Therefore λ is a pattern and by definition $\mathcal{F}$ follows the pattern λ in $\mathcal{D}$. Note that D_0 is the only element of $\mathcal{D}$ containing y. Since $y \in F_0$, we have that the only link of the chain $\mathcal{D}$ containing F_0 is D_0. Thus $\lambda(0) = 0$. Since D_m is the only element of $\mathcal{D}$ containing z, taking $j_0 \in \langle r\rangle$ such that $z \in F_{j_0}$, we have that $\lambda(j_0) = m$. Observe that $j_0 > 0$.

Define a function $\mu : \langle 2r - j_0\rangle \to \langle m\rangle$ in the following way.

$$\mu(i) = \begin{cases} \lambda(i), & \text{if } 0 \leq i \leq r, \\ \lambda(2r - i), & \text{if } r \leq i \leq 2r - j_0. \end{cases}$$

Since $\mu(2r - j_0) = \lambda(2r - (2r - j_0)) = \lambda(j_0) = m$, we have that μ is a well-defined pattern satisfying $\mu(0) = 0$ and $\mu(2r - j_0) = m$.

Choose a point $w \in C_m \setminus \text{cl}_X(C_{m-1})$.

Apply Theorem 17.11 to the chain C, the sets $A = \{x\}$, $B = \{w\}$ and the pattern μ. Then there exists a taut chain $\mathcal{G} = \{G_0, G_1, \ldots, G_{2r-j_0}\}$ of X that follows the pattern μ in C and satisfies $x \in G_0 \setminus \mathrm{cl}_X(G_1)$ and $w \in G_{2r-j_0} \setminus \mathrm{cl}_X(G_{2r-j_0-1})$.

Define $\mathcal{E} = \{E_0, E_1, \ldots, E_r\}$ as follows.

$$\begin{aligned} E_0 &= G_0, \ldots, E_{j_0-1} = G_{j_0-1}, \\ E_{j_0} &= G_{j_0} \cup G_{2r-j_0}, \\ E_{j_0+1} &= G_{j_0+1} \cup G_{2r-j_0-1}, \ldots, E_{r-1} = G_{r-1} \cup G_{r+1}, \\ E_r &= G_r. \end{aligned}$$

We check that $\mathcal{E}$ is a taut chain.

It is clear that the elements of $\mathcal{E}$ are open and their union contains the union of all sets G_i. Thus $\mathcal{E}$ covers X.

Since $x \in G_0$ and x does not belong to $\mathrm{cl}_X(G_1) \cup \cdots \cup \mathrm{cl}_X(G_{2r-j_0})$, we have that $x \in E_0 \setminus \mathrm{cl}_X(E_1)$.

Let $i \in \{1, \ldots, r\}$. Since $\emptyset \neq G_{i-1} \cap G_i \subset E_{i-1} \cap E_i$, we have that $E_{i-1} \cap E_i \neq \emptyset$.

Take $0 \leq k < l \leq r$ such that $\mathrm{cl}_X(E_k) \cap \mathrm{cl}_X(E_l) \neq \emptyset$. We check that $l = k + 1$.

If $l \leq j_0 - 1$, then $\mathrm{cl}_X(E_k) \cap \mathrm{cl}_X(E_l) \subset \mathrm{cl}_X(G_k) \cap \mathrm{cl}_X(G_l)$, so $l = k + 1$.

If $k \leq j_0 - 1 < l$, $\emptyset \neq \mathrm{cl}_X(E_k) \cap \mathrm{cl}_X(E_l) \subset \mathrm{cl}_X(E_0 \cup \cdots \cup E_{j_0-1}) \cap \mathrm{cl}_X(E_{j_0} \cup \cdots \cup E_r) = \mathrm{cl}_X(G_0 \cup \cdots \cup G_{j_0-1}) \cap \mathrm{cl}_X(G_{j_0} \cup \cdots \cup G_{2r-j_0})$, so $k = j_0 - 1$ and $l = j_0$.

Now, suppose that $j_0 \leq k < l$. Observe that

$$\mathrm{cl}_X(E_l \cup \cdots \cup E_r) = \mathrm{cl}_X(G_l \cup \cdots \cup G_{r-1} \cup G_r \cup G_{r+1} \cup \cdots \cup G_{2r-l-1} \cup G_{2r-l})$$

so this set is not intersected by $\mathrm{cl}_X(G_0 \cup \cdots \cup G_{l-2} \cup G_{2r-l+2} \cup \cdots \cup G_{2r-j_0}) \supset \mathrm{cl}_X(E_0 \cup \cdots \cup E_{l-2})$. Thus $k = l - 1$.

Finally, since $\mathcal{G}$ is a taut chain, if $j_0 < r$, then we have that $r + 1 \leq 2r - j_0$, so the connectedness of X implies that $\emptyset \neq G_r \setminus \mathrm{cl}_X((G_0 \cup \cdots \cup G_{r-1}) \cup (G_{r+1} \cup \cdots \cup G_{2r-j_0}) = E_r \setminus \mathrm{cl}_X(E_0 \cup \cdots \cup E_{r-1})$ and if $j_0 = r$, then $E_{r-1} = G_{r-1}$ and $E_r = G_r$. Thus $E_r \setminus \mathrm{cl}_X(E_{r-1}) \neq \emptyset$.

We have shown that $\mathcal{E}$ is a taut chain.

We show that $\mathcal{E}$ follows the pattern λ in C.

Let $i \in \langle r \rangle$.

If $0 \leq i < j_0$, then $E_i = G_i \subset C_{\mu(i)} = C_{\lambda(i)}$.

If $j_0 \leq i \leq r$, then $r \leq 2r - i \leq 2r - j_0$, $\mu(2r - i) = \lambda(i)$ and $E_i = G_i \cup G_{2r-i} \subset C_{\mu(i)} \cup C_{\mu(2r-i)} = C_{\lambda(i)}$. Thus $E_i \subset C_{\lambda(i)}$. This finishes the proof that $\mathcal{E}$ follows the pattern λ in C.

This finishes the proof of the claim.

We are ready to prove the inductive step. Suppose that we have constructed the chains $C_1, \ldots, C_n$ and $\mathcal{D}_1, \ldots, \mathcal{D}_n$ with the required properties.

By the claim, applied to the number $\frac{1}{2^{n+1}}$, the chains $\mathcal{C}_n$ and $\mathcal{D}_n$, and the points p and q, we have that there exist taut chains $\mathcal{E} = \{E_0, E_1, \dots, E_r\}$ of X and $\mathcal{F} = \{F_0, F_1, \dots, F_r\}$ of Y and there exists a pattern $\lambda : \langle r\rangle \to \langle m_n\rangle$ such that $\mathcal{E}$ follows the pattern λ in $\mathcal{C}_n$, $\mathcal{F}$ follows the pattern λ in $\mathcal{D}_n$, $p \in E_0 \setminus \mathrm{cl}_X(E_1)$, $q \in F_0 \setminus \mathrm{cl}_Y(F_1)$, and the mesh of $\mathcal{F}$ is less than $\frac{1}{2^{n+1}}$. Applying the claim again, now to the number $\frac{1}{2^{n+1}}$, the chains $\mathcal{E}$ and $\mathcal{F}$, and the points q and p, we obtain taut chains $\mathcal{C}_{n+1} = \{C_0^{(n+1)}, C_1^{(n+1)}, \dots, C_{m_{n+1}}^{(n+1)}\}$ of X and $\mathcal{D}_{n+1} = \{D_0^{(n+1)}, D_1^{(n+1)}, \dots, D_{m_{n+1}}^{(n+1)}\}$ of Y, and there exists a pattern $\delta : \langle m_{n+1}\rangle \to \langle r\rangle$ such that $\mathcal{C}_{n+1}$ follows the pattern δ in $\mathcal{E}$, $\mathcal{D}_{n+1}$ follows the pattern δ in $\mathcal{F}$, $p \in C_0^{(n+1)} \setminus \mathrm{cl}_X(C_1^{(n+1)})$, $q \in D_0^{(n+1)} \setminus \mathrm{cl}_Y(D_1^{(n+1)})$ and the mesh of $\mathcal{C}_{n+1}$ is less than $\frac{1}{2^{n+1}}$.

Since each element of $\mathcal{D}_{n+1}$ is contained in one of the chains $\mathcal{F}$ and the mesh of $\mathcal{F}$ is less than $\frac{1}{2^{n+1}}$, we have that $\mathcal{D}_{n+1}$ has mesh less than $\frac{1}{2^{n+1}}$.

Set $\lambda_{n+1} : \langle m_{n+1}\rangle \to \langle m_n\rangle$ given by $\lambda_{n+1} = \lambda \circ \delta$. Then λ_{n+1} is a pattern, $\mathcal{C}_{n+1}$ follows the pattern λ_n in $\mathcal{C}_n$ and $\mathcal{D}_{n+1}$ follows the pattern λ_n in $\mathcal{D}_n$.

This completes the inductive construction.

Now, we construct a homeomorphism $h : X \to Y$. Let d_X and d_Y be respective metrics for X and Y.

Take $x \in X$. Given $n \in \mathbb{N}$, choose $i_n \in \langle m_n\rangle$ such that $x \in C_{i_n}^{(n)}$ and choose a point $x_n \in D_{i_n}^{(n)}$. We check that the sequence $\{x_n\}_{n=1}^{\infty}$ is a Cauchy sequence in Y, so $\lim_{n\to\infty} x_n = h(x)$, for some $h(x) \in Y$. We also will check that h is well defined, bijective, continuous and $h(p) = q$.

Given $n \in \mathbb{N}$, $x \in C_{i_n}^{(n)} \cap C_{i_{n+1}}^{(n+1)} \subset C_{i_n}^{(n)} \cap C_{\lambda_n(i_{n+1})}^{(n)}$. So $|i_n - \lambda_n(i_{n+1})| \le 1$, $D_{i_n}^{(n)} \cap D_{\lambda_n(i_{n+1})}^{(n)} \neq \emptyset$ and $\mathrm{diameter}(D_{i_n}^{(n)} \cup D_{\lambda_n(i_{n+1})}^{(n)}) < \frac{1}{2^{n-1}}$. Since $x_n, x_{n+1} \in D_{i_n}^{(n)} \cup D_{i_{n+1}}^{(n+1)} \subset D_{i_n}^{(n)} \cup D_{\lambda_n(i_{n+1})}^{(n)}$, we have $d_Y(x_n, x_{n+1}) < \frac{1}{2^{n-1}}$. Given $m > n$, $d_Y(x_n, x_m) \le d_Y(x_n, x_{n+1}) + d_Y(x_{n+1}, x_{n+2}) + \cdots + d_Y(x_{m-1}, x_m) < \frac{1}{2^{n-1}} + \frac{1}{2^n} + \cdots + \frac{1}{2^{m-2}} < \frac{1}{2^{n-2}}$. From here, it follows that $\{x_n\}_{n=1}^{\infty}$ is a Cauchy sequence in Y, so $\lim_{k\to\infty} x_k = h(x)$, for some $h(x) \in Y$ and $d(x_n, h(x)) \le \frac{1}{2^{n-2}}$.

In order to see that $h(x)$ is well defined, suppose that for each $n \in \mathbb{N}$, we also choose $i_n' \in \langle m_n\rangle$ with $x \in C_{i_n'}^{(n)}$ and $x_n' \in D_{i_n'}^{(n)}$. Then $|i_n - i_n'| \le 1$ and $D_{i_n}^{(n)} \cap D_{i_n'}^{(n)} \neq \emptyset$. Thus $d_Y(x_n, x_n') < \frac{1}{2^{n-1}}$. This implies that $\lim_{n\to\infty} x_n = \lim_{n\to\infty} x_n'$. Hence h is well defined.

We show that h is continuous. Take $x \in X$. Let $\varepsilon > 0$. Take $m \in \mathbb{N}$, such that $\frac{3}{2^{m-2}} < \varepsilon$. Let $\{i_n\}_{n=1}^{\infty}$ and $\{x_n\}_{n=1}^{\infty}$ be as in the definition of $h(x)$. Then $U = C_{i_m}^{(m)}$ is an open neighborhood of x in X. Given $y \in U$, by the previous paragraph, for the definition of $h(y)$, we can choose $y_m \in D_{i_m}^{(m)}$. Then $d(x_m, y_m) < \frac{1}{2^m}$. By what we showed three paragraphs above, we have $d_Y(h(x), h(y)) \le d_Y(h(x), x_m) + d_Y(x_m, y_m) + d_Y(y_m, h(y)) < \frac{1}{2^m} + \frac{2}{2^{m-2}} < \frac{3}{2^{m-2}} < \varepsilon$. We have shown that if $y \in U$, then $d_Y(h(x), h(y)) < \varepsilon$. Therefore h is continuous.

To prove that h is one-to-one, suppose that $x, y \in X$ and $h(x) = h(y)$, let $\{i_n\}_{n=1}^{\infty}$, $\{x_n\}_{n=1}^{\infty}$, $\{i_n'\}_{n=1}^{\infty}$ and $\{y_n\}_{n=1}^{\infty}$ be as in the definition of $h(x)$ and $h(y)$, respectively. Then $\lim_{n\to\infty} x_n = \lim_{n\to\infty} y_n$. Given $n \in \mathbb{N}$, let $j \in \langle m_n \rangle$ be such that $h(x) \in D_j^{(n)}$. Then there exists a $k \in \mathbb{N}$ such that $x_k, y_k \in D_j^{(n)}$ and $n < k$. Since $x_k \in D_{i_k}^{(k)} \cap D_j^{(k)}$, we obtain that $|i_k - j| \leq 1$. Similarly, $|i'_k - j| \leq 1$. Thus $|i_k - i'_k| \leq 2$. Since $x, y \in C_{i_k}^{(k)} \cup C_{i'_k}^{(k)}$, we conclude that $d_X(x, y) < \frac{3}{2^k} < \frac{3}{2^n}$. Since this inequality holds for every n, we conclude that $x = y$. Therefore h is one-to-one.

To show that h is onto, it is enough to prove that $h(X)$ is dense in Y. Let $y \in Y$ and $\varepsilon > 0$. Choose $n \in \mathbb{N}$ such that $\frac{3}{2^n} < \varepsilon$. Take $j \in \langle m_n \rangle$ such that $y \in D_j^{(n)}$. Fix a point $x \in C_j^{(n)}$. Then, for the definition of $h(x)$, we may choose $x_n \in D_j^{(n)}$. We have shown that $d_Y(x_n, h(x)) < \frac{1}{2^{n-2}}$. Thus $d_Y(y, h(x)) \leq d_Y(y, x_n) + d_Y(x_n, h(x)) < \frac{1}{2^n} + \frac{1}{2^{n-2}} = \frac{3}{2^n} < \varepsilon$. So, $d_Y(y, h(x)) < \varepsilon$. We have shown that $h(X)$ is dense in Y. Therefore h is onto.

Given $n \in \mathbb{N}$, since $p \in C_0^{(n)}$ and $q \in D_0^{(n)}$ and we can choose $x_n = q$. Therefore $h(p) = \lim_{k\to\infty} x_k = q$.

This completes the proof of the theorem. □

Corollary 17.13 *The pseudo-arc is homogeneous.*

Corollary 17.14 *All hereditarily indecomposable chainable continua are homeomorphic.*

Corollary 17.15 *The pseudo-arc is the only chainable hereditarily indecomposable continuum.*

Corollary 17.16 *The pseudo-arc is homeomorphic to each of its non-degenerate subcontinua.*

Corollary 17.17 ([11, p. 346]) *The pseudo-arc is the only homogeneous chainable continuum.*

Proof Let X be a homogeneous chainable continuum. First we show that X has an end-point. Given $n \in \mathbb{N}$, let $\mathcal{C}_n$ be a proper covering chain of X with mesh less than $\frac{1}{n}$. Fix a point $q_n \in X$ such that the only link of $\mathcal{C}_n$ containing q_n is the first one. Let q be an accumulation point of the sequence $\{q_n\}_{n=1}^{\infty}$. Observe that q has the following property:

Property A. For each $\varepsilon > 0$ there exists a proper covering chain $\mathcal{C}$ of X such that $\text{mesh}(\mathcal{C}) < \varepsilon$ and $B(q, \varepsilon)$ contains the first link of $\mathcal{C}$.

The homogeneity of X implies that each point in X has Property A (Exercise 17.18).

Let C_1 be the first link of a proper covering chain $\mathcal{C}_1$ of X such that $\text{mesh}(\mathcal{C}_1) < 1$. Let $p_1 \in C_1$. Since p_1 has Property A, there exists a proper covering chain $\mathcal{C}_2$ of X such that $\text{mesh}(\mathcal{C}_2) < \frac{1}{2}$ and the first link C_2 of $\mathcal{C}_2$ satisfies that $\text{cl}_X(C_2) \subset C_1$. Proceeding in this way, it is possible to construct a sequence of proper covering chains $\{\mathcal{C}_n\}_{n=1}^{\infty}$ such that for each $n \in \mathbb{N}$, if C_n is the first link of $\mathcal{C}_n$, then $C_1 \supset \text{cl}_X(C_2) \supset C_2 \supset \text{cl}_X(C_3) \supset C_3 \cdots$, and $\text{mesh}(\mathcal{C}_n) < \frac{1}{n}$.

Let p be the unique point of X such that

$$\{p\} = \bigcap\{\text{cl}_X(C_n) : n \in \mathbb{N}\} = \bigcap\{C_n : n \in \mathbb{N}\}.$$

By Exercise 17.19, p is a terminal point. By Theorem 17.7, p is a final point of X. Since X is homogeneous, every point of X is final.

In order to finish the proof of this corollary, by Corollary 17.15 we need to show that X is hereditarily indecomposable. Suppose to the contrary that there exists a decomposable subcontinuum Y of X. Then $Y = A \cup B$, for some proper subcontinua A and B of Y. Therefore $A \setminus B \neq \emptyset \neq B \setminus A$. By the connectedness of Y, there exists a point $p \in A \cap B$. By definition, p is not a final point of X. This contradiction ends the proof of the corollary. □

17.6 Exercises

Exercise 17.18 Let X be a homogeneous chainable continuum. Prove that each point in X satisfies Property A of the proof of Corollary 17.17.

Exercise 17.19 Let X be a chainable continuum and $p \in X$. Suppose that for each $n \in \mathbb{N}$, there exists a proper covering chain C_n of X such that $\text{mesh}(C_n) < \frac{1}{n}$ and p belongs to the first link of C_n. Prove that p is a terminal point of X.

Exercise 17.20 Let X be a continuum, Y a subcontinuum of X and

$$C = \{C_0, C_1, \dots, C_n\}$$

a family of open subsets of X that covers Y. Suppose that $Y \cap C_0 \neq \emptyset$, $Y \cap C_n \neq \emptyset$ and C has the property that if $i, j \in \langle n \rangle$ and $C_i \cap C_j \neq \emptyset$, then $|i - j| \leq 1$. Thus for each $i \in \{1, \dots, n\}$, we have that $C_{i-1} \cap C_i \cap Y \neq \emptyset$.

Exercise 17.21 Prove that the sets A_0 and B_0 used in the proof of Lemma 17.8 are closed. (Hint: use Exercises 9.31 and 9.37(a).)

References

1. Alexandroff, P.S.: Über stetige Abbildungen kompakter Räume. Math. Ann. **96**, 555–571 (1927)
2. Anderson, R.D.: A characterization of the universal curve and a proof of its homogeneity. Ann. Math. **67**, 313–324 (1958)
3. Anderson, R.D.: One-dimensional continuous curves and a homogeneity theorem. Ann. Math. **68**, 1–16 (1958)
4. Anderson, R.D., Choquet, G.: A plane continuum no two of whose non-degenerate subcontinua are homeomorphic: an application of inverse limits. Trans. Am. Math. Soc. **10**(3), 347–353 (1959)
5. Andersen, R.N., Marjanović, M.M., Schori, R.M.: Symmetric products and higher dimensional products. Topol. Proc. **18**, 7–17 (1993)
6. Ayres, W.L.: Some generalizations of Scherer fixed-point theorem. Fund. Math. **16**, 332–336 (1930)
7. Banič, I., Goran, E., Kennedy, J.: The Lelek fan as the inverse limit of intervals with a single set-valued bonding function whose graph is an arc. Mediterr. J. Math. **20**(3), 159, 24pp. (2023)
8. Bellamy, D.P.: Indecomposable continua with one and two composants. Fund. Math. **101**(2), 129–134 (1978)
9. Bing, R.H.: A homogeneous indecomposable plane continuum. Duke Math. J. **15**, 729–742 (1948)
10. Bing, R.H.: Snake-like continua. Duke Math. J. **18**, 653–663 (1951)
11. Bing, R.H.: Each homogeneous nondegenerate chainable continuum is a pseudo-arc. Proc. Am. Math. Soc. **10**(3), 345–346 (1959)
12. Bing, R.H.: The elusive fixed point property. Am. Math. Month. **76**, 119–132 (1969)
13. Bing, R.H., Jones, F.B.: Another homogeneous plane continuum. Trans. Am. Math. Soc. **90**, 171–192 (1959)
14. Borsuk, K.: Sur les rétractes. Fund. Math. **17**, 152–170 (1931)
15. Borsuk, K.: Über eine Klasse von lokal zusammenhängenden Räumen. Fund. Math. **19**, 220–242 (1932)
16. Borsuk, K.: Problem 54. Colloq. Math. **1**, 332 (1948)
17. Borsuk, K.: On the third symmetric potency of the circumference. Fund. Math. **36**, 235–244 (1949)
18. Borsuk, K.A.: A theorem on fixed points. Bull. Acad. Polon. Sci. Cl. III. **2**, 17–20 (1954)
19. Borsuk, K., Ulam, S.: On symmetric products of topological spaces. Bull. Am. Math. Soc. **37**, 875–882 (1931)
20. Bott, R.: On the third symmetric potency of S_1. Fund. Math. **39**, 364–368 (1952)

A. Illanes, *Continuum Theory*, Universitext,
https://doi.org/10.1007/978-3-031-91011-1

21. Castañeda, E.: Embedding symmetric products in Euclidean spaces. In: Continuum Theory (Denton, TX, 1999). Lecture Notes in Pure and Applied Mathematics, pp. 67–79, vol. 230. Dekker, New York (2002)
22. Charatonik, J.J.: History of continuum theory. In: Aull, C.E., Lowen, R. (Eds.), Handbook of the History of General Topology, vol. 2, pp. 703–786. Kluwer Academic, Dordrecht (1998)
23. Charatonik, J.J., Charatonik, W.J.: Dendrites, XXX National Congress of the Mexican Mathematical Society (Spanish) (Aguascalientes, 1997). Aportaciones Material Communication., vol. 22, pp. 227–253. Sociedad Matemática Mexicana, México (1998)
24. Charatonik, J.J., Charatonik, W.J., Omiljanowski, K., Prajs, J.R.: Hyperspace retractions for curves. Diss. Math. **370**, 1–34 (1997)
25. Christenson, C.O., Voxman, W.L.: Aspects of Topology. Pure and Applied Mathematics, vol. 39, xi+517pp. Marcel Dekker, New York (1977)
26. Corona-Vázquez, F., Quiñones-Estrella, R.A., Sánchez-Martínez, J., Villanueva, H.: Embedding products into symmetric products of finite graphs. Topol. Appl. **241**, 162–171 (2018)
27. Curtis, D.: Growth hyperspaces of Peano continua. Trans. Am. Math. Soc. **238**, 271–283 (1978)
28. Curtis, D.W.: Application of a selection theorem to hyperspace contractibility. Can. J. Math. **37**, 747–759 (1985)
29. Curtis, D.W., Schori, R.M.: 2^X and $C(X)$ are homeomorphic to the Hilbert cube. Bull. Am. Math. Soc. **80**, 927–931 (1974)
30. Curtis, D.W., Schori, R.M.: Hyperspaces of Peano continua are Hilbert cubes. Fund. Math. **101**, 19–38 (1978)
31. de J. López, M.: Hyperspaces homeomorphic to cones. Topol. Appl. **126**(3), 361–375 (2002)
32. Dugundji, J.: An extension of Tietze's theorem. Pac. J. Math. **1**, 353–367 (1951)
33. Dugundji, J.: Topology. Reprinting of the 1966 Original, Allyn and Bacon Series in Advanced Mathematics, xv+447pp. Allyn/Bacon, Boston/London (1978)
34. Eberhart, C., Nadler, S.B., Jr.: Hyperspaces of cones and fans. Proc. Am. Math. Soc. **77**, 279–288 (1979)
35. Eilenberg, S.: Transformations continues en circonférence et la topologie du plan. Fund. Math. **26**, 61–112 (1936)
36. Engelking, R.: Wacław Sierpiński (1882–1969) his life and work in topology. In: Aull, C.E., Lowen, R. (eds.), Handbook of the History of General Topology, vol. 2, pp. 399–414. Kluwer Academic, Dordrecht (1998)
37. Esenin-Vol'pin, A.S.: Review 129. Referativnyĭ Zhurnal **1**, 25 (1955)
38. Fedorchuk, V.V.: Covariant functors in the category of compacta, absolute retracts, and Q–manifolds. Russian Math. Surv. **36**(3), 211–233 (1981)
39. Fedorchuk, V.V., Odintsov, A.A.: Theory of continua. I (in Russian). J. Math. Sci. **71**(2), 2329–2363 (1994). Itogi Nauki i Tekhniki, Algebra Topol. Geom. **29**, Algebra. Topology. Geometry, Vol. **29** (Russian), 63–119, Akad. Nauk SSSR, Vsesoyuz. Inst. Nauchn. i Tekhn. Inform., Moscow (1991)
40. Fugate, J.B.: Retracting fans onto finite fans. Fund. Math. **71**(2), 113–125 (1971)
41. Fugate, J.B.: Small retractions of smooth dendroids onto trees. Fund. Math. **71**(3), 255–262 (1971)
42. García-Máynez, A., Illanes, A.: A survey on unicoherence and related properties. An. Inst. Mat. Univ. Nac. Autónom. México **29**, 17–67 (1990)
43. Greenwood, S., Suabedissen, R.: 2-manifolds and inverse limits of set values functions on intervals. Discrete Contin. Dyn. Syst. **37**(11), 5693–5706 (2017)
44. Hagopian, C.L.: An update on the elusive fixed-point property. In: Pearl, E. (ed.), Open Problems in Topology II, pp. 263–277. Elsevier, Amsterdam (2007)
45. Hausdorff, F.: Grundzüge der Mengenlehre. Leipzig (1914)
46. Hausdorff, F.: Mengenlehre, Zweite, neubearbeitete Auflage. Walter de Gruyter, Berlin (1927)
47. Hoehn, L.C., Oversteegen, L.G.: A complete classification of homogeneous plane continua. Acta Math. **216**(2), 177–216 (2016)

48. Hoehn, L.C., Oversteegen, L.G.: A complete classification of hereditarily equivalent plane continua. Adv. Math. **368**, 107131 8pp. (2020)
49. Homma, T.: A theorem on continuous functions, (Volume numbers not printed on issues until Vol. 7 (1955)). Kōdai Math. Sem. Rep. **4**, 13–16 (1952)
50. Hurewicz, W., Wallman, H.: Dimension Theory, 9th edn. Princeton University Press, Princeton (1974)
51. Illanes, A.: Cells and cubes in hyperspaces. Fund. Math. **130**, 57–65 (1988)
52. Illanes, A.: The space of Whitney levels is homeomorphic to l_2. Colloq. Math. **65**(1), 1–11 (1993)
53. Illanes, A.: Hyperspaces which are products. Topol. Appl. **79**, 229–247 (1997)
54. Illanes, A.: The hyperspace $C_2(X)$ for a finite graph X is unique. Glas. Mat. Ser. III **37**(57), 347–363 (2002)
55. Illanes, A.: Finite graphs X have unique hyperspaces $C_n(X)$. Topol. Proc. **27**, 179–188 (2003)
56. Illanes, A.: A model for the hyperspace $C_2(S^1)$. Questions Answers Gen. Topol. **22**, 117–130 (2004)
57. Illanes, A.: Hyperspaces of Continua (Hiperespacios de Continuos) (Spanish). Aportaciones Matemáticas, Textos 28 (Nivel Medio). Sociedad Matemática Mexicana, México (2004)
58. Illanes, A.: Modelos de Hiperespacios (in Spanish). Aportaciones Matemáticas, Textos 31 (Nivel Medio). Sociedad Matemática Mexicana, México (2006)
59. Illanes, A.: A tree-like continuum whose cone admits a fixed-point-free map. Houston J. Math. **23**(2), 499–518 (2007)
60. Illanes, A.: A nonlocally connected continuum whose second symmetric product can be embedded in $\mathbb{R}^3$. Questions Answers Gen. Topol. **26**, 115–119 (2008)
61. Illanes, A.: A tree-like continuum whose hyperspace of subcontinua admits a fixed-point-free map. Topol. Proc. **32**, 55–74 (2008)
62. Illanes, A.: Fixed point property on symmetric products of chainable continua. Comment. Math. Univ. Carolin. **50**(4), 615–628 (2009)
63. Illanes, A.: A circle is not the generalized inverse limit of a subset of $[0, 1]^2$. Proc. Am. Math. Soc. **139**(8), 2987–2993 (2011)
64. Illanes, A.: Models of hyperspaces. Topol. Proc. **41**, 39–64 (2013)
65. Illanes, A.: Hereditarily non-weakly chainable continua in products of plane continua. Fund. Math. **239**(1), 19–27 (2017)
66. Illanes, A., de J. López, M.: Hyperspaces homeomorphic to cones, II. Topol. Appl. **126**(3), 377–391 (2002)
67. Illanes, A., Martínez-de-la-Vega, V.: Product topology in the hyperspace of subcontinua. Topol. Appl. **105**(3), 305–317 (2000)
68. Illanes, A., Martínez-de-la-Vega, V.: Models and homogeneity degree of hyperspaces of a simple closed curve. Rocky Mountain J. Math. **54**, 765–785 (2024)
69. Illanes, A., Nadler, S.B. Jr.: Hyperspaces: Fundamentals and Recent Advances. Monographs and Textbooks in Pure and Applied Mathematics, vol. 216. Marcel Dekker, New York (1999)
70. Illanes, A., Macías, S., Nadler, S.B., Jr.: Symmetric products and Q-manifolds. In: Geometry and Topology in Dynamics. Contemporary Mathematical Series of American Mathematical Society, vol. 246, pp. 137–141. American Mathematical Society, Providence (1999)
71. Illanes, A., Minc, P., Sturm, F.: Extending surjections defined on remainders of metric compactifications of $[0, \infty)$. Houston J. Math. **41**(4), 1325–1340 (2015)
72. Ingram, W.T.: A brief historical view of continuum theory. Topol. Appl. **153**, 1530–1539 (2006)
73. Ingram, W.T.: An Introduction to Inverse Limits with Set-Valued Functions. Springer Briefs in Mathematics, xii+86pp. Springer, New York (2012)
74. Ingram, W.T., Mahavier, W.S.: Inverse Limits: From Continua to Chaos. Developments in Mathematics, xvi+217pp., vol. 25. Springer, New York (2012)
75. Jones, F.L.: A History and Development of Indecomposable Continua Theory. Thesis (Ph.D.), Michigan State University, 206pp. (1971)

76. Jones, F.L.: Historia y desarrollo de la teoría de los continuos indescomponibles (in Spanish). [History and development of the theory of indecomposable continua], Dissertation, Michigan State University, East Lansing (2004). With an appendix containing a Spanish translation of "A brief history of indecomposable continua" [in Measure theory (Cincinnati, OH, 1994), 103–126, Dekker, New York, 1995] by Judy A. Kennedy. Aportaciones Matemáticas: Textos [Mathematical Contributions: Texts], 27. Sociedad Matemática Mexicana, México, 2004, vi+226 pp.
77. Kapuano, I.: Sur les continus linéaires (in French). C. R. Acad. Sci. Paris **237**, 683–685 (1953)
78. Kapuano, I.: Sur une proposition de M. Bing. C. R. Acad. Sci. Paris **236**, 2468–2469 (1953)
79. Keller, O.H.: Die Homoimorphie der kompakten konvexen Mengen in Hilbertschen Raum. Math. Ann. **105**, 748–758 (1931)
80. Kennedy, J.A.: A brief history of indecomposable continua. In: Continua (Cincinnati, OH, 1994). Lecture Notes in Pure and Applied Mathematics, pp. 103–126, vol. 170. Dekker, New York (1995)
81. Kinoshita, S.: On some contractible continua without fixed point property. Fund. Math. **40**, 96–98 (1953)
82. Knaster, B.: Un continu dont tout sous-continu est indécomposable. Fund. Math. **3**, 247–286 (1922)
83. Knaster, B.: Problèmes P 323 et P 324. Colloq. Math. **8**, 139 (1961)
84. Knaster, B., Kuratowski, C.: Problème 2. Fund. Math. **1**, 223 (1920)
85. Knaster, B., Kuratowski, K.: Sur les ensembles connexes. Fund. Math. **2**, 206–255 (1921)
86. Knaster, B., Kuratowski, K., Mazurkiewicz, S.: Ein Beweis des Fixpunktsatzes für n-dimensionale Simplexe. Fund. Math. **14**, 132–137 (1929)
87. Knill, R.J.: Cone, products and fixed points. Fund. Math. **60**, 35–46 (1967)
88. Kuratowski, K.: Une caractérisation topologique de la surface de la sphère. Fund. Math. **13**, 307–318 (1929)
89. Kuratowski, K.: Sur le problème des courbes gauches en Topologie. Fund. Math. **15**, 271–283 (1930)
90. Kuratowski, K.: Topology, Vol. II.. New edition, revised and augmented. Translated from the French by A. Kirkor, Academic Press, New-York-London, xiv+608pp. Pańﬆowe Wydawnictwo Naukowe Polish Scientific Publishers, Warsaw (1968)
91. Kuratowski, K.: Introduction to Set Theory and Topology. Containing a Supplement on "Elements of algebraic topology" by Ryszard Engelking. Completely revised second English edition. First edition translated from the Polish by Leo F. Boron. International Series of Monographs in Pure and Applied Mathematics, 352pp., vol. 101. Pergamon Press, Oxford-New York-Toronto; PWN-Polish Scientific Publishers, Warsaw (1972)
92. Levin, M., Sternfeld, Y.: The space of subcontinua of a 2-dimensional continuum is infinite dimensional. Proc. Am. Math. Soc. **125**, 2771–2775 (1997)
93. Lewis, W.: Continuous curves of pseudo-arcs. Houston J. Math. **11**(2), 225–236 (1985)
94. Lewis, W.: The pseudo-arc. Bol. Soc. Mat. Mexicana **5**(1), 25–77 (1999)
95. Lewis, W.: Characterizations of the pseudo-arc. Topol. Proc. **48**, 49–63 (2016)
96. Lewis, W.: Notes on the pseudo-arc (unpublished)
97. Lynch, M.: Whitney levels in $C_p(X)$ are ARS. Proc. Am. Math. Soc. **97**(4), 748–750 (1986)
98. Macías, S.: Fans whose hyperspaces are cones. Topology Proc. **27**, 217–222 (2003)
99. Macías, J.C.: On the n-fold pseudo-hyperspace suspension of continua. Glas. Mat. **43**(63), 439–499 (2008)
100. Macías, S., Nadler, S.B. Jr.: Absolute n-fold hyperspace suspensions. Colloq. Math. **105**, 221–231 (2006)
101. Mahavier, W.S.: Inverse limits with subsets of $[0, 1] \times [0, 1]$. Topol. Appl. **141**, 225–231 (2004)
102. Mańka, R.: The topological fixed point property—an elementary continuum-theoric approach. In: Fixed Point Theory and its Applications, vol. 77, pp. 183–200. Banach Center Publications, Polish Academy of Sciences (Nauk Matematycznych) (2007)

103. Mardešić, S.: Equivalence of singular and Čech homology for ANR-s, Application to unicoherence. Fund. Math. **46**, 29–45 (1958)
104. Martínez-de-la-Vega, V.: Dimension of n-fold hyperspaces of graphs. Houston J. Math. **32**, 783–799 (2006)
105. Martínez-de-la-Vega, V., Ordoñez, N.: Embedding hyperspaces. Topol. Appl. **159**, 2032–2042.
106. Mazurkiewicz, M.: Problème 14. Fund. Math. **2**, 286 (1921)
107. Menger, K.: Allgemeine Räume und Cartesische Räume. Erste Mitteilung. Proc. Sect. Sci. Koninklijke Akademie van Wetenschappen te Amsterdam **29**, 476–482 (1926); Zweite Mitteilung: Über um fassendste n-dimensionale Mengen, 1125–1128
108. Moise, E.E.: An indecomposable plane continuum which is homeomorphic to each of its non-degenerate sub-continua. Trans. Am. Math. Soc. **63**, 581–594 (1948)
109. Molski, R.: On symmetric products. Fund. Math. **44**, 165–170 (1957)
110. Mostovoy, J.: Lattices in $\mathbb{C}$ and finite subsets of a circle. Am. Math. Montly **111**, 357–360 (2004)
111. Nadler, S.B. Jr.: Continua whose cone and hyperspace are homeomorphic. Trans. Am. Math. Soc. **230**, 321–345 (1977)
112. Nadler, S.B. Jr.: Hyperspaces of Sets: A Text with Research Questions. Monographs and Textbooks in Pure and Applied Mathematics, vol. 49. Marcel Dekker, New York (1978)
113. Nadler, S.B. Jr.: Continuum Theory: An Introduction. Monographs Textbooks Pure and Applied Mathematics, vol. 158. Dekker, New York (1992)
114. Naranjo-Murillo, J.A.: Examples concerning means on dendroids of generalized type N. Topol. Appl. **301**, 107532, 23pp. (2021)
115. Oversteegen, L.G., Tymchatyn, E.D.: On hereditarily indecomposable compacta. In: Geometric and Algebraic Topology, pp. 407–417, vol. 18. Banach Center Publications, Warszawa (1986)
116. Peano, G.: Sur une courbe, qui remplir toute une aire plane. Math. Ann. **36**, 157–160 (1890)
117. Rosenholtz, I.: Another proof that any compact metric space is the continuous image of the Cantor set. Am. Math. Month. **83**(8), 646–647 (1976)
118. Sierpiński, W.: Sur une courbe cantorienne qui contient une image continue de toute courbe donnée. Comptes Rendus de l'Académie de Paris **162**, 629–632 (1916); *Mat. Sb.* **30** (1916), 267–287 (Russian); French traslation in Waclaw Sierpiński, Oeuvres choisies, vol. II, PWN Warszawa 1974, 107–119
119. Sierpiński, W.: Sur une condition pour qu'un continu soit une courbe jordanienne. Fund. Math. **1**, 44–60 (1920)
120. Thomas, E.S. Jr.: Monotone decompositions of irreducible continua. Dissert. Math. **50**, 1–74 (1966)
121. Tucker, A.: The parallel climbers puzzle: a case study in the power of graph models. Math. Horizons **3**(2), 22–24 (1995)
122. Tuffley, C.: Finite subset spaces of S^1. Algebr. Geom. Topol. **2**, 1119–1145 (2002)
123. Vidal Escobar, P.I.: Los hiperespacios de continuos desde el punto de vista de sus modelos geométricos. Bachelor's Thesis, Director: Enrique Castañeda, Universidad Autónoma del Estado de México (2009)
124. Wojdysławski, M.: Sur la contractilité des hyperspaces de continus localment connexes. Fund. Math. **30**, 247–252 (1938)
125. Wojdysławski, M.: Rétractes absolus et hyperspaces des continus. Fund. Math. **32**, 189–192 (1939)
126. Wu, W.: Note sur les produits essentiels symétriques des espaces topologique, I. C. R. Acad. Sci. **16**, 1139–1141 (1947)
127. Yoneyama, K.: Theory of continuous sets of points. Tôhoku Math. J. **12**, 43–158 (1917)

Index

A. Illanes, *Continuum Theory*, Universitext,
https://doi.org/10.1007/978-3-031-91011-1

U

V

W

Y

Z

MIX
Papier aus verantwortungsvollen Quellen
Paper from responsible sources
FSC® C105338

If you have any concerns about our products,
you can contact us on
ProductSafety@springernature.com

In case Publisher is established outside the EU,
the EU authorized representative is:
Springer Nature Customer Service Center GmbH
Europaplatz 3, 69115 Heidelberg, Germany

Printed by Libri Plureos GmbH
in Hamburg, Germany